AF775026

Problem Solving Approaches for Maintaining Operational Excellence in Process Plants

Problem Solving Approaches for Maintaining Operational Excellence in Process Plants

Joseph M. Bonem
Engineering Consultant
ExxonMobil Chemicals, US

Nattapong Pongboot
Project Leader
Refinery Catalyst Testing Group at Avantium

Wiroon Tanthapanichakoon
Engineering Consultant and CEO
Global R&D Co. Ltd., Thailand

Second Edition

AIChE
The Global Home of Chemical Engineers
WILEY

Copyright © 2025 by the American Institute of Chemical Engineers, Inc. All rights reserved.

A Joint Publication of the American Institute of Chemical Engineers and John Wiley & Sons, Inc.

Published by John Wiley & Sons, Inc., Hoboken, New Jersey.
Published simultaneously in Canada.

No part of this publication may be reproduced, stored in a retrieval system, or transmitted in any form or by any means, electronic, mechanical, photocopying, recording, scanning, or otherwise, except as permitted under Section 107 or 108 of the 1976 United States Copyright Act, without either the prior written permission of the Publisher, or authorization through payment of the appropriate per-copy fee to the Copyright Clearance Center, Inc., 222 Rosewood Drive, Danvers, MA 01923, (978) 750-8400, fax (978) 750-4470, or on the web at www.copyright.com. Requests to the Publisher for permission should be addressed to the Permissions Department, John Wiley & Sons, Inc., 111 River Street, Hoboken, NJ 07030, (201) 748-6011, fax (201) 748-6008, or online at http://www.wiley.com/go/permission.

Trademarks: Wiley and the Wiley logo are trademarks or registered trademarks of John Wiley & Sons, Inc. and/or its affiliates in the United States and other countries and may not be used without written permission. All other trademarks are the property of their respective owners. John Wiley & Sons, Inc. is not associated with any product or vendor mentioned in this book.

Limit of Liability/Disclaimer of Warranty: While the publisher and author have used their best efforts in preparing this book, they make no representations or warranties with respect to the accuracy or completeness of the contents of this book and specifically disclaim any implied warranties of merchantability or fitness for a particular purpose. No warranty may be created or extended by sales representatives or written sales materials. The advice and strategies contained herein may not be suitable for your situation. You should consult with a professional where appropriate. Further, readers should be aware that websites listed in this work may have changed or disappeared between when this work was written and when it is read. Neither the publisher nor authors shall be liable for any loss of profit or any other commercial damages, including but not limited to special, incidental, consequential, or other damages.

For general information on our other products and services or for technical support, please contact our Customer Care Department within the United States at (800) 762-2974, outside the United States at (317) 572-3993 or fax (317) 572-4002.

Wiley also publishes its books in a variety of electronic formats. Some content that appears in print may not be available in electronic formats. For more information about Wiley products, visit our web site at www.wiley.com.

Library of Congress Cataloging-in-Publication Data Applied for:

Hardback ISBN: 9781394207152

Cover Design: Wiley
Cover Image: © Joseph M. Bonem

Set in 9.5/12.5pt STIXTwoText by Straive, Chennai, India

Contents

Preface *xi*

1 **Focus of Book** *1*
1.1 Introduction *1*
1.2 Metrics and KPIs *2*
1.3 Finding Hidden Problems *3*
1.4 Experience-Based Solutions *4*
1.5 Achieving and Maintaining Operational Excellence *5*

2 **How to Achieve and Maintain Operational Excellence** *7*
2.1 What is Operational Excellence? *7*
2.2 What is the Value of Operational Excellence? *8*
2.3 What are the Limitations to Achieving and Maintaining Operational Excellence? *11*
2.4 Achieving and Maintaining Operational Excellence *11*

3 **Initial Considerations** *15*
3.1 Introduction *15*
3.2 An Electrical Problem *17*
3.3 A Coffeemaker Problem *18*
3.4 Applications of Concepts to Plant Problem Solving *21*
3.5 Limitations to Problem Solving in Process Plants *24*

4 **Successful Plant Problem Solving** *29*
4.1 Introduction *29*
4.2 Finding Problems with a Daily Monitoring System *29*
4.3 Solving Problems with a Disciplined Learned Problem-Solving Approach *36*
4.4 Determining the Optimum Technical Depth *42*
4.5 Using the Directionally Correct Hypothesis Approach *47*

5 Examples of Plant Problem Solving *49*
5.1 Industrial Examples *49*
5.2 Polymerization Reactor Example *49*
5.3 Application of the Disciplined Problem-Solving Approach *51*
5.4 Lessons Learned *55*
5.5 Multiple Engineering Disciplines Example *56*
5.6 Application of Disciplined Problem-Solving Approach *57*
5.7 Lessons Learned *62*
5.8 A Logical, Intuitive Approach Fails *63*
5.9 Lessons Learned *65*
Nomenclature *65*

6 Development of Working Hypotheses *67*
6.1 Introduction *67*
6.2 Areas of Technology *67*
6.3 Formulating Hypotheses Via Key Questions *68*
6.4 Beauty of a Simplified Approach *71*
6.5 Verification of Proposed Hypotheses *71*
6.6 One Riot—One Ranger *73*

7 Application to Prime Movers *75*
7.1 Introduction *75*
7.2 Kinetic Systems *75*
7.3 Pump Calculations *78*
7.4 Centrifugal Compressor Calculations *80*
7.5 Displacement Systems *82*
7.6 Displacement Pump Calculations *85*
7.7 Calculations for Positive Displacement Compressors *87*
7.8 Problem-Solving Considerations for Both Systems *89*
7.8.1 Compressors *89*
7.8.2 Pumps *91*
7.9 Example Problem 7.1 *92*
7.10 Lessons Learned *98*
7.11 Example Problem 7.2 *99*
7.12 Lessons Learned *103*
7.13 Example Problem 7.3 *104*
7.14 Example Problem 7.4 *104*
7.15 Lessons Learned *109*
7.16 Example Problem 7.5 *109*
7.17 Example Problem 7.6 *111*
7.18 Lessons Learned *118*
7.19 Example Problem 7.7 *118*
7.20 Lessons Learned *124*
7.21 Example Problem 7.8 *125*
7.22 Lessons Learned *131*
Nomenclature *131*

8 Application to Plate Processes *135*
8.1 Introduction *135*
8.2 Fractionation with Sieve Trays *135*
8.3 Problem-Solving Considerations for Fractionating Towers *139*
8.4 Development of Theoretically Sound Working Hypotheses *140*
8.5 Problem-Solving Reboiler Circuits *142*
8.6 Example Problem 8.1 *145*
8.7 Lessons Learned *154*
Nomenclature *154*

9 Application to Kinetically Limited Processes *155*
9.1 Introduction *155*
9.2 Kinetically Limited Models *155*
9.3 Limitations to the Lumped Parameter Approach *158*
9.4 Guidelines for Utilization of this Approach for Plant Problem Solving *159*
9.5 Example Problem 9.1 *160*
9.6 Lessons Learned *167*
9.7 Technique for Estimating Polymer–Volatile Equilibrium *168*
9.8 Example Problem 9.2 *169*
9.9 Lessons Learned *174*
9.10 Example Problem 9.3 *175*
9.11 Lessons Learned *183*
Nomenclature *184*

10 Application to Unsteady State *187*
10.1 Introduction *187*
10.2 Approach to Unsteady State Problem Solving *188*
10.3 Example Problems *189*
10.4 Problem 10.1 *189*
10.5 Lessons Learned *196*
10.6 Example Problem 10.2 *197*
10.7 Lessons Learned *204*
10.8 Final Words *205*
Nomenclature *206*

11 Application to Other Plant Improvements *209*
11.1 Introduction *209*
11.2 Debottlenecking Reactors *210*
11.2.1 Kinetics *210*
11.2.2 Heat Removal *211*
11.2.3 Stepwise Procedure *211*
11.2.4 Real Life Example *212*
11.2.5 Macro Heat Removal *212*

11.2.6 Micro Heat Removal *213*
11.3 Real-World Hydraulic Debottleneck *216*
11.3.1 Introduction *216*
11.3.2 Always Account for Non-idealities *217*
11.3.3 A Quick Guide for a Successful Hydraulic Test Run *218*
11.3.4 A Recommended Workflow for an Effective Hydraulic Revamp *220*
11.3.4.1 Utilize Equipment/Device's Characteristics *220*
11.3.4.2 Use a Calibrated Hydraulic Model *221*
11.3.5 Maximize Diesel Production by Revamping the Draw-off System *222*
11.3.6 A New Feed Preheater for Energy Saving *224*
11.3.7 Increase Wash Water Flow to Reduce the Corrosion Rate *227*
11.3.8 Improve Diesel Recovery from a Crude Distillation Column *228*
11.3.9 Other Considerations *230*
11.4 Debottlenecking By Improving Operating Procedures *232*
11.4.1 Improving Water Clarification *232*
11.4.1.1 Introduction to Water Clarification *232*
11.4.1.2 Sand Filtration Basics *233*
11.4.1.3 Evaluate Backwash Performance *235*
11.4.1.4 Operational Issues Related to Inefficient Backwash *236*
11.4.1.5 Field Techniques for Backwash Tuning *238*
11.4.1.6 A Real-World Example *241*
11.4.1.7 Other Tips *242*
11.5 Trust Creating a Disaster—Heat Exchanger Corrosion from Improper Cooling Water System Operation *243*
11.5.1 Background *243*
11.5.2 Penny Wise, Pound Foolish—Misconception of Cooling Water System Operation *245*
11.5.3 Take Ownership *251*
Nomenclature *251*

12 Applications of Novel Process Engineering Fundamentals to Plant Problem Solving *255*
12.1 Introduction *255*
12.2 Novel Approaches to Plant Problems *255*
12.3 Mostly used Engineering Fundamentals to Solve Plant Problems *256*
12.4 Application of New Engineering Fundamentals to Plant Maintenance Problems (Example Problem 12.1) *257*
12.5 Application of the Disciplined Problem-Solving Approach *258*
12.6 Lessons Learned *260*
12.7 Tank Roof Raising for Maintenance Example *260*
12.8 Application of the Disciplined Problem-Solving Approach *261*
12.9 Lessons Learned *263*
Nomenclature *264*

13 Verification of Process Instrumentation Data *265*
13.1 Introduction *265*
13.2 Data Verification Via Technical Resources *265*
13.3 Flow Measurement *269*
13.4 Temperature Measurement *272*
13.5 Pressure Measurement *272*
13.6 Level Measurement *273*
13.7 Data Verification Via Human Resources *275*
13.8 Example Problems *275*
13.9 Example Problem 13.1 *276*
13.10 Lessons Learned *280*
13.11 Example Problem 13.2 *280*
13.12 Lessons Learned *284*
13.13 Example Problem 13.3 *284*
13.14 Example Problem 13.4 *286*
13.15 Lessons Learned *290*
Nomenclature *290*

14 Successful Plant Tests *291*
14.1 Introduction *291*
14.2 Ingredients for Successful Plant Tests *292*
14.3 Pretest Instrument and Laboratory Procedure Evaluation *292*
14.4 Statement of Anticipated Results *293*
14.5 Potential Problem Analysis *295*
14.6 Explanation to Operating Personnel *297*
14.7 Formal Post-Test Evaluation and Documentation *298*
14.8 Examples of Plant Tests *299*
14.9 Example Plant Test 14.1 *299*
14.10 Lessons Learned *301*
14.11 Example Plant Test 14.2 *302*
14.12 Lessons Learned *307*
14.13 More Complicated Plant Tests *307*
14.14 Other Uses for Plant Tests *308*
14.15 Key Plant Tests Considerations *308*

15 Utilization of Commercially Available Simulation Tools *309*
15.1 Process Simulation and Modern Chemical Engineering *309*
15.2 Breaking Down the Problem *311*
15.3 Green Field Problem Example *313*
15.3.1 Situation *313*
15.3.2 Input Parameter *313*
15.3.3 Output Parameter *313*
15.3.4 Design Parameters *314*
15.4 Brown Field Problem Example *315*

15.4.1 Input Parameter *316*
15.4.2 Output Parameter *316*
15.4.3 Design Parameters *317*
15.5 Do Not Gamble with Physical Properties for Simulations *317*
15.6 Examples—Effects of Equation of State on the Required Compression Power and Cooling Duty *318*
15.6.1 Input Parameter *318*
15.6.2 Output Parameter *318*
15.6.3 Design Parameters *318*
15.7 Be Skeptical with your Initial Design Assumptions *319*
15.8 Obtaining a High-Quality Plant Data for your Process Model *320*
15.9 Verifying your Plant Data *323*
15.10 Example—Heat Balance of Heavy Gas Oil Pumparound *324*
15.11 Reconciling your Data *325*
15.12 Example—Hydrocracking Catalyst Testing *326*
15.13 Model Calibration *331*
15.14 Process Simulation as a Training Tool *336*
Nomenclature *336*

16 Putting it Altogether *339*
16.1 Introduction *339*
16.2 Do Not Forget to Use Fundamentals *339*
16.3 Example Problem 16.1: Do Fundamental Processes Developed in the United States Translate to Europe? *340*
16.4 Lessons Learned *348*
16.5 Example Problem 16.2: An Embarrassing Moment *349*
16.6 Lessons Learned *355*
16.7 Example Problem 16.3: Prime Mover Problems are not Always What They Appear To Be *356*
16.8 Lessons Learned *364*
16.9 Example Problem 16.4: The Value of a Potential Problem Analysis *365*
16.10 Lessons Learned *371*
16.11 Example Problem 16.5 *371*
16.12 Lessons Learned *377*
Nomenclature *378*

17 A Final Note *379*

Appendix Conversion Factors from English Units to CGS Units *381*

References *383*
Index *385*

Preface

Whenever one starts the process of writing a book, consideration must be given to the important question—Does this book create any value added for society?

This book answers that question with a resounding YES. The book addresses the question of how can the Operational Excellence of process plants (refineries, gas plants, and chemical plants) be improved and maintained at this higher level of excellence. Mastering the concepts in this book will improve the efficiency, profitability, and safety of process plants. While the concepts are focused on process plants, the concepts are general enough that they can be applied to almost any industrial operation. In order to show how the general concepts are put into detailed application, the examples are real-life illustrations from process plants.

This book makes frequent use of an earlier book by the editor who is also one of the authors of this book. The previous book was entitled *PROCESS ENGINEERING PROBLEM SOLVING: Avoiding "The Problem Went Away But It Came Back" Syndrome.* The previous book and this book are focused on a calculation-based approach to problem solving. This approach uses first principles of engineering and is based on fundamental concepts. However, it also recognizes the time pressures that most operational problems create. As such, the fundamental approach utilizes relatively simple first principles that are theoretically correct. The previous book has been expanded by defining what "Operational Excellence" means, by indicating the incentive for achieving this level of performance, by summarizing an approach to achieving Operational Excellence, by providing insight into the use of available software programs, and by including additional example problems. The focus remains on using a structured methodical calculation-based problem-solving approach in order to achieve and maintain Operational Excellence.

While this book is not a management book, it does make the assumption that management has set as a priority improving/maintaining Operational Excellence and that management will provide the necessary resources to implement the concepts recommended in this book. These additional resources will increase the expense side of the P/L, BUT they will more than pay for themselves by reduced operating costs, increased production capability (revenue), and increased safety.

The book can be thought about as four sections. The initial section deals with a general approach to problem solving that can be used in all areas of life (work, home, hobbies, or volunteer activities). The second section deals with application of this approach to process operating plants, the third section deals with successful plant tests including confirmation of data, and the last section covers concluding comments.

In preparing any book like this, the need to use personal pronouns is almost mandatory. The use of personal pronouns helps to make the book more readable and minimizes the more complex ways of expression. Even though I have worked with very talented female engineers, operators, and mechanics, it was my choice to use the personal pronoun "he" to cover all gender possibilities.

September 12, 2024

Joe Bonem
Ft. Worth Texas

1

Focus of Book

1.1 Introduction

The book is focused on what technical concepts and calculation-based approaches are required to improve and maintain Operational Excellence. The types of problems that are envisioned are either chronic operating problems or operating problems for which the problem solver has no experience-based solution. The example problems discussed are for the most part real-life problems that actually occurred, and solutions were those that were developed from a careful analysis and calculations. These problems and solution concepts are enhanced by a limited discussion on setting metrics and uncovering hidden problems.

While if experience-based solutions are available, they can be of great value. However, not every problem encountered has been experienced before by the problem solver, and as such these problems are rarely solved with experience-based solutions. In fact, the attempt to solve them based on extrapolated experience often results in delaying the true solution. For example, an impurity problem associated with ethylene used in a polyethylene process cannot be extrapolated to a similar impurity problem associated with propylene used in a polypropylene process. While this seems obvious, time pressures associated with poor operating performance often create the management decree to "Do something even if it is wrong."

In reading Sir Arthur Conan Doyle's late 1880s fiction about the famous detective Sherlock Holmes and his companion Dr. Watson, I see a parallel to current industrial problem solving techniques. Watson often quickly jumped to conclusions based on a few facts and his experience. On the other hand, Holmes took a thoughtful and methodical approach and used all the available data to obtain the right solution. Holmes' approach being methodical appeared to be too slow, but it avoided the endless detours caused by jumping to conclusions. A quote from Sherlock in the novel *The Valley of Fear* illustrates this well—"The temptation to form premature theories upon insufficient data is the bane of our profession." For

Problem Solving Approaches for Maintaining Operational Excellence in Process Plants, Second Edition.
Joseph M. Bonem, Nattapong Pongboot, and Wiroon Tanthapanichakoon.
© 2025 the American Institute of Chemical Engineers, Inc. Published 2025 by John Wiley & Sons, Inc.

the complicated or chronic process problem, the successful problem-solver is a modern-day Sherlock Holmes.

1.2 Metrics and KPIs

Before one decides that a problem exists, there must be a deviation from some sort of standard or metric. These metrics are often referred to as key performance indicators (KPIs). This terminology KPIs will be used in the book to refer to the anticipated value of a key variable. An Operational Excellence deviation is one that is significantly different than the anticipated value of a key operating variable or lumped parameter. The key variable or lumped parameter can be as simple as pressure of a key utility or as complicated as a reaction rate constant. Before one determines that a key operating parameter has significantly deviated from the target value, the target value must be known. The reported analysis of blood of a medical patient uses this concept well when the technician reports actual values and the expected range.

The first step in problem-solving is to determine values for the KPIs. The approach of employing a lumped parameter should be utilized. For example, the lumped parameter for a heat exchanger is the well-known heat transfer coefficient. This includes temperatures of both hot and cold sides, flow rates of both hot and cold materials, and thermal characteristics of the materials in a single value. A similar approach could be used for a reaction rate constant. The calculation of this lumped parameter will provide a value of this KPI at any point in time, which can be compared to the theoretical value. In the case of a heat transfer coefficient, the theoretical value is the design coefficient. This comparison can then be expressed as a percent of the theoretical value. A similar approach can be used for equipment such as pumps, where the pump curve serves as the theoretical data source. For example, a key pump operating at the design flow rate, but below the theoretical head, could be flagged as an operating problem using the design head as the theoretical KPI.

Regardless of the terminology (metrics or KPIs), the most important thing is how the theoretical value is determined. It can be determined by design calculations such as heat/material balances, equipment specifications, or pilot plant/laboratory results. They can also be set by actual plant performance. It is often true that process plants can be operated at 10% or more above design capacity for extended periods of time. This capability should be used to determine new KPIs rather than using design calculations.

These KPIs must be theoretically obtainable, achievable (with often significant effort), and sustainable. There is often a temptation to set the KPIs too low so that they can be easily obtained. This temptation must be avoided if the goal is to achieve Operational Excellence. The successful military mission or airplane journey represent examples that are theoretically possible, achievable, and sustainable.

Now considering this terminology in more detail:

- Theoretically obtainable KPIs are those set by the physical limits of the operating unit. These are metrics that would be achieved if there were no limits caused by failures, errors, or uncontrollable events.
- Achievable KPIs are those that can be achieved with existing equipment and available raw materials.
- Sustainable KPIs are those that can be achieved every day of the year. Again, referring to the airline industry and assuming that one is traveling from Tokyo to DFW (Dalllas-FT. Worth International Airport) with an **exceptionally** strong tail wind, the flight may be as much as an hour shorter than planned. However, this does not mean this shorter flight time is sustainable day after day.

KPIs can include things such as operating rates, reaction kinetics/rate constants, heat transfer in critical exchangers, tower performance, and/or dryer performance. To achieve improved Operational Excellence, these KPIs should be established during good plant operations. The approach to establishing them during good operation frequently at greater than design capacity will often create concerns about the number of chronic problems that are uncovered. This uncovering of what are referred to as hidden problems should be considered a valuable source of potential operational improvements. The other extreme of setting metrics so low that they are easily achieved is not the focus of improved Operational Excellence. While setting KPIs too low does reduce the number of perceived operational problems, it does limit the KPIs or however metrics are measured to levels that can be easily achieved. A quote from Frederick Bonfils, the founder of The Denver Post, is certainly appropriate for the manufacturing industry—"There is no hope for the satisfied man." It is only by setting high standards and tight KPIs that Operational Excellence can be improved.

1.3 Finding Hidden Problems

Comparing these theoretical KPIs to actual operation is a method for finding "hidden problems". Finding these problems is one of the keys to improving Operational Excellence. These hidden problems exist for a multitude of reasons. They can be understood as operating problems that have become accepted and hence ignored as "just the way that it is." Since these problems are accepted, the necessary analysis to fully understand and to eliminate the problem is not done. Some typical areas are as follows:

- It is often difficult to convince a process control engineer to make a programming change in a process that is computer-controlled. This results in less than excellent adaptive solutions.
- In processes that are not computer-controlled, operators at shift change often "tweak" set points to their personal preferences. While the process will operate at either set point, one of the set points will be less than optimum.
- Often a process engineer does not understand mechanical equipment such as pumps and compressors. This lack of understanding limits the ability to make any potential recommended changes that might improve process operations. The process will operate, but not at an optimum.
- There is confusion between using standard deviations for control and using them for defining if a situation is likely a real problem. The well-known $\pm 3\sigma$ has become a benchmark for process control. However, for operating problems, problem-solving should begin before that point. This concept along with the risks is described in a later chapter.

These hidden problems or ones that are and have been blatantly in the open for either short or extended periods of time are referred to as chronic problems, and solving them is what the book covers. These operating problems are almost always related to some aspect of technology. This technology could be equipment technology or process technology. Unfortunately, attempts to solve these problems are often based on operating experience rather than an understanding and analysis of the problem. They are defined as chronic because these attempts have failed to resolve the problem. Many of the example problems included in this book fit the definition of chronic operating problems. They were long-standing problems with many trial-and-error attempted solutions. The attempted solutions were based on limited experiences, intuition, or guesses. The time period to attempt these multiple solutions almost always exceeds the time that would have been required for a more in-depth analysis of the problem and equipment involved at an early stage. For unsuccessful solutions that require new equipment, the pain associated with the failure is especially damaging.

1.4 Experience-Based Solutions

Operating problems where there is no experience-based solution are an increasing significant problem in today's environment. This is due to several factors given as follows:

- In developed countries, there is a loss of experience in the technical, operational, and mechanical areas. Thus, even in a process that has been in operation for many years, these chronic problems have no apparent solutions because of this loss.

- This loss of experience is made even more noticeable in developed countries where technical staff have been reduced with the belief that experienced operation and mechanical personnel can handle all situations.
- Developing countries have similar problems compounded by the fact that they are just now developing experience.
- Educational institutions in both developed and developing countries are continuing to emphasize research and the utilization of computer algorithms as opposed to the understanding of calculation-based problem solving or design.
- In both types of countries, there is an increased movement of technically trained personnel into business, management, marketing, and supply chain activities.

Looking for proof that these factors are negatively influencing Operational Excellence, consideration could be given to increased Force Majeures, major fires/safety incidents, supply chain disruptions, and derating of industrial plants.

1.5 Achieving and Maintaining Operational Excellence

While this book discusses the setting of operating metrics and finding hidden problems, the main focus is a problem-solving technique for maintaining plant Operational Excellence along with real life examples. A five-step procedure including theoretical-based calculations is proposed. The calculations can be described as fundamental and using first principles. They also recognize the urgent need that management often feels when the problems impact the bottom line. These pressures are often extended to those working the problem. Thus, the calculations are those that can be done relatively quickly while still remaining fundamentally accurate. The example problems given in the book are real-life problems that were solved by the techniques, and the calculations are shown.

The authors of the book have over 100 years of industrial experience and have seen how an emphasis on setting good operating metrics and having an accurate and fast response problem-solving technique can improve Operational Excellence, which leads to improved profitability. As indicated in the preface, the content of the book can be thought of as having four sections. The initial section deals with a general approach to problem-solving that can be used in all areas of life (work, home, hobbies, or volunteer activities). The second section deals with the application of this approach to process operating plants, the third section deals with successful plant tests including confirmation of data, and the last section covers concluding comments.

The initial section (preface through Chapter 4) of the book includes items such as

- Operational Excellence—definition, value, limitations, and achieving/maintaining this level of performance.
- Components of a fast response problem-solving approach.
- The five-step problem-solving technique that works in process plants.

The second section (Chapters 5 through 14) contains examples of application of the five-step approach to process plants and process equipment, development of working hypotheses, and verification of plant instrument data. The second section also deals with solutions to problems where a previous analysis has confirmed a need for new equipment or enhanced operating procedures. This section provides comments on the utilization of plant tests as a mechanism to test a working hypothesis (Step 4 of the five-step approach).

The third section (Chapter 15) deals with simulation tools and the building of "digital twins" to simulate plant operations. An accurate digital twin can be used for predicting yields and plant operations as well as for training operating and technical personnel.

The last section (Chapter 16) is a summary along with some additional more complicated examples of the application of the five-step approach. A final note is provided in Chapter 17.

2

How to Achieve and Maintain Operational Excellence

2.1 What is Operational Excellence?

While the focus of this book is maintaining Operational Excellence, it is necessary to know both what Operational Excellence is and how to achieve that level of operation. Operational Excellence is a well-known phrase to those involved in many aspects of the worldwide society. In one form or another, it seems to be present from the academic world to heavy industry. While there are many books focused on achieving Operational Excellence by good management practices, there are few that discuss what technical approaches are required to both achieve and maintain this level of operations.

In order to provide the necessary background, this chapter provides a brief summary of the aspects of Operational Excellence as follows:

- The meaning of the term.
- The value of achieving this level of operations.
- Current limitations to achieving and maintaining Operational Excellence.

As a starting point, what is meant by the term "Operational Excellence"? In understanding what is meant by this term, one might use dictionary terminology, cliches, organization goals, or mission statements. Some examples of these different ways to define the term are as follows:

- *Dictionary*: The idea is that in a company with Operational Excellence, operations would be superior to others operating the same facilities or doing the same work.
- *Cliches*: Some of the more common ones are – "Be all that you can be," "Don't leave anything on the court or track," "Give it all you got," or "If a job is worth doing at all it is worth doing well."

Problem Solving Approaches for Maintaining Operational Excellence in Process Plants, Second Edition.
Joseph M. Bonem, Nattapong Pongboot, and Wiroon Tanthapanichakoon.
© 2025 the American Institute of Chemical Engineers, Inc. Published 2025 by John Wiley & Sons, Inc.

- *Organization goals*: "Be the highest quality producer" or "Reduce utility consumption by 10%."
- *Mission statements*: These are often organization goals with fancy wording.

Regardless of lofty ambitions expressed in these examples, they all fail to achieve a satisfactory level of Operational Excellence because they are not specific enough and do not include measurable KPIs, or in the case of organizational goals such as increased profits and/or reduced costs, they do not provide specific tactics to allow this goal to be achieved. Well-defined KPIs on a select number of key variables will provide these tactics for strategic organization goals.

While each individual company will have to define exactly what these key variables are, this book takes the approach that any deviation greater than ±5% from these selected KPIs should be investigated as a means to improve Operational Excellence. A positive deviation should be considered. For example, a calculated heat transfer coefficient that shows greater than the design value may indicate that a critical exchanger has more capacity than design and that the process could be operated at a higher rate. Conversely, the apparent higher coefficient might be due to an instrument error. Either way, additional evaluation would be of value.

2.2 What is the Value of Operational Excellence?

There is no doubt that developing and communicating these KPIs will require a significant amount of time and effort. This leads to the logical question of what is the incentive for developing these target KPIs. Developing these KPIs with a method described in Chapter 1 is key to improved operations. The value of Operational Excellence can be measured in terms of improved economics, enhanced product quality, improved safety, and greater employee morale and engagement.

Industry/business is concerned about making and/or increasing profits. This can be achieved by increasing net revenue and/or decreasing costs. Profits can always be increased by raising prices if all other things are constant. Since this book is primarily associated with operations, only increased sales or decreased costs will be considered a means to improve profits. Sales can be increased by either increased production or increased market share.

When considering revenue, the focus is on net revenue. This can be defined as the difference in increased revenue less increased operating costs associated with an increase in production. The marginal increase in production associated with Operational Excellence will generate more profit on a unit basis than the average profit on the average production. For example, a typical polyolefin plant might be generating a profit of 10 cents/pound based on a sales price of 50 cents/pound. By difference, the total operating cost is 40 cents/pound. Of this operating cost, 20% is fixed and thus does not increase if production is increased. The remainder of the costs are variable and consist of feed, chemicals, catalysts, and utilities. Based on these values, the profit on the marginal increased production is 18 cents/pound or 80% higher than the average cost prior to production increase.

The source of this increased production can be due to a scenario where the operating plant is running at an alleged maximum capacity. It can also be due to enhanced product quality, which allows a company to obtain a higher market share during times when production is limited by demand. In either of these scenarios, Operational Excellence techniques can be utilized to increase production and net revenues. It is highly likely that the effort required to improve operations will be more than justified by the increase in profits (net revenue).

In the case of reduced operating costs that are associated with Operational Excellence, the incentive is not as easy to generalize. The goal of reducing operating costs can be associated with a reduction in demand. In this case, there is a desire to keep the variable cost (cents/pound) at reduced production as close as possible to that at full rates. In the case of a polyolefin plant where 80% of the costs are variable, there are at least two major limits to reducing costs as production is reduced. These are as follows:

- The plant is often not designed for a significant turndown.
- Plant operators find it easier to operate at conditions that are not optimum. An illustration from domestic situations is that it is easier to leave the lights on than turn them off and have to turn them on again.

As an example of the first limit described above, consider the cycle gas compressor of a gas-phase polyolefin plant, and assume that this compressor is supplied with a fixed-speed electric motor. In this process, the exothermic heat of reaction is removed by the circulating gas. With this type of compressor design, there is a minimum flow rate that must be maintained to avoid the surging of the compressor. This limitation can be mitigated by the utilization of a variable-speed motor or steam turbine which will reduce the surge point of the compressor. However, while this design may mitigate the problem, there are other limitations associated with the gas rate. At reduced production rates, the minimum gas rate will likely be set by the fluidization rate requirements or the necessary superficial velocity to achieve good individual particle heat removal rather than overall heat load on the reactor.

In addition to the case of reduced production due to demand limit, there is a case where at full rates, it is desirable to reduce variable costs for closer approach to a theoretical limit. This will generally be a situation where operating parameters have drifted away from design values. If engineering calculations confirm the validity of the design values, then the second limitations described above are the limiting factor. This situation is usually driven by the operating personnel's negative experience such as gas-phase reactor fouling as gas rates were reduced. This limitation can be overcome by careful technical analysis and strong management. While this example is highly simplified, it illustrates the difficulty in achieving an operational goal of maintaining a constant variable cost as production rates are decreased. It will always be necessary to consider both engineering details and possible previous experience in order to be successful in this cost reduction goal.

There are real, but somewhat less tangible benefits associated with Operational Excellence. As indicated earlier, these intangible benefits are in the areas of improved product quality, safety, and greater employee job satisfaction and engagement. Product quality can be both a tangible and intangible benefit. Thus, it shows up in both areas. These areas are intertwined and are all related to the smoother operation of a process that has a high degree of Operational Excellence. This interdependence of these attributes is shown in Figure 2.1.

As indicated in this figure, Operational Excellence results in more steady-state operations, minimal shutdowns, minimal rate reductions due to process problems, and minimal upsets. In addition, this improved operation reduces the temptation to take on risks that might be associated with undesirable plant shutdowns. These risks will eventually lead to a safety incident. The figure also includes a feedback loop where improvements in product quality, safety, and job performance also improve Operational Excellence. If the impact of improvements, whether tangible or somewhat intangible, is communicated adequately to the workforce, this feedback loop will be very strong. Conversely, if the communication is not done or not done effectively, this feedback loop will be weak or nonexistent. This feedback loop is present because there is almost always a sense of pride in the operations, mechanical, and technical workforce when the process they are responsible for is operating well. This leads to a desire to make even more improvements.

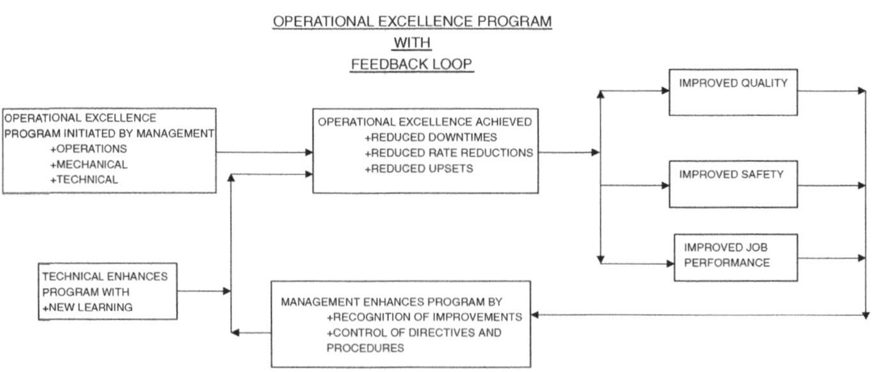

Figure 2.1 Operational excellence program with feedback loop.

Conversely, poor operations or lack of good communications can lead to an attitude of nonchalance and poor job performance. Verbal exchanges can often give a clue that this is occurring. With comments such as "management doesn't care about this process," "No one pays any attention to what I say," or "This place is going downhill at a fast pace." A happy engaged worker that feels safe can be a positive influence on Operational Excellence.

2.3 What are the Limitations to Achieving and Maintaining Operational Excellence?

Since Operational Excellence does not occur automatically, an analysis of the limitations to obtaining this excellence will be of value. While each company has somewhat different limitations, there are four general areas that tend to be present in all areas of the world whether the areas can be classified as "developed nations" or "developing nations." These are as follows:

- Limited experience
- Changing academic focus
- Increasing complexity and size of industrial processes
- Emphasis on cost reduction

These areas are discussed in more detail in Chapter 3.

2.4 Achieving and Maintaining Operational Excellence

Given the industrial environment that many operating facilities are surrounded by today, what are the steps to achieve and maintain Operational Excellence. These steps must touch on both technical and management issues. Since this book deals primarily with technical issues, the need for strong management is only briefly considered.

The division between technical and management varies from company to company. One often used division is that the technical workforce is responsible for developing operating procedures/directives and solving both chronic and complicated operating problems. On the other hand, management is responsible for the assuring that operating directives/procedures are both adequately developed and followed.

As discussed in Chapter 1, the means for measuring Operational Excellence is carefully developing, following, and reacting to KPIs. The initial step in achieving Operational Excellence is to develop these metrics for the specific process.

While KPIs measure how close a given process is to Operational Excellence, they do not provide a means to achieve this desired level of performance. To achieve the target KPIs requires well-defined operating procedures and operating directives. It is also important to provide sufficient technical detail so that operating and mechanical personnel can understand the importance of the directives and procedures. The differentiation between operating directives and procedures is that operating directives are used to determine how a process should operate at steady state. Operating procedures include both what steps are required to reach steady state (start-up procedures, for example), procedures to conduct batch operations, and procedures associated with interim operations (such as replacing the catalyst in reactors). These procedures and directives should be consistent with the KPIs discussed earlier.

Even with well-developed and established directives and procedures that are being well followed, there will always be external events that create problems that must be solved. Some examples of these external factors are as follows:

- *Changes in supplier feedstock*: While most process plants have feedstock specifications, sometimes, a change of suppliers or a change in the supplier's operations will introduce an impurity that is not specified.
- *Changes in climatic conditions*: A sudden change in ambient temperature may introduce a change in gas density in a process drum. This may cause an occasional process upset. While the frequency is minimal, the result maybe a chronic operating problem.
- The marketing needs to operate a process for an extended period without a periodic downtime to inspect equipment may create the failure of a critical piece of equipment.
- The expansion of a process fails to deliver as promised because the limitation of a minor piece of equipment was not considered by the contract engineering company.

These are examples that the techniques described in this book would have prevented and/or quickly eliminated if they did occur.

Since this book is focused on technical problem solving, a reasonable question is what attributes are required to achieve excellence in this goal and thus assist in maintaining Operational Excellence. The following attributes are mandatory to obtain excellence in the area of problem solving:

- Technical personnel should have a good understanding of the process and process equipment before beginning problem solving. This likely means that initially, the problem solver will spend time learning the process and/or a specific piece of equipment.

- The problem solver will not only react to problems brought to them but also be proactive and seek to find potential problems and hidden problems. The problem solver does this daily or more frequently monitoring KPIs. The problem solver will know and understand the KPIs and how to monitor them.
- In solving problems, technical personnel will normally be recommending new directives/procedures or new equipment. These recommendations should be based on a firm technical basis including a potential problem analysis associated with proposed changes or new equipment. This requirement is illustrated by the cliché—do not create another problem while solving the current problem.
- The problem-solver will assess the problem in a calm and deliberate fashion. This is in contrast with problem-solving sessions where the loudest expression is the chosen path.
- There will always be time pressures associated with problem solving. It is important to establish a good compromise between expediency and accuracy.

A thoughtful examination of these attributes will lead to the conclusion that while meetings are necessary to table problems, develop understanding of problems, and to set priorities, the real problem solving of chronic and complicated problems occurs as more understanding is developed and as thoughtful calculations or careful examination of the data is done.

As indicated in Chapter 1, the remaining chapters are dedicated to the techniques of problem solving in process plants. Besides the techniques, examples are given to illustrate the use of these techniques.

3

Initial Considerations

3.1 Introduction

Chapters 1 and 2 discussed the generalized approach to achieving and maintaining Operational Excellence including the existence of a fast response problem-solving approach. This chapter and subsequent chapters focus on this fast response approach. These chapters cover questions such as—What is this approach? What are the limitations? And are there real-life examples of this approach? This chapter specifically focuses on limitations to implementing this fast response problem-solving approach and the guidelines for a successful problem-solving organization. The purpose of this chapter is as follows:

- To illustrate and discuss the importance of fully understanding the problem.
- To stress the importance of a theoretical first-principles approach.
- To provide examples of real-world inadequate problem solving.
- To elaborate on the limitations to successful plant problem solving discussed in Chapter 2.

Problem solving is an area that is found throughout all activities of daily life. It tends to take place in two mindset modes. There is the intuitive or instinctive reactionary mode, which has also been called "gut feel." Then, there is the methodical reasoning approach, which is usually based on theoretical considerations and almost always on calculations.

Either of these approaches has its place in real-world problem-solving activities. The intuitive reactionary person will respond much faster to a problem. Their response is usually based on experience. That is, the problem solver has seen the same thing before or something very similar and remembers what the problem solution was. However, if what is occurring is a new problem or is somewhat different, this approach may well lead to an incorrect problem solution. The methodical reasoning person will not be able to react to problems quickly but will usually

Problem Solving Approaches for Maintaining Operational Excellence in Process Plants, Second Edition.
Joseph M. Bonem, Nattapong Pongboot, and Wiroon Tanthapanichakoon.
© 2025 the American Institute of Chemical Engineers, Inc. Published 2025 by John Wiley & Sons, Inc.

obtain the correct problem solution for complicated problems much faster than the intuitive reactionary person who must develop several aborted "gut feel" solutions. In the real world, we are all a mixture of these types of problem-solving personalities. The effective problem solver will know when to choose and which mode to use for any situation that the problem solver confronts.

An example of how two people with these different mindsets will react can be found in the most unlikely places. For example, on a golf course, the cry of "Fore" will illicit different responses. The person responding based on intuition or instinct will immediately cover his head and crouch. This will reduce the probably that the errant golf ball hits a sensitive body part. The person responding based on methodical reasoning will begin to assess where the cry came from, where the ball might be coming from, and reach a conclusion as to where it might land. Obviously, in this case, reacting based on intuition or instinct is a far superior mode of operating. There could be many more examples from the sports world where reacting in an intuitive fashion yields far superior results than reacting in a methodical reasoning manner. **However**, essentially, all of these examples will be experience-based. People who are reacting successfully in an intuitive mode know what to do because they have experienced the same or very similar situations.

Similar things happen in industrial problem solving. Experienced people (engineers, mechanics, or operators) react instinctively because they have experienced similar events. This instinctive approach serves people well in handling emergency situations or making decisions during a start-up. As a rule, the person who tends to respond based on methodical reasoning and calculations rarely can react fast enough to be of assistance in an emergency or if quick action is required in a start-up situation. The exception to this rule is the engineer who has designed the plant and has gone through calculations to understand what will happen in an emergency or start-up. In effect, he has gained the experience through calculations, as opposed to actual experience.

The experience necessary to conduct problem solving in the real world does not always exist. In addition, while the need for a quick response when solving industrial problems is real, there is not always an emergency or need to take immediate action. Thus, the methodical reasoning approach is often the desirable mode of operating. The three components of this methodical reasoning approach are as follows:

- A systematic step-by-step procedure. This includes the three essential problem-solving skills (daily monitoring system, disciplined problem-solving approach, and determining optimum technical depth).
- A good understanding of how the equipment involved works.
- A good understanding of the specific technology involved.

Before discussing problem solving in industrial facilities, two examples from everyday life are discussed. It often aids learning to discuss things that are outside the scope of the original thrust of the teaching. The two examples from everyday life discussed below will be helpful in understanding the difference between intuitive problem solving and those based on methodical reasoning.

3.2 An Electrical Problem

While trimming bushes with an electric hedge trimmer, a laborer accidentally cut the extension cord being used to power the trimmer. He had been using an electrical outlet in a pump house located approximately 70 feet from the main house. The only other use for 110-V electricity in the pump house was for a small clock associated with the water softener. The laborer found another extension cord and replaced the severed cord. However, when he plugged it in and tried to turn on the hedge trimmer, it did not have any power. He then had to report the incident to the homeowner. The homeowner checked the panel mounted circuit breakers. None of them appeared to be tripped. However, to be sure, he turned off the appropriate circuit breaker and reset it. However, power was still not restored to the outlet in the pump house. To make sure that the replacement extension cord was not the problem, the homeowner plugged another appliance directly into the electrical outlet in the pump house. It did not work either. The homeowner then concluded that the electric outlet had been "blown out" when the cord was cut. He replaced the electric outlet. However, this still did not provide power to the equipment. When the homeowner rechecked the circuit breaker, he noticed that a ground fault indicator (GFI) in a bathroom in the main house was tripped. Resetting this GFI solved the problem.

While the homeowner believed that in this particular house every GFI protected a single outlet, it is not unheard of to protect more than a single outlet with a GFI. It seemed surprising that the GFI in a bathroom also protected an outlet in the pump house 70 ft away. The homeowner then recalled that at some point in the past, he had noticed that the small clock in the pump house was about 2 hours slow. This clock was always very reliable. In retrospect, he remembered that at about the same time that the clock lost 2 hours, this particular GFI had tripped during a lightning storm and had not been reset for a few hours. Thus, it became obvious that the accidental cutting of the extension cord had caused the GFI to trip rather than tripping the circuit breaker or "blowing out" the electrical outlet. The failure to correctly identify the problem cost the homeowner a small amount of money for the electrical outlet and a significant amount of time to go to town to purchase the outlet and then install it.

Note that the homeowner's intuitive approaches were all real possibilities. That is, the circuit breaker could well have tripped, the replacement extension cord could have had an electrical break in it, or the electrical outlet could have failed when the original extension cord was cut. His problem solving just did not go into enough detail to solve the problem quickly. Several lessons can be learned from this experience. While it seemed to be a simple problem that could be easily solved based on the homeowner's experience, the intuitive approach did not work. A more systematic approach based on methodical reasoning might have improved results as follows:

- Consideration would have been given to the possibility that GFIs can protect more than one outlet. The distance between the GFI and the electrical outlet would not be a consideration. The homeowner did not fully understand the technology.
- A voltmeter would have been used to check that power was available coming to the electrical outlet. If power was not available coming to the outlet, the "blown outlet" hypothesis would be invalid. A systematic approach was not used.
- In addition, a systematic approach would have raised the question of whether the clock losing 2 hours could be related to the lack of power at the electrical outlet.

3.3 A Coffeemaker Problem

A housewife experienced problems with a coffeemaker overflowing about half of the time when she made either a flavored or decaffeinated coffee. The coffee and coffee grounds would overflow the top of the basket container and spill all over the counter. The coffeemaker performed flawlessly when regular coffee was used. A sketch of the coffeemaker is shown in Figure 3.1. When the coffeemaker is started, water is heated, and the resulting steam provides a lifting mechanism to carry the mixture of water, steam, and entrained air into the basket where the ground coffee resides. The hot water dissolves the coffee, and it flows into the carafe. The coffeemaker is fitted with a cutoff valve that causes the flow out of the basket to stop if anyone pulled the carafe out while the coffee is still being made.

The housewife asked her husband, a graduate engineer, to determine what was wrong. The housewife's engineer husband examined the problem by first convincing himself that his wife was following directions to make the coffee. He then carefully examined the equipment, especially the cutoff valve. He concluded that somehow the cutoff valve was restricting the liquid flow whenever decaffeinated or flavored coffee was being made. That is, the incoming flow of hot water and steam was greater than the flow out of the valve. This would cause the level in the

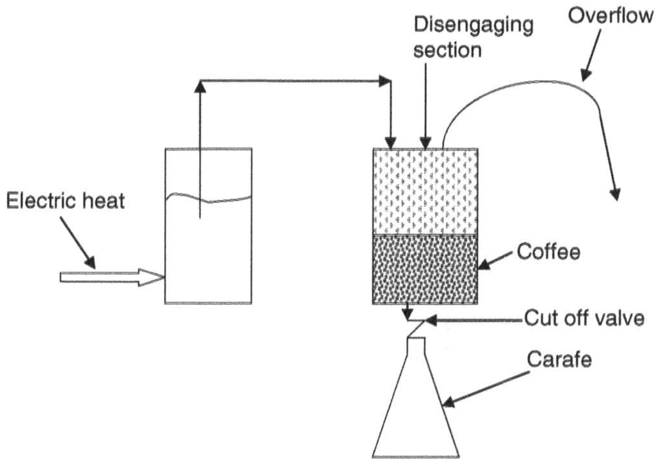

Figure 3.1 Coffee pot schematic.

container to build up and run over. The problem solution seemed relatively simple. He removed the valve and made a sign that said "Do not remove carafe until coffee is finished brewing." He felt a surge of pride in not only solving the problem but also in that he prevented a future problem by providing instructions to prevent someone from pulling out the carafe. The next time that one of the suspect coffees was made, the container did not overflow. He then announced that the problem was solved.

Unfortunately, the glow of successful problem solving did not last long. The next time that flavored coffee was made, the problem reoccurred—that is, the coffee and grounds overflowed the top of the basket container. The engineer then began a more detailed investigation of the problem including understanding the technology for making flavored and decaffeinated coffee. He discovered that when decaffeinated coffee was being made, a surface-active material was utilized. This surface-active material was mixed with the coffee to extract the caffeine. He extrapolated from this and theorized that when flavored coffee was being made, a surface-active material was used to evenly distribute the flavor to the coffee. Once he understood the difference in the coffeemaking processes, he theorized that residual amounts of the surface-active material being left on the coffee reduced the surface tension of the hot water and coffee and allowed it to foam up in the container and out over the sides onto the counter.

Since the amount of the residual surface-active material would vary slightly from batch to batch, it was theorized that only the batches of either flavored or decaffeinated coffee that contained greater than a critical level would cause an overflow. After studying this theory, the engineer decided that the problem solution would

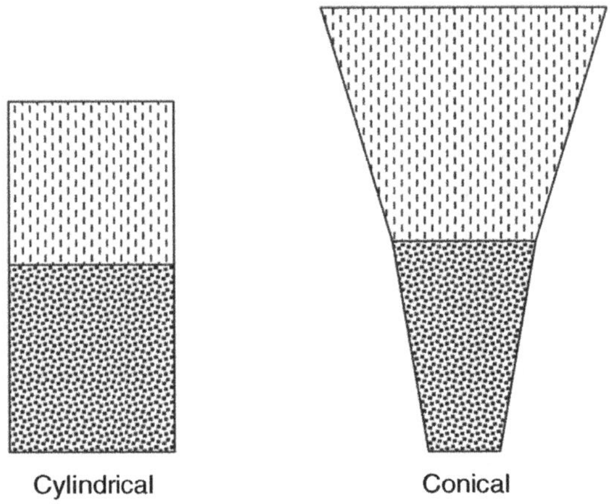

Figure 3.2 Basket comparisons.

be to obtain a coffeemaker that had a basket container with a different design. The problem coffeemaker had a small cylindrical-shaped basket. A new coffeemaker with a large conical design basket was purchased. The comparison of the two baskets is shown in Figure 3.2. It was theorized that the large conical design would provide a reduced upward velocity of the foaming material, and this would allow more disengaging of the vapor trapped in the foam and that this would reduce the probability of the overflow occurring. The purchase of this coffeemaker eliminated the problem completely.

Several lessons can be learned from this problem-solving exercise. The intuitive hunch that coffee was not flowing through the valve as fast as hot water coming into the basket made logical sense. However, no logical explanation was provided for why this only happened with flavored or decaffeinated coffee. Any theory that has the phrase "for some reason" is suspect and is an indication of an incomplete problem analysis. An incomplete problem analysis almost always has a portion of the analysis that is very logical. However, it is imperative that the entire analysis be logical. Another error was that in formulating the hypothesis, the engineer assumed that only liquid water and solid coffee existed in the container. He overlooked the fact that steam vapors and entrained air were always carried into the container with the hot water. The presence of steam and air would provide a mechanism for creating a frothy mixture when the surface-active chemical was present. The example also illustrates the need for the following:

- A systematic approach—a systematic approach as will be described later in this book would have eliminated the incomplete hypothesis that suggested the outlet valve was a restriction on only certain grades of coffee.
- A good understanding of how the equipment works—if the engineer had understood how the coffeemaker worked, he would not have assumed that only a

liquid was present along with the coffee in the container. He would have recognized that both steam and air were carried over into the container along with the hot water.
- A good understanding of the technology involved—the fact that decaffeinated and flavored coffee performed differently than regular coffee should have been an indication to the engineer that he needed to examine the difference in the coffee-making technology.

These relatively simple examples of how successful problem solving requires a more detailed analysis than simple logic and/or intuition are meant to set the stage for the next section that deals with limitations to industrial problem solving. While industrial problems are almost always more complicated than those described in this section, the same problem-solving approach needs are present.

3.4 Applications of Concepts to Plant Problem Solving

While the concepts discussed earlier definitely apply to plant problem solving, one may question whether plant problem solving is engineering. This section touches on the question—"Is problem solving really engineering?" If one defines engineering as dictionaries do—"The science of making practical application of knowledge in any field," we must conclude that problem solving is truly engineering. Often, engineers may conclude that problem-solving is not truly engineering because of the following:

- Engineering is defined in such narrow terms that only "design work" appears to be engineering. Engineers often consider that they are not using any of the material that they learned in college to solve problems.
- Intuition and "gut feel" have replaced thorough analysis as a preferred tool for problem solving. As indicated earlier, there are certain times when this experience-based problem solving is the preferred route. **However**, it does not work for chronic or complicated problem solving.
- Considerations of "optimum technical depth" are not well understood.

The later chapters will consider the engineering approach to problem solving of chronic or complicated industrial problems. These problems illustrate the application of first principles as well as the need to understand both the equipment and the technology being investigated.

It is also important to understand why a course in engineering problem solving is of value. The following two examples are included to illustrate the importance of understanding fundamentals in solving real-world industrial problems caused by poor operation in one case and by improper design in another case.

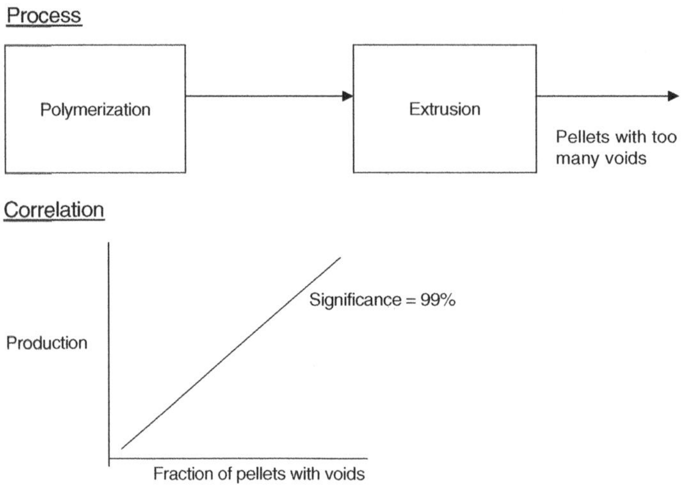

Figure 3.3 An example of improper problem solving.

In a typical industrial problem, a customer is unhappy with the appearance of the plastic pellets being received from his supplier. Specifically, the pellets have visual discontinuities that are described by the customer as "voids." A simplified statement of the problem is shown in Figure 3.3.

As shown in this figure, the process where the pellets are manufactured consists of two sections (polymerization and extrusion). A strong correlation was developed between the pellet appearance (fraction of pellets with voids) and the polymerization production rate. The problem solver recommended that the production rate be reduced to solve the "multi-void" problem. The solution to the problem (reducing production rate) is at best only a short-range solution. Figure 3.3 shows a simplified description of the process and the correlation developed by the problem solver.

This solution cannot be considered a lasting solution because of the following:

- The basic cause of the voids was not considered.
- The solution required a severe economic penalty (it might have worked for one problem, but it created another one). In most process industries, the limited profits are made at production rates above 75% or 80% of capacity.
- Since the basic cause of the voids was not discovered, the problem will likely reoccur even at the reduced production rates as operating changes are made in the polymerization section.

Design errors often require problem solving to provide correct and rapid solutions to quickly improve operations. These improvements require an understanding of both design details and fundamental chemical engineering. A simple illustration of this is shown in Figure 3.4. As shown in this figure, a

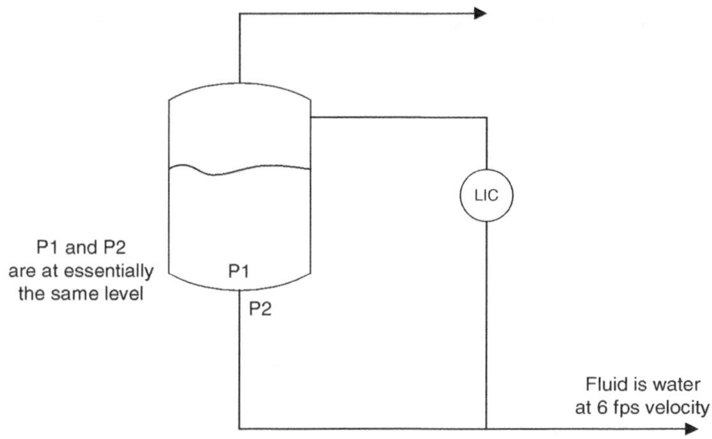

Connecting the level instrument in the process line as shown will result in the measured level reading being 0.5 feet lower than actual.

This is based on Bernoulli's theorem

$$dP/D + dV^2/2g + dZ = 0$$

where

dP = difference in pressure
D = density of liquid
dV^2 = difference in liquid velocities squared
g = gravitational constant
dZ = difference in liquid height

At base level the pressure at the level instrument will be less than the same pressure in the drum as follows:

$$(P2 - P1)/62.4 + (36 - 0)/64.4 = 0$$
$$P1 - P2 = 34.9 \text{ lbs/ft squared}$$

This is equivalent to 0.5 ft in measurement of level this ignores the friction loss in the line and nozzle

Figure 3.4 An example of improper level design.

last-minute addition of a level instrument was designed to use the liquid line as the source of the higher-pressure connection. During the design phase, Bernoulli's theory could have been used to prove the need for providing a connection on the drum. The actual design using the liquid line resulted in the level indication being lower than the actual level. Failure to use the same fundamentals resulted in the problem solver concluding that the incorrect level reading was caused by condensation in the vapor connection and piping. An inspection of the drum indicated that the vapor connection and associated piping were steam traced and insulated. Thus, there could be no condensation in the vapor line.

3.5 Limitations to Problem Solving in Process Plants

The previous examples are typical of much of the improper problem solving that occurs in many industries in today's hectic fast-paced society. It also illustrates why a course in engineering problem solving is of value.

When considering these examples and the need for a more efficient problem-solving approach, one must question: What are the limitations to using this approach? Chapter 2 provided a general list of limitations to achieving and maintaining operational excellence. These four general areas have been expanded and are discussed as they apply to problem solving in the following paragraphs.

- *Limited experience*: In the developed world, the average experience of the workforce is decreasing as more experienced operations, mechanical and technical resources die or retire. In the developing world, the new world of process technology is often replacing the industrial world of low technology associated with piece work. Either of these situations results in a workforce that has minimal experience with the process operations world. This minimal experienced workforce must start with a basic understanding and build experience-based knowhow. For example, take the rather simple case of an experienced operator observing a pump that is not performing and is making an unusual noise will conclude the pump is "vapor locked." An experienced engineer making the same observation will analyze the situation using calculations and the pump curve and conclude that the NPSH (net positive suction head) required is greater than the NPSH available. He might then recommend increasing the liquid level in the drum feeding the pump. Without experience, it is unlikely that either conclusion would be reached and it is likely that the pump would be unnecessarily taken out of service for mechanical repairs.
- *Changing academic focus*: The modern chemical engineering curriculum, while providing an excellent theoretical foundation in some areas, often fails to adequately stress the application of fundamentals. The changing academic focus is necessary due to the widespread use of computers, process design simulation programs, and the opening of the biochemical branch of chemical engineering. However, these additional courses reduce the amount of time available for academic instruction in the basic understanding of how process equipment works. The second factor is that the academic instructors are focused on doing research rather than designing equipment. Thus, an engineering graduate is equipped to pursue an advanced degree, which leads to a research career or a career in process control/computer simulation, but has little knowledge of how basic process equipment works. As a result, he is poorly equipped to provide assistance in a program to develop or maintain operational excellence. This situation often

encourages engineering principles being ignored or inadequately applied and results in a problem-solving mode that might be referred to as "shooting from the hip." This may be completely appropriate in some limited situations. However, the purpose of this book is to address the chronic problem that is only slightly wounded by the "shoot from the hip" technique. This failure to utilize fundamental engineering skills is often combined with the failure to use a methodical structured approach to problem solving. Without these components, the problem solver might quickly do as described earlier when presented with the situation shown in Figure 3.4 and assess the problem as condensation in the vapor side of the level instrument or another erroneous hypothesis. Another possible example of failing to use a methodical approach and understand fundamentals is a fractionation column that is performing poorly. The failure to use a methodical approach could lead one to hypothesize that the tower had a plugged tray, and that was the cause of the high-pressure drop. In fact, the problem might well be associated with a change in internal vapor and liquid loading, buildup of an impurity that boils between the light key and heavy key, foaming caused by a trace impurity, or improper assumptions regarding the tower loading point. This change in focus is often compounded by the fact that operating facilities often do not perform as would be indicated by theoretical calculations. For example, an actual pressure drop measurement may be much higher than calculated. A pressure survey with a pressure gage might show that a plug existed at a single spot in the line.

- *Increasing complexity and size of industrial processes*: Modern day process plants are large and complex. For example, a relatively simple process such as propylene purification has evolved from fractionation, followed by a drying process to remove water to a process incorporating "heat pump fractionation" and more complicated conversion steps to remove impurities to the parts per billion concentrations. In addition, the plant sizes have increased significantly. Thus, there is even more emphasis on solving problems quickly and correctly. Similar trends are present in the technology for the production of other chemicals such as para-xylene and polyolefins. These trends are also true in other areas of the process industry. There does not seem to be any change in these trends. This creates an environment that leads to several confusing factors. One of these is that the problem is usually more complicated than first described. Typical initial problem descriptions might consist of statements such as "It won't work as designed" or "It won't work unless you modify it to ..." If either of these problem descriptions is accepted exactly as stated, the problem solver is likely doomed to failure. In order to practice true problem solving, he must use a disciplined approach that involves **writing out** an accurate description of the problem that **does not** include a problem solution. This is necessary to avoid ignoring data and "jumping to conclusions." Conflicting data are always present and can take many forms. Some examples are verbal description of eyewitnesses can disagree, laboratory data may be in disagreement with physical factors, instrumentation or even other laboratory data, and/or instrumentation/computer data may be in conflict with other sources of data. Variable interaction can be a source of confusion in modern day plants. This results in difficulty in isolating the real problem-causing variable and/or strong correlations between dependent variables. (While this type of strong correlation may be of interest, it rarely results in solving problems.) Besides a high degree of variable interaction, there is also a high degree of interaction between the various engineering disciplines. Thus, what appears to be an obvious mechanical engineering problem often has its true roots in chemistry and/or chemical engineering. The converse is also true. Modern day plants often involve longer residence times because of the incentive to "push the process" to higher efficiency or higher purity. Problem solving with long residence time equipment requires the use of a dynamic model. Unfortunately, when faced with the need for a dynamic model, the problem solver will often take one of two unsatisfactory approaches. He will give up on the basics and say "It's too complicated." Since the dynamic model is truly required to solve the problem, the problem solver now must take an approach that can be characterized as "guesswork." The other extreme is that he will begin the development of an elaborate technically correct model that will rarely be finished in time to be of any assistance. Either of these approaches overlooks the fact that there are ways to build simple technically correct dynamic models. These simple models will contain assumptions. However, the model even with these assumptions will still be of sufficient accuracy to solve most industrial problems.

- *Emphasis on cost reduction*: The ever-present emphasis on cost reduction has resulted in reduced staffing. This reduced staffing along with the reduced level of experience results in excessive amounts of trial-and-error solutions by inexperienced personnel. The existence of this can be found in the number of operating directives/procedures as well as problem solutions that must be rewritten or reworked because the time to do the necessary technical calculations was deemed to be unacceptable. Certain well-known phrases such as "Why is there never time to do it right, but always time to redo it" or "Do something even if it is wrong" are indications that the technical depth is not sufficient. This emphasis on cost reduction and the subsequent reduction in forces leads to a multitude of problems. There might not be time to see the whole picture. There are often verbal clues that this is happening such as—"That's a mechanical problem" or "The laboratory is wrong again." While these statements may be valid, they are often indications that the problem solver is excluding essential pieces of data. It should be noted that someone using the methodical approach is less vulnerable, but still subject to this limitation. A separate issue caused by cost reduction might be an overdependence on history. While a historical database is a mandatory prerequisite for successful problem solving, the database should be used to define deviations rather than a repository of answers. The statement "The last time that this happened, it was due to. ..." must always be tested by data analysis.

As described earlier, Figure 3.3 shows a typical industrial problem. Several of the limitations discussed above are apparent. The problem was certainly complex, in that it could be caused by conditions in either the polymerization or the extrusion area. There appears to be a lack of both a methodical approach and an inadequate application of engineering principles. In addition, while only a limited amount of data is present in Figure 3.3, the problem solution appears to be only historically based. There is no evidence that a hypothesis was developed and tested with a plant test. Was the problem solver seeing the entire picture? For example, was the independent variable polymerization production rate or extrusion rate? Was the independent variable production rate or residence time (the inverse of production rate)? Perhaps, the confusion of the problem solver is illustrated by the figure, which shows the "voids" on the x-axis, which is normally reserved for the independent variable.

In this chapter, we have shown examples of failed problem solving in both home and industrial situations. In addition, we have explored some of the limitations to industrial plant problem solving. For many, the limitations are all too real. They know the limitations, but not the solution. Chapter 4 covers the five steps of successful problem solving in an industrial situation. These steps will be of value for any given environment from chemical/refinery operations to the volunteer organization.

4

Successful Plant Problem Solving

4.1 Introduction

Before beginning a discussion on how one conducts successful engineering problem-solving, perhaps a definition of the activity is appropriate. Engineering problem-solving is defined as the application of *engineering principles* to allow *finding*, *defining*, and *solving* plant-operating problems in an expedient and complete fashion. The *finding* and *defining* phases of problem-solving are often ignored, considered obvious, or unimportant. However, these phases prevent small problems from growing into large problems and allow the problem-solving phases to be done in an expedient fashion. Finding the problem involves sorting through the mass of laboratory and process data to uncover deviations that may only be 1 sigma from normal, but which have the potential to grow into large deviations. Defining the problem involves developing a quantitative description of the problem specifications.

Successful engineering problem solving will always involve the following:

- A *daily* monitoring system.
- A *disciplined* (not intuitive) and *learned* (not inherited) engineering problem-solving approach.
- The ability to distinguish between problems requiring technical problem solving and those only requiring an expedient answer. The ability to determine how detailed a technical analysis should be is also required to efficiently solve plant process problems. This is referred to later as *optimum technical depth*.

4.2 Finding Problems with a Daily Monitoring System

In order to successfully find and define problems, the problem-solver must obtain and maintain a historical database. The database can be maintained by

Problem Solving Approaches for Maintaining Operational Excellence in Process Plants, Second Edition.
Joseph M. Bonem, Nattapong Pongboot, and Wiroon Tanthapanichakoon.
© 2025 the American Institute of Chemical Engineers, Inc. Published 2025 by John Wiley & Sons, Inc.

Table 4.1 Sources of historical data.

Source	Managerial Objective				
	Minimize Routine Work	Maximize Finding Hidden Problems	Maximize Trend Spotting	Maximize Variable Retention	
				Volume	Key[a]
Computer data storage	X			X	
Computer or hand graphs			X		X
Delta data graphs[b]		X	X		X
Communication with hourly workers		X			
Visual observation of field equipment		X			

a) The concept of key variable retention involves retaining the graphs or delta data graphs of only the key variables, whereas volume retention involves a data source that relies on maintaining values of every variable.

b) "Delta Data Graphs" are the difference between actual values and a theoretical or established value. An example of such a plot is shown in Figure 4.1.

using several different sources. The *Managerial Objective* will also be important. The *Managerial Objective* is defined as the goal that management has defined for the particular process. This goal will vary depending on the age of the process, technical staffing of the location, and the value added by the process, to name a few. Table 4.1 shows a grid of both *Managerial Objectives* and sources of data.

As an example in the use of this table, assume that a well-established process is producing a commodity chemical. As a general rule, commodity chemicals have a low value added. That is, the difference between the product revenues and the cost of production is very small. The management might elect to staff so that the technical organization could only respond to established significant problems. Thus, the *Managerial Objective* might be characterized as minimizing routine work and maximizing variable retention. In this case, the number of process variables to be retained would be maximized. As shown in Table 4.1, computer data storage would be the desired source of data to fit this objective. If a problem developed, the problem solver could then go back and use the stored data to attempt to resolve the problem. The problem solver might find this difficult due to the vast amount of data that must be analyzed. In addition, the data sources entitled communication with hourly workers and visual observation of field equipment would likely not be

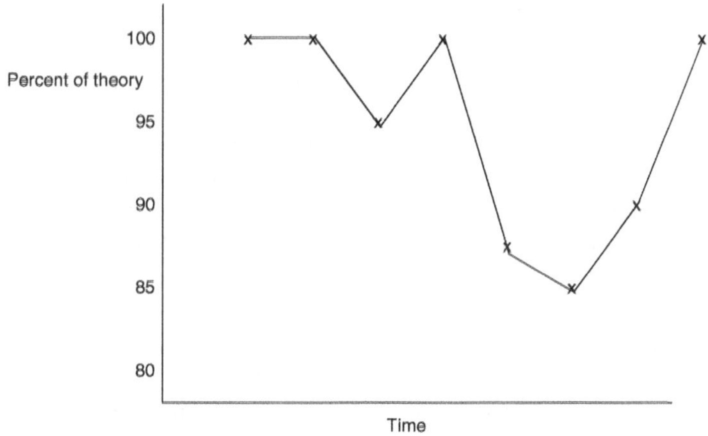

Figure 4.1 Essential variable (reaction kinetics) percent of theory versus time.

available since people's memory might have faded and changes might well have occurred in the field equipment.

On the other hand, if the process being considered is an unproven process and/or is a high value-added process, management might elect the objective of maximize finding of hidden problems. in this case, the problem-solver would use delta data plots, communications with hourly workers, and visual observations of field equipment as their data sources. It is likely that the main source of historical data would be the trend graphs or delta data graphs. Of course, in this case, the computer would still be used to store all process variable data. However, it would not be the primary source of data for the problem-solver. While this objective allows for finding problems quickly, it likely will require more technical staffing.

In the two cases cited above, there are implicit assumptions. In the case where the *Managerial Objective* is Minimizing Routine Work and Maximizing Variable Retention, the implicit assumption is that essentially all process problems that occur can be readily solved without a detailed problem analysis. In the case where the *Managerial Objective* is Maximize Finding of Hidden Problems, the implied assumption is that essentially all problems will require a detailed problem analysis.

If graphs are to be used in any of the cases shown in Table 4.1, they should be drawn, reviewed, and monitored on a daily basis. To monitor the process by preparing these graphs only once a week defeats the purpose of finding problems or spotting trends.

This daily monitoring system should be designed to allow the problem-solver to monitor process variables by incorporating several variables into process models that summarize the operation of each section of the process. Figure 4.1 could be used to express reaction kinetics as a percent of theoretical. Significant deviations from 100% are indicative of process impurities, catalyst contamination, or inaccurate process measurements. Thus, with only a glance at the figure, it is possible to assess the status of the reactor section of the process.

Even a cursory look at Figure 4.1 will raise the question of when does the problem solver declare that a problem has occurred. Is it the first drop in kinetics or the second? For each variable that is graphed, there should be a limit that if the actual value increases above or decreases below will indicate to the problem-solver that it is likely that a problem is occurring. This will be discussed later and is referred to as the concept of a "trigger point."

One source of historical data requiring elaboration is discussions with operating, mechanical, and laboratory hourly personnel. Their observations may be highly subjective but at the same time very meaningful. For example, discussions with laboratory personnel revealed that a standard Millipore filter test used to determine the level of solid contamination in a hydrocarbon caused "fusing" (melting together of the two parts) of the Millipore filter container. This plastic container was known to be inert to the hydrocarbon, and "fusing" had never been encountered before. Based on the laboratory technician's comment that the container was "fusing," an investigation was initiated. This investigation showed that the hydrocarbon was contaminated with methanol. The plastic used in the Millipore apparatus was soluble in methanol. Small amounts of methanol would cause the melting point to be reduced and the "fusing" of the two parts of the container. The determination that the hydrocarbon was contaminated with methanol provided a strong clue for developing a hypothesis for determining the source of the known hydrocarbon contamination.

In this day and time, with the multiple means of "nonpersonal" data acquisition techniques, the communication flow from hourly workers to professionals must be cultivated and nourished primarily by the professional.

The observation of field equipment is done by walking through the process plant and both looking at and listening to the equipment to detect any differences since the last walk-through. For example, a loud noise that appears to be emanating from a process vessel might be indicative of condensation of vapor inside of the drum. A problem-solver on a walk-through may observe a new sample connection, which on closer examination may appear to be installed in such a fashion that it will not give a representative sample. These observations by themselves may not be problems, but they are sources of data that can be considered as other problems are detected. The problem-solver should make notes on anything that seems different. These notes will provide data with a time stamp that can be used for future references.

It is inadequate to only record data and collect observations. The examination of the data can best be done with "trigger points." "Trigger points" are limiting values of either laboratory analyses, instrument readings, or computed variables. If the variable being monitored is outside of these limits, the *successful* problem-solver will declare that a problem exists and begin to solve the problem. It should be emphasized that the *successful* problem-solver will *find and define* the problem well before it becomes a major problem. The finding and defining the problem is the first step toward problem resolution. Resources may not be available to resolve the problem completely; however, management will recognize that a problem has been uncovered. The "trigger point" approach is similar to that used by the medical profession. Medical and laboratory tests such as blood pressure, cholesterol level, and hemoglobin levels are used to spot minor problems before they become major problems.

One of the most important things to recognize in setting trigger points is that there is a difference between using statistics to control a process and using statistics to find problems. Control statistics require that the process be greater than 3 sigma from target before changes are made. This implies that there is a 99+ percent confidence level that there has been a real change in the process, as opposed to process variability. The successful problem-solver cannot wait until he is 99+ percent confident that there is a process problem. For example, very few car owners wait until they are 99+ percent confident that they have an automobile problem before they begin a problem-solving activity.

Trigger points can be set for the following different types of variables:

- *Theoretical/laboratory/pilot plant demonstrated*: Each of these variables would have a "lumped parameter constant" (to be discussed later) that can be calculated from plant data. These constants can then be compared to similar constants demonstrated in the laboratory or pilot plant or that can be developed from theory. Examples of these are reaction rate constants (demonstrated in the laboratory or pilot plant) or fractionation tower tray efficiencies (demonstrated by theoretical calculations).
- *Plant demonstrated*: These include variables that are equipment-related such as production, purity, slurry concentration, or additive controllability. They can only be demonstrated in a commercial size plant where full-scale equipment is utilized.
- *Vendor demonstrated or guaranteed*: These will be almost exclusively equipment items. These variables will include items such as highly specialized valves, volatile removal equipment, or heat exchange equipment.

Statistical techniques can be utilized to set "trigger points." For example, in a process demonstrated to have a catalyst efficiency of 5000 with a 1 sigma of 200, a low "trigger point" of 4900 would be ludicrous. Conversely, a "trigger point" of 4400 would cause many problems to be ignored. Obviously, determining a meaningful standard deviation is mandatory if the "trigger point approach" is to be utilized.

It is highly likely that in an industrial process, the standard deviations of essential variables are not well known. Rather than doing elaborate laboratory statistical studies, a more expedient approach will involve developing approximate standard deviations and allowing the daily process monitoring to help determine the real commercial standard deviation.

In a new process, there is great value in setting the standard deviation on the low side and attempting to explain as many deviations as possible. The tightening of the standard deviation for a new process will cause the maximum number of problems to be uncovered, while the management's attention is focused on getting the new process operational and adequate resources are available to solve problems. The opposite approach of having a large standard deviation will result in an apparent good start-up followed by a multitude of problems 6–12 months later. These problems that occur in 6–12 months were actually present during the start-up as small problems that went undetected due to the large standard deviation being utilized.

While problem-solvers generally think in terms of negative deviations (failure to achieve a target), positive deviations must also be considered. For example, a critical heat exchanger that had a known heat transfer coefficient of 120 BTU/hr-ft^2-°F suddenly began operating with a coefficient of 150. An investigation is warranted to determine what had happened to cause an apparent new baseline. This investigation of positive deviation will often lead to new and improved operating procedures.

While the actual setting of trigger points depends on the process as well as the individual company, Table 4.2 shows some suggested ones. It should be recognized that each of these is based on a statistical approach once a standard deviation has been developed or approximated.

In summary, the key concepts in the use and the defining of trigger points are as follows:

- They should be based on statistics and theoretical values where possible.
- The criterion for declaring that a problem exists is different from the criterion for taking control action.
- The criterion for declaring that a problem exists will be a function of severity of the problem. In addition, the point on the learning curve for a specific process should be considered.
- Positive deviations must always be considered.

Another area that will require attention in a full-scale commercial unit is that some problems can be caused by transient process upsets. An adequate explanation of these upsets will usually require extrapolation to the steady-state condition. For example, an impurity is present for only 30 minutes in the feed to a reactor with 3 hours residence time and causes the conversion to drop 2%.

Table 4.2 Suggested trigger points.

Magnitude of Variable on Profits	Trigger Point; Sigma	Probability of Type 1 Error[a]; Percent
Very significant (or new process)[b]	1	30
Moderate	2	5
Insignificant	3	<1

a) A type 1 error is the probability that a problem would be declared when no problem really existed. For example, if the trigger point criterion is set at 1 sigma, there is a 30% probability that a declared problem is really just a normal fluctuation in the process. *However*, there is a 70% probability that a real problem exists.

b) Items that could cause a very serious upset (plant shutdown) should be evaluated using a *one-sided* test against a criterion of being 60% or less sure that you are right. Thus, a trigger point of only 0.3 sigma could be important.

To determine the seriousness of this feed impurity will require extrapolation to steady-state conditions. The approach to developing a simplified dynamic model is discussed in Chapter 10.

While the "daily monitoring system" has been discussed primarily in process engineering terms, it can also be used for following mechanical equipment. One of the essential areas that can be followed is "mean time between failures." This is the time that a piece of equipment is in service before it fails. Good records will allow determining whether there is any difference in failure history of a piece of mechanical equipment. In many process plants, there is a strong relationship between the process and mechanical equipment. The problem-solver should be careful that he does not exclude events that are occurring in the plant because they are not strictly in his area of training. For example, a decrease in mean time between failures of a mechanical seal may be related to the presence of very small particles in the seal flush fluid. The presence of these particles may be related to some change in process conditions.

The implementation of an effective Daily Monitoring Program can be established using the information discussed above along with the following guidelines.

- Pick 6–10 essential variables and graph (computer or hand) them on a *continuous daily basis* using delta graphs and theoretically determined target values. Combine as many variables as is justified based on theory into single graphs. For example, the graph shown in Figure 4.1 combines variables such as catalyst efficiency, reactor residence time, production rate, reactor temperature, and reactor pressure into one graph.
- Establish positive and negative trigger points for each variable. Compare the actual value to the trigger points on a daily basis.

- Obtain hourly worker comments daily and follow-up on any unusual observations.
- Visually observe equipment in the field at least weekly.
- Store the essential variable plots so that this information can be easily accessed.

4.3 Solving Problems with a Disciplined Learned Problem-Solving Approach

A *Disciplined Learned Problem-Solving Approach* is a technique that allows determining if the problem really occurred and specifying the problem in quantitative terms, as well as resolving the problem accurately and quickly. The approach discussed here *differs significantly* from traditional problem-solving courses. The approach discussed in this book emphasizes using techniques for verifying that the problem really occurred. Many problems presented to the problem-solver are either not real problems or are radically different than first described.

In addition, it emphasizes the need to use engineering principles in formulating a hypothesis to explain the problem. In the void problem described earlier in Chapter 3, the relationship between voids and production rate is an idea or vision. A scientifically correct hypothesis would be developed by exploring the following logic path along with appropriate calculations. The steps in this logic path are as follows:

1) Voids are caused by immiscible volatiles.
2) These volatiles are present due to either or a combination of excessive *immiscible* volatiles in the feed or from a steam leak, poor mass transfer, and/or lack of residence time in the dryer.
3) At this point, additional data could be collected and hypotheses could be developed that explain the data and observations associated with the voids on this specific grade.

The approach in this book also emphasizes that any hypothesis must be confirmed with a plant test, calculations, or by making "directionally correct changes." A successful plant test is one that conclusively proves or disproves the hypothesis. The concept of confirming a hypothesis by making "directionally correct changes" will be discussed later. The approach in the book emphasizes that a problem solution must not create new problems.

The *Disciplined Learned Problem-Solving Approach* consists of the following five steps:

Step 1: *Verify* that the problem actually occurred.
Communications in an operating environment are almost always second or third hand and often highly "garbled." The problem-solver must have a means to reduce the confusion at this point.

Step 2: *Write* out an *accurate* statement of what problem you are trying to *solve.*

Answers to the following questions may be helpful:

- What happened?
- When did it happen?
- Where did it happen?
- What was the magnitude of the problem?
- What else happened at the same time or shortly before?
- What actions are you planning?

Step 3: *Develop* a *theoretically sound* working hypothesis that explains as many specifications of the problem as possible.

Step 4: *Provide* a mechanism to test the hypothesis.

Step 5: *Recommend* remedial action to *eliminate* the problem without creating another problem.

Perhaps the problem verification phase is the most overlooked part of this five-step procedure. The problem of the day often arrives at the problem-solver so jumbled that the best approach is to go direct to the "horse's mouth." For example, by talking to the operator who is having an equipment-related problem or to the laboratory technician who got a "strange" result, the engineer can find out directly and exactly what was observed. He will often find that the real problem is considerably different than described in an e-mail that he received. Problem verification may also take the form of data verification. While this is the subject of a later discussion, it should be noted here that application of engineering principles can often eliminate a problem by determining the alleged problem was only a defective instrument. For example, an engineer sent to investigate the poor operation of a 40-psig steam desuperheater found that the measured steam temperature was below the temperature of 40 psig saturated steam, and yet water did not appear to be present in the steam. Since this is a theoretically impossible situation, he began to investigate the accuracy of the instrumentation. He determined that the steam temperature instrument had been incorrectly calibrated.

The person directly involved with the problem can usually be helpful in the problem specification phase (Step 2). However, his knowledge base will normally not allow him to formulate technically sound hypotheses, although he will normally do so. At this point, it is very important to focus in on the activities of Step 2 (writing an accurate specification of the problem). While this description does not have to be a formal document, *shortcuts or even shorthand to allow getting to the answer quickly will be counterproductive*. The problem statement should be as short as possible while still including pertinent data. There is great value in writing out the problem specification using a structured approach. The structured approach provides a means to uncover gaps in the data. In addition, the writing process forces one to clarify data and thinking.

Table 4.3 Problem specification example.

SHORT TITLE OF PROBLEM ________________________
DESCRIPTION OF EVENT (make sure that Step 2 is utilized to provide a complete problem description)
____
____
____
____
____
____
____
HOW WAS THE PROBLEM DISCOVERED? (Was it by data plotting, operator discussion ...?)
____
____
____
____
PRELIMINARY PROBLEM ASSESSMENT
COST OF PROBLEM (HIGH, MODERATE, OR LOW) ______________________
IS IT AN OPERATING OR TECHNICAL PROBLEM? ___________________
IS THERE AN OBVIOUS IMMEDIATE FIX? ___________________________
IF YES, WHAT IS THE PROBABILITY OF SUCCESS? ___________________
IF NO, WHAT AMOUNT OF EFFORT IS INVOLVED IN PROVIDING A FIX? ____
ARE YOU ACTIVELY WORKING ON THIS PROBLEM? _______________

Table 4.3 shows an example of a problem statement format that could be used. The key part of this format is the problem statement (description of the event). The other parts of this format may or may not be of value depending on the organization needs.

The purpose of this format is to provide a simplified communication tool between the problem-solver and different managerial layers and to provide a format to allow the problem-solver to both state the problem in problem-solving terms and assess the severity and solution difficulty of the problem.

While this form is only presented to serve as an example, there are two important concepts involved in using this or similar forms. The form should be kept as simple as possible. In addition, the tendency of management to review and edit all documents must be avoided. It should be remembered that this form is only a device to advise the management of the status of problem-solving activity in the problem-solvers' realm of responsibility. Bureaucracy or any type of editing will be counterproductive and will often reduce the desire of the problem-solver to use this technique.

The development of a theoretically sound hypothesis (Step 3) to explain the problem is an essential concept to allow industrial problems to be *eliminated.* A cause–effect relationship does little good unless the cause can be eliminated or understood. For example, in the "void problem" discussed earlier, reducing the production rate only masks the problem rather than eliminating the problem. An example of a theoretically sound hypothesis for this problem is as follows:

> "There is a condensate leak causing water to flow from the steam side of the indirect dryer to the polymer side. This water is trapped in the pores of the polymer flakes. It is not removed when it is heated in the extruder because the extruder does not have a vent. As the particle is cooled, the water condenses forming a second phase in the polymer particle. This second phase is what causes the discontinuity in appearance."

This hypothesis must be tested against plant data, but it could explain both the appearance of voids and the sensitivity to rate. As the production rates and heat input requirements are increased, the steam pressure on the dryer would have to increase in order to provide the temperature driving force necessary to provide more heat input. This increase in steam pressure would create more leakage potential.

Chapter 6 provides more information on how hypotheses can be formulated. Development of theoretically correct hypotheses will involve the application of engineering principles as described in the following paragraphs.

Unit operations and/or equipment design calculations can be used to formulate hypotheses associated with pump or compressor motor overloading. For example, changes in the pump or compressor horsepower requirements that occur as the composition changes might be used to determine why a motor overloads. Another example could be the calculation of the amount of the condensate produced from a steam turbine, which then might be used to show that a steam trap was being overloaded, resulting in poor performance of a heat exchanger.

Unsteady-state accumulation calculations that allow for analyzing a process in a dynamic mode could be used for determining how fast propane builds up in a polypropylene process. These calculations could also be used to determine how many displacements of a system are required to achieve a given degree of cleanliness during a transition or start-up operation.

Mass and energy balances can be used for analyzing steady-state or dynamic operations. Examples of the use of these balances for dynamic operations are how hot would the wall of a reactor become if heat transfer failed? Or how long could a process operate without cooling water?

The development of a theoretically correct working hypothesis is mandatory to reduce the unlimited number of hypotheses to a few that make sense. *Problem Solving that is not based on theoretically sound hypotheses will degenerate into unstructured brainstorming*. Unstructured brainstorming quickly becomes a contest to determine who can generate the greatest number of hypotheses (sound or unsound).

As the next step (Step 4) is addressed, the definition of a successful plant test must be considered. A successful hypothesis test is often thought of as one that proves the hypothesis is correct. However, disproving a proposed hypothesis is as valuable as proving one. Therefore, the definition of a successful hypothesis test is a test that either proves or disproves the proposed hypothesis *conclusively*. A failed hypothesis test is simply one that is *inconclusive*. Whether the test proves or disproves the hypothesis, the results of the test must be documented. Even for a test that disproves the hypothesis, documentation is important. This will avoid redoing the test later. This subject is covered in greater detail in Chapters 11 and 14.

The mechanism to test the hypothesis can consist of a plant test of new operating conditions, an increased data collection frequency and/or new data, a series of calculations, or a temporary "mechanical fix."

Regardless of which mechanism is selected to test the hypothesis, a great deal of salesmanship will be required to obtain the necessary cooperation from all parties that are involved. The first meeting where the hypothesis test is proposed may be the first encounter with the individual who originally uncovered the problem. Regardless of whether this is true or not, the carefully prepared problem statement and a statement of the theoretically correct working hypothesis will be very beneficial at this point. These two documents, along with the proposed hypothesis test, will provide an outline of the following:

- What problem are you trying to solve?
- What is the working hypothesis?
- How do you plan to prove/disprove the hypothesis?

The mode to a successful hypothesis test often lies in the hands of the hourly personnel. If the hypothesis is to be demonstrated by a plant test or by any technique that involves the hourly workforce, the need to communicate the goals of the test must not be overlooked. This will also be an opportunity to explain the theoretically correct working hypothesis to the hourly workforce. This pretest communication is an excellent opportunity to teach and train, as well as getting support for the test. A test that fails allegedly because "The operator did not want it to succeed" usually indicates that there was inadequate communication with the operator. The successful problem-solver will always have the paradigm that the hourly workforce wants the test to be successful also. Post-test communication is also of value. Such items as the test results, the conclusions, and future plans will help ensure future positive results.

While a plant test is a typical approach to testing a hypothesis, increased data collection can also be used as a test mode. While this appears to be an obvious statement, the presence of highly specialized techniques for obtaining additional data should be considered. Examples are as follows:

- Temperature measurements using infrared detectors can be used to either supplement and/or confirm existing instrumentation.
- High-speed data acquisition devices can often be of benefit to determine the exact sequence of events.
- A specialty designed venturi flowmeter can be used to detect the presence of two-way flow in a pipe.
- X-ray pictures of equipment can be used to confirm the presence of a plugged downcomer or a damaged fractionation tray.
- Qualitative laboratory tests can be used to confirm the presence of an element that could only be present if an "O-ring" were failing.

The important consideration is to use all the resources available and "think outside the box" to allow the proposed hypothesis to be *conclusively* tested.

While it is similar to a plant test, a temporary mechanical fix can be used to provide a test of a hypothesis. This approach provides a circuitous route to proving/disproving a hypothesis. In cases where a plant test is undesirable or would require an excessive amount of time, a temporary mechanical fix may allow confirming the hypothesis. The problem-solver will use logic prior to the mechanical fix to specify what criteria would be required to demonstrate that the hypothesis was correct. The logic might be stated as "If the hypothesis is true and the proposed mechanical fix is made, the following will be observed __________." If the anticipated results are obtained, the hypothesis is confirmed with some degree of certainty. However, there is always the possibility that the hypothesis was wrong and the mechanical fix, while providing the anticipated results, did so because of a different hypothesis than proposed.

Once a proposed hypothesis has been demonstrated to be true, the problem must now be *eradicated* (Step 5). The three keys to Step 5 (*recommend* remedial action to *eliminate* the problem without creating another problem) are as follows:

- In order to avoid creating another problem with the solution to the initial problem, a thorough potential problem analysis should be conducted. A potential problem analysis is a technique for visualizing what problems might occur if the recommended solution is implemented. Once a potential problem is discovered, consideration can be given to eliminating or ameliorating the problem via preventative or contingency actions. *This analysis should include safety aspects also.*

- Make sure that the problem solution is the simplest one that will work. Keeping things simple is even more important in problem solving than in plant design. Make sure that in attempting to provide a perfect solution, that the solution complexity does not create a trap.
- Make allowances for "follow-up." New operating or maintenance techniques will require a great deal of "nagging," "hand-holding," and "coddling" to keep them from being forgotten or ignored.

4.4 Determining the Optimum Technical Depth

Any discussion of optimum technical depth is meant to apply to those process problems that require serious engineering considerations. There are large numbers of process plant problems that are best solved by intuitive judgment and experience-based know-how, as discussed in Chapter 1. Some of these manifest themselves in emergency and start-up situations. In those situations, there is no question of *optimum technical depth*. Things must be done quickly with little time for introspective analysis.

However, for those problems that require engineering analysis, there will always be a question of the required technical depth. For the purpose of this discussion, *optimum technical depth* can be defined as "the ability to compromise between expediency and thoroughness in order to solve a process problem in a minimum amount of time." This definition is shown schematically in Figure 4.2.

This is an exceptionally difficult area to quantify. It will vary greatly from company to company. Even in the same company, different divisions appear to have different standards. A few definitions are required before proceeding further.

Confidence level is defined as the probability that the recommended problem solution will completely eliminate the problem without creating another problem. There are two different confidence levels to be considered. One is the required confidence level as suggested by the management. The other is the probable confidence level as assessed by the problem-solver. In order to avoid misunderstandings, the probable confidence level should be greater than the required confidence level.

Project execution time is the amount of time required from the time that the problem-solver begins to work on the problem until the problem is solved. This involves time for data collection, data analysis and implementation of operating changes, or installation of mechanical equipment. Obviously, this can be as short as a few days to as long as several months.

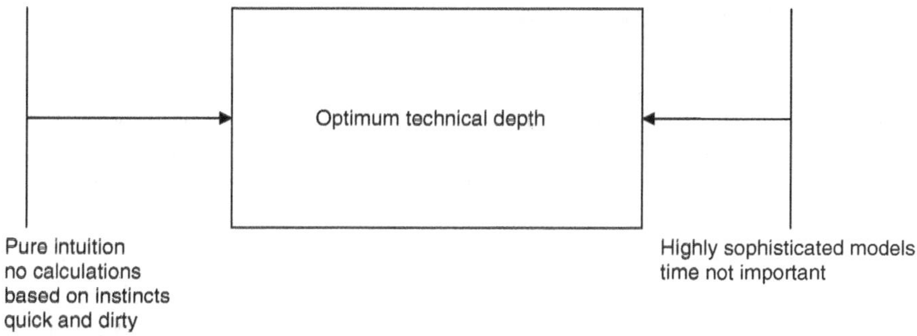

Figure 4.2 Optimum technical depth schematic.

In spite of the difficulty in quantifying optimum technical depth, the problem-solver should give consideration to this variable prior to initiating problem solving efforts. Some suggested guidelines that may help in quantifying the optimum technical depth are provided in the following paragraphs.

The *probable* confidence level that the problem solution is correct is directly proportional to the technical depth involved in the problem-solving activity. For example, a pressure drop calculation that assumes the length of the line is about 200 ft is much less accurate than a calculation based on line measurement and a count of the number of fittings.

The *required* confidence level in an industrial environment is much lower than in an academic or research environment. The courtrooms are filled with examples from the pharmaceutical and medical research fields of alleged inadequate required confidence levels. In an industrial environment, where product liability is not an issue, the daily cost of the process problem often dictates the need for a lower required confidence level. The exception to this is where safety or product liability is involved. In these cases, there is a need to have a high degree of confidence that the urgency to solve a process problem does not create a product liability or safety-related problem.

The *required* confidence level is directly proportional to the cost of the solution and/or the execution time of the solution. Often, the solution to a process problem involves the engineering and construction of additional facilities. This can require a period of 12–48 months depending on the complexity of the design. These facilities can often be very costly. If the chosen problem solution will require additional facilities, the problem-solver should have a great deal of confidence that the revisions will result in a true problem solution. On the other hand, there are problem solutions that require minimal cost and can be installed quickly. These will require a lower degree of confidence prior to installation.

The *required* confidence level is also directly proportional to the cost of the problem. That is, the required confidence levels for solutions to costly problems are higher than those for the less costly problems. In an industrial environment, the costly problems also get the greatest visibility. That is, they get more management attention and as such require a higher degree of confidence in the problem solution. The less costly problems that require a long execution time or involve a large expense for equipment also require a high confidence level in the chosen solution. The less costly problems that can be solved by a quick low-cost fix do not require as high a confidence level.

Unfortunately, the very expensive problems often require a detailed technical analysis. Since they are expensive, they create a lot of pressure to develop a quick fix. Rather than doing the required technical analysis, the problem-solver often submits to the temptation to “try something.” He then finds himself spending some of his limited amount of time implementing the “something” multiple times rather than doing a detailed technical analysis.

Another aspect of assessing the optimum technical depth involves estimating the project execution time. For typical engineering projects, most industrial companies have well-established project execution times. However, for problem solving, these engineering and construction guidelines will likely not be applicable. The problem-solver is the best-equipped person to make an estimate of the work that needs to be done and the amount of time and resources that will be required. Once the number of man-days for completion of the project is known, the manning can be estimated. As a very general rule, if the estimated project execution time exceeds 3 months, it will be desirable to increase the manning so that the execution time can be reduced to 3 months or less.

There is always a minimum degree of confidence that is acceptable regardless of the cost of the problem, cost of project execution, or length of project execution. It would seem that one should be at least 70% confident of the problem solution before it is proposed.

These concepts are illustrated in Figure 4.3. In this figure, the required and probable confidence levels are shown on the *y*-axis. The *x*-axis is the project execution time and/or project cost. Thus, as the project execution time and/or the project cost increase, the probability of success also increases. Two levels of required confidence are shown (80% and 90%). The required confidence level is set by discussions at the start of the project. It should be realized that these discussions may well start off with some generalized statements such as “Just get it right” or “Do something quick.” These can be translated into required confidence levels that provide some idea of what degree of certainty is required before a recommendation is presented. Obviously, these are very subjective evaluations. However, the time that it takes to do them will eliminate future disappointment when recommendations are presented. This effort will also allow the problem-solver to indicate to management whether the project can be accomplished with the required confidence level in the time and cost constraints that have been determined.

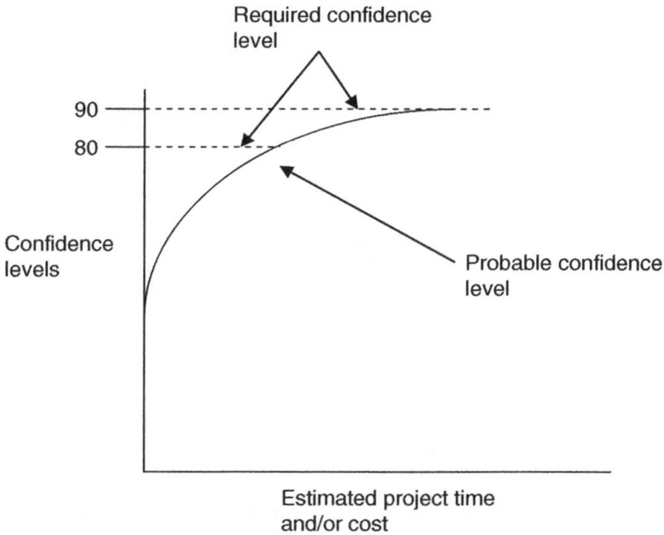

Figure 4.3 Probable and required confidence levels.

As indicated earlier, the goal of any problem-solving exercise is to obtain a correct solution to the process problem in the minimum amount of time. It may be possible to do this by either

- A detailed analysis that leads to one unique solution with a high probable confidence level that exceeds the required confidence level.
- A multitude of attempts to solve the problem. Each of these attempts will likely have a probable confidence level less than the required confidence level. However, because each of these attempts is done sequentially, the problem will eventually be solved. It should be noted that this concept still requires a technical analysis to confirm that each attempt to solve the problem is a theoretically correct hypothesis. The technical analysis is not a detailed analysis, and therefore each attempted solution has a low probable confidence level.

It is possible that the problem-solver will have to consider both of these execution approaches. Management may indicate that the required resources are not available or that the required execution time is too long for the alternative with a high probable confidence level (detailed analysis). In this situation, the problem-solver can help make an execution plan decision by providing management with his best assessment of the alternative execution strategy. One way of doing this is shown in Figure 4.4. The hypothetical example shown in this figure presents two approaches to solving the same problem. In this example, there is one unique solution to the problem. In one approach, this solution can be reached

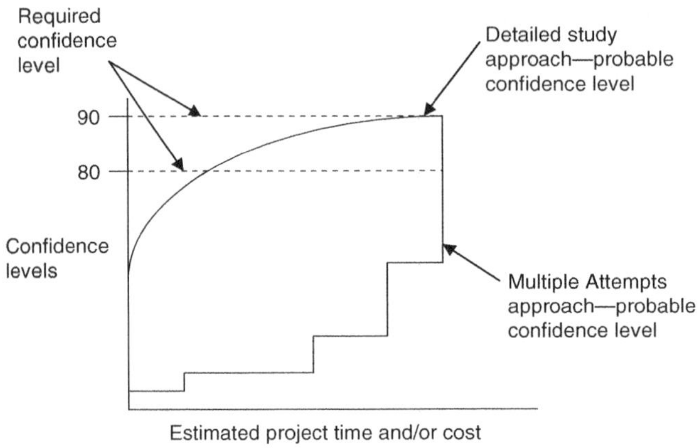

Figure 4.4 Detailed study approach compared to multiple attempts.

with a 90% probable confidence level with detailed study. The *x*-axis is the cost and/or the length of time required to reach this solution. The *y*-axis represents the probable and required confidence levels. In the first approach (detailed study), the required confidence level is reached by a single path. The more detailed the study is, the higher is the probability of success. In the alternative approach that uses multiple attempts to solve the problem, the required confidence level is reached, but only after several failed attempts.

In this example, the times/costs have been adjusted so that the 90% confidence level is reached after the same expenditure of cost and/or time for both approaches. This is not the normal chain of events. Normally, the detailed study will allow reaching the required confidence level in less time and at a lower cost. Figure 4.4 also shows what happens in the hypothetical example if the required confidence level is reduced to 80%. In this case, the detailed analysis will allow reaching this level faster and/or at a lower cost. This is more typical of industrial problem solving. It is realized that this is a highly theoretical example, but it is given to illustrate the thinking that should be done in assessing the best approach to reach a final successful problem solution in a minimum amount of time and/or cost.

Once the solution path can be agreed upon, it can be used to steward the progress of the project using the confidence level versus time/cost type of relationship that was developed to determine the project execution approach. For example, in Figure 4.4, if the detailed study route was chosen as the project execution strategy, the problem-solver should feel about 85% confident in the approach when the project is about 50% complete. This is true whether the project is building a mathematical model or installation of new equipment. Progress reports would consist of status of the project, as well as an indication of the probability of success. This accountability will allow for any necessary midcourse corrections.

In the second approach of multiple attempts, the relationship can also be used to steward progress. Management reports in this case will likely consist of reports of failed attempts and number of trials remaining. This reporting of multiple failed trials often leads to loss of management confidence in the problem working process.

While it may seem that the above considerations are not worth the effort required, it should be realized that whether this type of thinking is quantified or not, it is going on as any problem-solving exercise is being considered. The proof that such thinking is present is found in statements such as "We have to try something quickly," "That approach is nice, but it takes too long," or "Let's just put in a treatment bed." When faced with these criticisms, the wise problem-solver will attempt to define a cost/timing and confidence level analysis of the various approaches. The above technique is one approach to try to quantify this analysis.

The approach described here can be criticized as attempting to quantify items that are so very subjective that they cannot be quantified. However, it should be realized that whether these items are quantified or not, they are always resident in the minds of management and the problem-solver. By attempting to quantify such areas as required confidence level, probable confidence level, and execution times, better decisions on approaches and resource allocation will result.

4.5 Using the Directionally Correct Hypothesis Approach

There are times that the problem-solver will be faced with a problem where he believes that there is a hypothesis that appears to be correct, but because of lack of calculation techniques or lack of technology correlations, the hypothesis cannot be proven with calculations. If the problem-solver truly has a "directionally correct hypothesis" in mind, this knowledge may by itself lead to an effective problem solution. This approach is different than the trial and error approach where calculations could be made but are not because of the time or costs involved. This "directionally correct hypothesis" approach assumes that if one can make a small and low-cost (in either time or money) change to an independent variable, the impact of this change will by itself either prove or disprove the working hypothesis. The change must be small enough so that no other parts of the process are impacted, but large enough to be confident that the impact if any on the dependent variable is statistically significant. As a general rule, there must be at least a 75% confidence level that the independent variable does affect the dependent variable. The sources of these directionally correct hypotheses are likely to be experience-based. The problem-solver is likely to have experience with a similar process or similar piece of equipment. In his experience, a problem was solved by a change in an independent variable. He thinks that a similar change will solve the current problem. The hypothesis could be tested by making small and low-cost changes in the process. Since this small change would be expected to make only a small change in the dependent variable, it would be necessary for the process to continue operating in the new mode until sufficient data were accumulated to statistically prove or disprove the theory. It should be noted that this concept still requires a technical analysis and a theoretically correct hypothesis.

An example of this approach is an engineer that has previous experience indicating that there should be a direct correlation between a quality attribute of a product and a plant operating variable. His experience is based on a similar but not identical process. However, in the existing process, there is a large amount of scatter in the correlation of the product attribute and the plant operating variable such that there is only a 60–70% confidence level that there is any correlation at all. Increasing the plant operating variable is the "directionally correct" approach to increase the product attribute. As a test, the plant operating variable can be increased a small amount to determine if there is any change in the product attribute. Over an extended period of time, sufficient data will be collected to show whether or not there is a statistically significant difference in the quality attribute.

A similar approach to this is the need for an efficient means to test hypotheses that have been developed by the detailed analysis methods discussed earlier. It will often be desirable to test hypotheses in a low-cost fashion with minimal disruptions to plant operations. In a case study to be discussed later, two alternative approaches were available to test a hypothesis. One of the approaches would require 215 days at reduced operating conditions to provide a 90% confidence level that the hypothesis was correct. The alternative approach required no reduction in operating conditions, could be implemented immediately, and was very inexpensive. Thus, it met the criteria of being low cost and having a minimal impact on the process. It still required 215 days to provide a 90% confidence level that the hypothesis was correct. However, during these 215 days, no reduction in operating conditions was required.

5

Examples of Plant Problem Solving

5.1 Industrial Examples

In an industrial environment where the emphasis is usually on increased productivity, doubts about the validity of this technique will always be present. Typical questions are as follows:

- Does this technique really work?
- On what kind of problems can it be used for?
- Is it really possible in an industrial environment to use engineering calculations as opposed to intuitive problem solving?

In an attempt to answer these questions, the following examples are presented. The three examples are all actual real-life examples. All of the examples come from the polymer industry. Polymer manufacturing problems are often the most difficult to solve, and one of the author's primary experiences is in this area. Two of the examples were solved successfully. The first example requires only process engineering skills, and the problem solution emphasizes the need for a *daily monitoring system*. The second example requires both minimal knowledge of statistics and mechanical engineering as well as process engineering skills. The last example is presented to illustrate the problem of inadequate intuitive problem solving.

5.2 Polymerization Reactor Example

At 02:00 hours on April 2, one of the six continuous polymerization reactors experienced a temperature runaway. That is, the reactor temperature increased exponentially from a normal temperature of 150–175 °F in a 30-minute period. Polymerization is an exothermic reaction that generates a significant amount of heat for each pound of polymer produced. The heat of reaction is removed by

Problem Solving Approaches for Maintaining Operational Excellence in Process Plants, Second Edition.
Joseph M. Bonem, Nattapong Pongboot, and Wiroon Tanthapanichakoon.
© 2025 the American Institute of Chemical Engineers, Inc. Published 2025 by John Wiley & Sons, Inc.

circulating cooling water. Polymerization reaction rates generally double with every 20 °F increase in temperature. When the reactor in question reached 175 °F, the reaction was terminated by injection of a quench agent. All the other reactors were operating normally.

The temperature control system on the reactor was such that an increase in temperature caused an immediate increase in the cooling water supply flow. It was known that a small increase in the catalyst rate occurred right before the temperature began increasing. However in the past, catalyst rate increases of this magnitude only resulted in a slight temperature increase. Following this slight increase, the reactor temperature very quickly returned to normal as the cooling water control system responded. The heat exchanger that is used to remove the heat of polymerization is periodically removed for cleaning. On April 1, the exchanger seemed to be "ok."

A simplified sketch of the equipment and various data is shown in Figure 5.1.

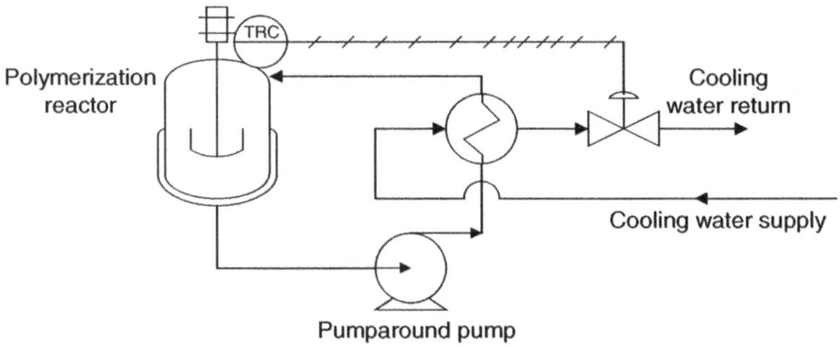

Data midnight

Temperatures

Cooling water
In 90
Out 120

Pumparound liquid
In 150
Out 142

Flow rates, pph

Cooling water 195,000
Reactor slurry pumparound 2,000,000

Other information

The valve on the cooling water is 95% open.

Technology information

Reaction heat generated = $K\,e$ ^(−11000/T)

where K is a constant that contains monomer concentration, catalyst concentration, reactor volume and heat of reaction.

$K = 3.9(10^{\wedge 14})$,
T is in Rankin

The specific heat of the reaction fluid = 0.5 BTU/lb-F.

Figure 5.1 Reactor schematic.

At this point, the problem solver is faced with at least three questions:

1) What should be done to return the reactor back to service?
2) What caused the episode?
3) What can be done to prevent it from recurring in the future?

The first of these questions can be handled by a combination of good operating practices (clean out the reactor) and intuitive problem solving (the exchanger should be cleaned). However, the last two can best be approached by the problem-solving techniques discussed in the previous chapters.

5.3 Application of the Disciplined Problem-Solving Approach

Step 1: *Verify* That the Problem Actually Occurred
While on first glance there may not seem to be a need to do this step, the problem solver made a cursory review of all variables to confirm that the reaction really was terminated due to a "temperature runaway." He found that all temperature instruments indicated an increase in temperature. In addition, the pressure on the reactor also increased.

Step 2: *Write* Out an *Accurate* Statement of What Problem You Are Trying to *Solve*
In this example, the problem that should be solved was twofold – what caused the episode? In addition, what can be done to prevent it from recurring in the future? The problem solver developed the following problem statement.

"Temperature control was lost in the polymerization operation on April 2. This loss of control occurred at about 0200 following a very small increase in the reactor temperature caused by a slight increase in catalyst flow. This loss of control occurred on only one of six reactors all operating at the same charge rate on the same feedstock. The reactor had to be removed from service and cleaned prior to restarting polymerization. There was no mechanical or utility failure on the reactor in question. The weather turned slightly warmer on March 30. Once the reactor temperature began increasing it rose exponentially from 150 °F to 175 °F in an extended period (30 minutes). Determine what caused this loss of control, and once the cause has been determined, develop recommendations to prevent this problem from recurring."

Step 3: Develop a *Theoretically Sound* Working Hypothesis That Explains the Problem
Several possible hypotheses could be proposed, and the problem statement could eliminate all but one, as shown in Table 5.1.

Table 5.1 Hypothesis conclusions.

Hypothesis	Why It Can Be Eliminated
Recirculation pump stopped	"No mechanical failure"
Pumparound exchanger plugged	"No mechanical failure"
Cooling water supply lost	"No utility failure"
Catalyst activated by feedstock	"Only occurred on one reactor"
Heat generated > heat removal capability	Not eliminated

Thus, a theoretically sound working hypothesis developed by the problem solver was as follows:

"The temperature runaway was caused by the fact that the rate that heat generation increased with temperature was greater than the rate that heat removal increased with temperature."

Or mathematically:

$$dQ_g/dT > dQ_r/dT \tag{5.1}$$

where

$dQ_g/dT =$ the rate that heat generation increases with temperature

$dQ_r/dT =$ the rate that heat removal increases with temperature

This working hypothesis would predict a loss of temperature control. In addition, since the rate of reaction increased with temperature, this hypothesis would also predict an exponential increase in temperature.

Step 4: Provide a Mechanism to Test the Hypothesis

While testing a hypothesis often involves experimental work, using fundamentally correct engineering calculations can also test hypotheses. In this case, experimental work would involve the risk of another loss of reactor temperature control. Thus, the problem solver used engineering calculations as the best approach to testing the hypothesis. These calculations are shown below:

Hypothesis

$$dQ_g/dT > dQ_r/dT \tag{5.1}$$

Engineering calculations

$$Q_g = K^* e^{(-11{,}000/T)} \tag{5.2}$$

Equation (5.2) is a typical Arrhenius equation for polymerization. Note that the gas constant, R, has been incorporated into the numerical exponent constant.

In equation (5.2):

K = a constant that depends on monomer and catalyst concentrations, reactor volume, and heat of polymerization. A typical value for this specific process and operating conditions is $3.9(10^{14})$

T = absolute temperature, °R

Equation (5.2) can be differentiated with respect to the absolute temperature, T, to give the rate that heat generated increases with respect to temperature, as shown in equation (5.3).

$$dQ_g/dT = (K^*11,000/T^2)^* e^{(-11,000/T)} \tag{5.3}$$

The rate that heat is removed from the reactor can be represented by a typical heat removal equation shown in equation (5.4).

$$Q_r = U^*A^* \ln \Delta T \tag{5.4}$$

where

U = the exchanger heat transfer coefficient

A = the exchanger surface area

$\mathrm{Ln}\Delta T$ = the log temperature difference between the polymerization slurry side and the cooling water side

Since as noted in Figure 5.1, the cooling water flow is almost at a maximum (valve is 95% open), the average water temperature will not decrease. Therefore, as a first approximation:

$$\mathrm{Ln}\,\Delta T = (T - T_w) \tag{5.5}$$

$$Q_r = U^*A^*(T - T_w) \tag{5.6}$$

where

$(T - T_w)$ = the difference between the average reactor temperature and the average cooling water temperature

In the initial few minutes, the average cooling water temperature will remain constant. Thus, differentiation of equation (5.6) gives equation (5.7).

$$dQ_r/dT = U^*A \tag{5.7}$$

By substituting actual values into equation (5.3), equations (5.8) and (5.9) can be developed.

$$dQ_g/dT = 3.9(10^{14})^*(11,000/(610)^2)^*(e^{(-11,000/610)}) \tag{5.8}$$

$$dQ_g/dT = 170,000\ \mathrm{BTU/hr\text{-}^\circ F} \tag{5.9}$$

U^*A can be estimated from the midnight values shown in Figure 5.1 using equation (5.10).

$$U^*A = Q_r/\ln \Delta T \tag{5.10}$$

where

$$Q_r = 5.75(10^6)\ \text{BTU/hr}$$
$$\text{Ln}\Delta T = 40$$

Therefore,

$$dQ_r/dT = U^*A = 144,000\ \text{BTU/hr-}^\circ\text{F} \tag{5.11}$$

As indicated earlier, the hypothesis was that—"The temperature runaway was caused by the fact that the rate that heat generation increased with temperature was greater than the rate that heat removal increased with temperature" or mathematically $dQ_g/dT > dQ_r/dT$. Since the calculated value of dQ_g/dT (170,000 BTU/hr-°F) exceeds the calculated value of dQ_r/dT (144,000 BTU/hr-°F), the hypothesis was proved with calculations.

Step 5: Recommend Remedial Action to Eliminate the Problem without Creating Another Problem

The required remedial action developed by the problem solver consisted of providing operating procedures to ensure that the rate of heat removal always increases faster than the rate of heat generated. Mathematically, this can be expressed as follows:

$$dQ_r/dT > dQ_g/dT \tag{5.1}$$

To be conservative, a 10–20% safety factor should be included. Thus,

$$dQ_r/dT > 1.1^*dQ_g/dT \tag{5.12}$$

From equation (5.9) $dQ_g/dT = 170{,}000$ BTU/hr-°F.
Thus,

$$dQ_r/dT > 187,000\ \text{BTU/hr-}^\circ\text{F} \tag{5.13}$$

$$\text{or }\ UA > 187,000\ \text{BTU/hr-}^\circ\text{F} \tag{5.14}$$

Therefore, to prevent future occurrences, he specified that the exchanger should be removed from service whenever the "*UA*" drops below 187,000 BTU/hr-°F.

Since "*UA*" could be easily calculated (see equation (5.10)), it became one of the "key variables" that was plotted and monitored on a daily basis. This could be done using the plant process control computer or by hand-plotting.

Some may question the need to actually calculate a "*UA*" value since the narrative indicates that the cooling water was close to a maximum flow rate. While this fact should have been a "red flag" warning to both operations and technical personnel, there is value in being as precise as possible.

It should also be noted that the calculated value of dQ_g/dT depends on both reaction rate and reaction temperature. If the reaction rate increases (larger value of "K") or the reactor temperature decreases (larger slope of rate versus temperature relationship), the value of dQ_g/dT will increase. This will cause the minimum value of "UA" to increase.

A potential problem analysis might reveal that the main potential problem was the degree of conservativeness used to evaluate the heat removal capacity required. A study of the variability of the rate of reaction would reveal whether the 10–20% above the dQ_g/dT factor was sufficient. The proposed solution is certainly a simple solution. However, follow-up will be difficult because it will involve requiring that operations remove a "perfectly good" exchanger from service for cleaning to avoid future episodes of temperature runaways.

5.4 Lessons Learned

The value of being as quantitative as possible is actually twofold. The daily monitoring of a numerical value allows for planning of an exchanger downtime for cleaning as opposed to an "unplanned cleaning," which will always occur at the most inopportune time. If the heat transfer capability were followed on a daily basis in a numerical fashion, the exchanger could be removed from service for cleaning during periods of line downtimes for other mechanical reasons or for downtimes associated with a reduction in sales volumes. In addition, the subjective observation that the cooling water flow was close to a maximum may depend on climatic conditions. These can change rapidly. Thus, what appeared to be a situation where the exchanger had "plenty of capacity" changed quickly as the ambient temperature changed. If the value of the heat transfer coefficient or a comparable value had been calculated, there would be a minimal effect of climatic conditions.

Frequently, in situations like this, the problem solver that is under time pressures will often just participate in doing whatever is necessary to get the equipment back into service. The question of what should be done to prevent the same or a similar problem from happening in the future is not considered. In this particular case, as the problem solver investigated the problem in detail, they uncovered a new area and a new technique that would prevent future temperature runaways.

While the approach that the problem solver used allowed solving the problem and developing a system to prevent future problems, it would have been better to have a more methodical approach to developing a theoretically correct working hypothesis (Step 3). The approach to developing this working hypothesis can be enhanced by a list of questions that will stimulate theoretically correct creative thinking. This list of questions will be given in Chapter 6.

5.5 Multiple Engineering Disciplines Example

This example illustrates the value of a disciplined problem-solving approach when dealing with people or organizations that appear to have fixed positions based on sound logic, but inadequate data. In addition, it also illustrates the advantages for making simple changes to test hypotheses.

A process plant using a rotary filter shown in Figure 5.2 was plagued by excessive downtimes caused by tears of the screen cloth. The fine mesh screen cloth covered the rotating metal drum. It was held in place against the metal drum by retainers. These excessive downtimes occurred on only one out of three rotary filters. These filters were thought to be operating under essentially the same conditions.

Whenever the screen cloth would tear, solids would enter the liquid stream, causing a shutdown of critical equipment and a resulting shutdown of the plant. After each screen cloth tear, the screen cloth and rotating metal filter drum were carefully examined. The examinations showed that the metal drum would be scratched. There was not any apparent reason for the scratches. That is, there was no residual that could have caused the scratches. Solids would be present between the cloth and the drum. This was not surprising since the cloth was torn, and it was known that solids had passed into the filtrate. The cloth would be torn

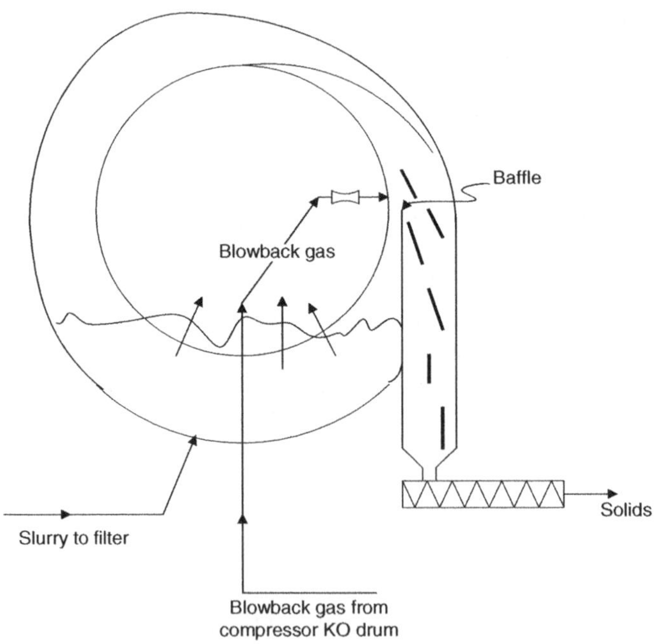

Figure 5.2 Rotary filter schematic.

in a circumferential manner, with most of the tears and drum scratches occurring in the middle 60–70% of the rotating drum.

Even after this careful observation of the filter, no consensus conclusions were reached concerning the failure. In fact, several heated arguments developed with several fixed positions being taken. The mechanical engineers believed that the hard solid polymer particles were cutting the cloth. They believed that these polymer particles were so small that they leaked through the cloth and around the cloth retaining facilities. The process engineer believed that some hard metallic part of the filter was rubbing against the cloth and the metal drum. This would cause the cutting and failure of the screen cloth, letting large amounts of solids into the filtrate, as well as scratching the metal drum. He thought that the baffle, which isolated the solids from the filtrate, might be the part of the filter that was rubbing against the cloth and drum. However, he had no explanation for how this might happen since there was acceptable clearance between the baffle and rotating drum.

Since there were people in the research organization who were experienced in this process, they were also called for assistance. They believed that there was liquid in the blowback gas and that this liquid was cutting the cloth. This would allow solids to enter the filtrate and also to scratch the rotating metal drum.

Faced with such a diversity of opinions and minimal data, the problem solver approached the problem using the five-step approach discussed earlier. He made a decision to obtain as much data as possible from all sources.

5.6 Application of Disciplined Problem-Solving Approach

Step 1: Verify That the Problem Actually Occurred

In this example, there was no doubt that the problem actually occurred. However, there was considerable question whether the problem was worse than in previous years. That is, problem verification consisted of considering if there had been a change in the frequency of screen cloth tears.

A review of mechanical records indicated the following, as shown in Table 5.2. Obviously, a problem existed. It should be noted that without detailed mechanical records (daily monitoring), quantifying the extent of the problem would have been impossible.

A further review of what changed between the past and current data revealed that the filtration temperature on this filter was increased from 130 to 170 °F. There was a significant advantage to operating at the higher filtration temperature so that returning to the previous process conditions was not a satisfactory solution to the problem.

Table 5.2 Mechanical history.

Time Period	Mean Time Between Failures, Days	Type of Tear
Past data	43	Horizontal along the tension rods that held the cloth in place
Current data (all runs)	16	Circumferential
Current data excluding the very short runs	25	Circumferential

Step 2: *Write* Out an *Accurate* Statement of What Problem You Are Trying to *Solve*

The following problem statement was written by the problem solver:

"There has been a significant increase in screen tearing frequency that occurred on only one filter of three filters. This increase appeared to occur at the same time as the filtration temperature was raised. In addition, to a reduction in mean time between screen failures (increased frequency), the nature of the screen failure changed. Previous failures were fatigue failure caused by the cloth being weakened during flexing while being held in place by the tension rods. The current failure is a catastrophic circumferential failure. The current failure is also characterized by scratch marks on the metal drum. Determine the cause for the significant change in screen tearing frequency. In addition, recommendations should be made for what changes are necessary to eliminate this problem."

Step 3: Develop a *Theoretically Sound* Working Hypothesis That Explains the Problem

Since the new failure mode appears to be related to the increase in filtration temperature, the following three hypotheses were developed:

1) The screen cloth is decomposing at the higher temperatures.
2) The baffle (see Figure 5.2 and Figure 5.3) is expanding due to thermal growth and bowing into the filter cloth and metal drum.
3) The rotating drum is deforming at the higher temperatures, causing poor distribution of blowback gas. The poor distribution causes an increase in blowback gas in the middle of the drum, which then blows the filter cloth into the baffle, causing the cloth to tear.

Of these hypotheses, only the single hypothesis of baffle expansion could account for *both* screen cloth tears and scratches on the metal drum. The baffle position required to cause the observed failures is shown in Figure 5.3.

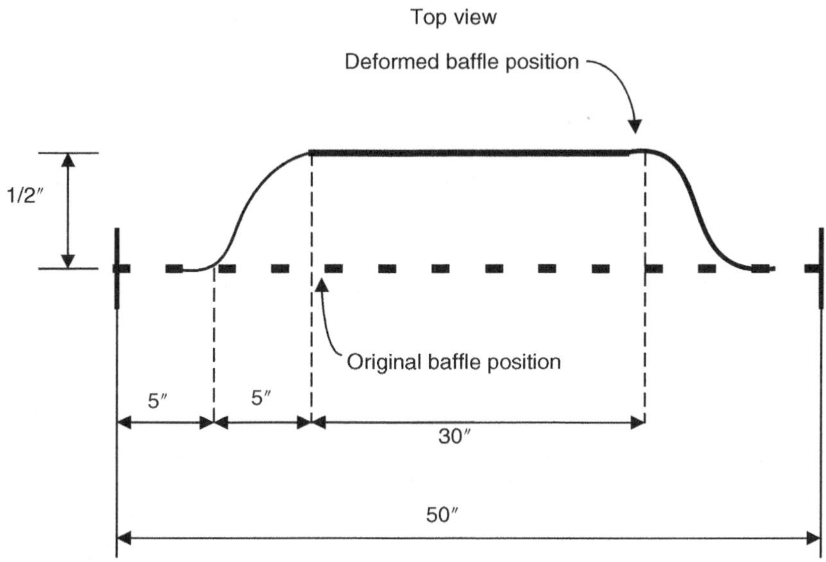

Figure 5.3 Hypothetical baffle deformation.

Step 4: Provide a Mechanism to Test the Hypothesis

This hypothesis was tested by calculations of the thermal growth of the baffle as shown below:

Given:

Original baffle length = 50 in.

Original distance between the baffle and rotary drum = 1/2 in.

Coefficient of linear expansion = 0.000011 in./in.-°F

A typical relationship relating length to temperature is as follows:

$$I_t - I_o = I_o{}^{*}.000011^{*}dt \tag{5.15}$$

where

l_t = the baffle length at the new temperature

l_o = the original baffle length

dt = the change in temperature, °F

Considering the baffle shape shown in Figure 5.3, the baffle would only have to grow 0.05″ to cause it to bow into the rotating drum. Thus, the new baffle length would be 50.05″. The increase in temperature that would cause this amount of growth was calculated using equation (5.15), as shown below in equation (5.16).

$$dt = 0.05/(50^{*}.000011) = 90^{\circ}\mathrm{F} \tag{5.16}$$

As indicated by these calculations, the baffle would be expected to grow sufficiently to expand into the metal drum and screen if the differences between the filtration temperature and ambient temperature exceeded 90 °F. Thus, an increase in filtration temperature from 130 to 170 °F significantly increased the probability of the baffle bowing into the drum. The fact that the baffle could also bow away from the drum without any particular consequences explained why failures did not always occur when the temperature difference approached 90 °F.

Two alternatives were available for further testing this hypothesis. The filtration temperature could be reduced back to the level of previous operations. A second possibility was that a mechanical constraint could be provided to cause the baffle to always bow away from the drum.

Since hot filtration was desirable for the process, reducing the filtration temperature would only be permissible for testing. In addition, the testing period would have to provide a high degree of confidence that the problem was caused by the higher filtration temperature, while involving a minimum amount of time.

An analysis was made to determine the minimum amount of time required to give a 90% confidence level that returning to the lower filtration temperature could eliminate the problem. The basic statistical data developed for this analysis are shown in Table 5.3.

In addition, the statistical approach used by the problem solver to determine how to proceed with Step 4 is defined in the following paragraphs.

If the filtration temperature is returned to the lower value, the mean time between failures will increase to the previous value (43 ± 26). It will be assumed that the values of the mean and standard deviation will be the same as "before hot filtration." The experimental test of returning to the lower filtration temperature needs to be accomplished in the minimum amount of time (minimum number of experimental runs). The large values of the standard deviation

Table 5.3 Statistical data.

Period	Mean Time Between Failures, Days	Runs
Before hot filtration	43 ± 26	31
After hot filtration	16 ± 10	26

(26 and 10) mean that more than a single test run will be required. Thus, the minimum number of runs to prove that there has been a significant statistical improvement from "hot filtration" at the 90% confidence level needs to be determined. The minimum number of runs can be determined by using a statistical comparison of the means of hot filtration and of the results after the process is returned to the lower-temperature filtration. When statistics are used to compare two values of means, a two-sided test is involved. This comparison of means will produce a numeric value such as "$A \pm B$," where "A" is the difference in the means and "B" will depend on the standard deviations and the actual number of runs to be used in the statistical test. To conclude that there is a real difference between hot filtration and cold filtration in the experiment, both possible values of the term "$A \pm B$" must be positive. Since the value of "B" decreases as the number of experimental runs increases, the minimum number of experimental runs will be the number that produces a positive value for both sides of the statistical test. An iterative procedure is required to develop the final answer. The final iteration is shown below:

Assumed number of runs at reduced temperature = 5

$$\mathrm{SE} = \left(s_1^2/n_1 + s_2^2/n_2\right)^{1/2} = (100/26 + 676/5)^{1/2} = 11.79 \tag{5.17}$$

$$1/\varphi = (1/\varphi_1)^* \left(s_1^2/n_1/\left(s_1^2/n_1 + s_2^2/n_2\right)\right)^2 + (1/\varphi_2) {}^* \left(s_2^2/n_2/\left(s_1^2/n_1 + s_2^2/n_2\right)\right)^2 \tag{5.18}$$

$$1/\varphi = (1/26)^*(3.85/139)^2 + (1/5)^*(135.2/139)^2 \approx 0.2 \tag{5.19}$$

where

SE = standard error for comparing the two means

s = standard deviation of the two samples

n = the number of runs in each sample

φ = the degrees of freedom in each sample (number of runs)

From statistical tables for "t" distribution (two-sided at 90% confidence level):

$$\mu = 2.01$$

$$\begin{aligned} \therefore \text{Difference in mean time between failures} \\ &= 43 - 16 \pm 2.01^* 11.79 \\ &= 3 \text{ to } 50 \text{ days} \end{aligned} \tag{5.20}$$

When only four runs at the lower temperature are assumed, the calculated difference in mean time between failures is −1 to 55. Thus, four runs are not sufficient to provide conclusively significant data. However, when assuming that there are five runs at the lower temperature, both sides of the statistical test (3 and 50) are positive. This represents the minimum number of runs that are required. At this point, the problem solver could say with a 90% confidence that returning to the lower-temperature filtration would eliminate the high frequency of screen tears. However, it would be difficult to accurately define the anticipated advantage for returning to the lower temperature due to the large standard deviation. More experimental runs would be required to narrow the range of anticipated benefits.

An estimated period of time for these five runs would be 5*43 or 215 days. Thus, after 215 days at the lower-temperature filtration (90% confidence level), the problem solver could say that returning to the previous temperature conditions will return the average screen cloth life to 43 days. *However*, this does not conclusively prove or disprove the working hypothesis. It only proves or disproves the effect of filtration temperature on the average screen cloth life. There might be another possible hypothesis that explains the problem. In addition, the test does not lead to any acceptable problem solution since it was desirable to operate at hot filtration.

The other alternative testing procedure (adding a mechanical constraint to ensure that the baffle always bowed away from the drum) was easy to do and provided a good mechanism to test the "baffle bowing" hypothesis, while allowing continued operation at higher filtration temperatures. However, it involved "political risks" since the addition of the mechanical constraint had "never been done this way before."

Step 5: Recommend Remedial Action to Eliminate the Problem without Creating Another Problem

Selecting the remedial action in this example was a strong function of how the hypothesis was tested. If the process conditions were modified to allow returning to the lower-temperature filtration for 215 days, there would be a tendency to recommend staying at the lower-temperature operation as a problem solution. Note that since this was undesirable from process considerations, it would not be an acceptable recommendation.

The alternative technique of mechanically constraining the baffle so that it always bowed away from the drum would provide both a testing procedure and a permanent solution. Thus, after 215 days, Steps 4 and 5 could both be considered complete. This was the alternative that the problem solver recommended for the test and for the permanent solution. The potential problem analysis focused on how to make sure that sufficient tension was applied to the mechanical constraint to ensure that the baffle bowed away from the filter.

5.7 Lessons Learned

As indicated in the problem description, the initial thinking was that all of the filters were operating at essentially the same conditions. Thinking of this kind is almost always present in any problem-solving activity. It is only when one dedicates sufficient time to analyze data that it will almost always be found that the initial reaction of "essentially at the same conditions" or "no process changes were made" is found to be incorrect.

Oftentimes, minor mechanical changes such as adding the baffle brace will provide simple solutions to complex problems. However in this case, a potential problem analysis of the proposed remedial action missed the possibility that the mechanical constraint might fall off the baffle due to corrosion, vibration, or metal fatigue. If this happened, the device would likely go with the polymer. This indeed happened and caused failure of downstream equipment and some contamination of the polymer with metal. If this problem had been uncovered in a potential problem analysis, preventative action could have consisted of using a "backup nut" on the constraining device and/or ensuring that the bolt and clamp were made out of corrosion-resistant materials. This illustrates the importance of potential problem analyses. Often, the problem solver is so intent on moving into the execution phase of a problem solution that he does not give adequate consideration to potential problem analyses. This phase deserves as much attention as developing a problem solution.

Besides the engineering advantage for using the disciplined problem-solving approach in this example, there is also a psychological advantage. Once people take a fixed position on any subject, it is almost impossible to change their mind without sound data. Often, it takes more sound data to change a person's opinion than it would have if the data had been obtained prior to the development of a working hypothesis. Any of the initial positions described by the different engineering disciplines could be *partially* supported with logic. However, it was only after the problem solver uncovered as much data as possible and developed a hypothesis based on these data that a theoretically correct working hypothesis was developed. Normally, the application of the proposed approach will significantly narrow the hypotheses down to one or two that can then be tested in Step 4 of the five-step procedure.

5.8 A Logical, Intuitive Approach Fails

A customer complaint was received at the manufacturing location of a highly regarded supplier of a baled elastomer (12″ × 28″ × 7″). The customer alleged that he had received some green bales of the product in a recent shipment. The bales were normally of yellow color. The process for manufacturing the elastomer was about 10 years old, and a similar problem had never been encountered. When confronted with this complaint, the Operating Department Head used his problem-solving technique and developed the following problem statement:

> "The customer complaint has been investigated. We have not made any significant changes in our operation in 10 years except for the use of 'magic markers'. Our operators have started carrying 'magic markers' to mark equipment that requires maintenance at the next downtime. We believe that one of these markers must have fallen from one of their pockets into the extruder. The subsequent fracture and dispersion of the material caused several bales to have a green appearance."

This intuitive logical approach overlooked several areas that a more structured procedure would have uncovered. There was no verification that the problem really occurred. It would have been valuable for the customer to send a sample of the material that he received. The problem definition was incomplete. Consideration of other questions such as the following would have been helpful in a better problem statement.

- What shift did the problem occur on?
- Did other customers notice the problem?
- How many bales were green?
- Did the problem occur on all lines?
- Did the laboratory retained samples on the problem date, previous dates, or subsequent dates show green color?
- Were there really no operational changes?
- Were the bales green when they were boxed?

In addition to problem definition failures, the hypothesis was not tested against any type of theory. For example, how much of a green magic marker would be required to turn a single yellow bale into a green color? There was no way to estimate the concentration of the green magic marker in the bale since there was no knowledge of the number of green bales. No mechanism was provided to test the hypothesis. The hypothesis could have easily been tested by dropping a green magic marker into the extruder. This would only provide a one-sided test. If the bales did not turn green, the test would be successful, in that it would prove that the hypothesis was incorrect. However, if the bales did turn green, it would be necessary to consider the hypothesis in more detail. For example, the following questions should be considered:

- How many bales turned green from a single magic marker?
- Did this correspond to the number of green bales that the customer observed?

Subsequent to the manufacturing manager writing the customer a letter of apology for a magic marker falling into the extruder, several other complaints on the same subject were received from different customers. The continued customer complaints led to the formation of a multidiscipline problem-solving team that determined after a lengthy investigation that the green color was associated with an obscure change in the makeup water used for the polymer–water slurry system.

5.9 Lessons Learned

This example indicates how a failure to adequately develop a problem statement can lead to an embarrassing faulty problem solution. If the problem statement had been fully developed using some of the questions shown above, it is highly likely that the problem would have been recognized as systemic rather than as an individual isolated accident. The failure to adequately define the problem then cascaded to the point that only a simplified logical approach to problem solving was used. If the problem had been recognized as a systemic problem initially, the problem-solving team would have been formed at a much earlier point in time.

This example also illustrates how what appears to be an operation that has not made any changes can be impacted by subtle and obscure changes in utilities. The normal assumption that "water is water" was not true in this case.

Nomenclature

A	the exchanger surface area
dQ_g/dT	the rate that heat generation increases with temperature
dQ_r/dT	the rate that heat removal increases with temperature
dt	the change in temperature, °F
K	a constant that depends on monomer and catalyst concentrations, reactor volume, and heat of polymerization
$\text{Ln}\Delta T$	the log temperature difference between the polymerization slurry side and the cooling water side
l_o	the original baffle length
l_t	the baffle length at the new temperature
n	the number of runs in each sample
s	the standard deviation of the two samples
SE	the standard error for comparing the two means
T	the absolute temperature, R
$(T - T_w)$	the difference between the average reactor temperature and the average cooling water temperature
U	the exchanger heat transfer coefficient
φ	the degrees of freedom in each sample (number of runs)

6

Development of Working Hypotheses

6.1 Introduction

This book does not deal with the multitude of methods for generating potential working hypotheses. Many of these methods put almost complete emphasis on accurate problem statements. The implicit assumption is that if the problem can be well defined in sufficient detail, that the problem solution will be apparent. In complex process plants, multifaceted problems can rarely be solved with this simple approach.

This chapter presents an approach for the development of theoretically sound working hypotheses based on consideration of a series of questions. These questions will require an analysis of the data in an introspective fashion. It is highly unlikely that the approach described in this chapter will provide only one possible problem solution. Thus, this chapter and subsequent chapters also deal with how the large number of possible working hypotheses can be narrowed down using logic and most importantly one's technology training.

6.2 Areas of Technology

As indicated earlier, two kinds of technologies are utilized in process plants. These are process-related technologies and equipment-related technologies.

Process technologies are technologies that deal with the specific process, for example, a polymerization process, an isomerization reactor process, or a distillation process. The process-specific technologies may include items such as reaction rate kinetics, polymer product attributes, or relative volatility data. Each of these processes will have specific technology details that must be well known and understood before attempting to do any process technology-related problem solving.

Equipment-related technologies are technologies that are valid for any specific piece of equipment regardless of the process technology where it is

Problem Solving Approaches for Maintaining Operational Excellence in Process Plants, Second Edition.
Joseph M. Bonem, Nattapong Pongboot, and Wiroon Tanthapanichakoon.
© 2025 the American Institute of Chemical Engineers, Inc. Published 2025 by John Wiley & Sons, Inc.

being used. Examples of these are details and calculations associated with pumps, compressors, and distillation towers. In addition, most kinetically limited processes (heat transfer, for example) can be generalized in terms so that a hypothesis can be developed regardless of the specific technology. This approach to kinetically limited processes is described in detail in Chapter 9.

6.3 Formulating Hypotheses Via Key Questions

The primary thrust of Chapters 7 through 10 is using the five-step problem-solving procedure discussed in Chapter 4. The equally important parts of having a daily monitoring system and determining the optimum technical depth were adequately covered in Chapter 4.

The emphasis in this chapter is on formulating and verifying *theoretically sound working hypotheses*. Formulating theoretically sound working hypotheses deals with an in-depth thinking process that requires the problem solver to utilize his engineering training to develop a hypothesis. The in-depth thinking process will rarely be done in meetings. It often requires data analysis, literature research, and/or "one on one" discussions with experts in the field or those that can serve as a source of data. This in-depth thinking process often involves consideration of the potential questions shown below to help define the cause of the problem. Examples are given for each question. These examples are not meant to be an inclusive list but are given only to amplify the specific question. The questions are given in the order of priority. Obviously, this priority will depend on the specific problem and the specific process. These questions also assume that Steps 1 (verify that the problem actually occurred) and 2 (*write* out an *accurate* statement of what problem you are trying to *solve*) have been completed.

1) Are all operating directives and procedures being followed?—An inspection of operating conditions at most process plants will show that deviations from procedures and directives are occurring. These may or may not be related to the problem under study. While it is important to verify that all procedures and directives are being followed, a small deviation should not be deemed to be the source of the problem, unless there is a theoretically sound working hypothesis for explaining how the deviation is causing the problem.
2) Are all instruments correct?—Incorrect flowmeters may result in reaction rates being different than expected, fractionation separation being below design levels, a pump or compressor appearing to be operating "off the curve," or failure to adequately strip an impurity from a polymer. A level instrument being wrong could result in a reduced or increased reaction rate, which would manifest itself as a less than or greater amount of reaction.

Heat and material balances are exceptionally good tools and can often be used to answer this question.

3) Are the laboratory results correct?—If the problem under study is related in any way to laboratory results, confirming that the laboratory results are correct is a high priority. This confirmation can require review of the procedures as written, review of the procedures as performed, and review of the chemicals being used. For example, the results of an extraction procedure can be greatly altered if cyclohexane is used as the solvent when the procedure called for the use of normal hexane.
4) Were there any errors made in the original design?—The high priority given to this possibility is due to the need to assess this question early in the problem-solving activity prior to doing a large amount of work in other areas. This assessment can be made based on experience with the process and the duration for which it has been in operation. The probability that original design errors are causing the operating problem decreases with the age of the process. However, one should not assume that just because a process has been in operation for several months that it is free of design errors. A new operating condition or new product grade may expose design errors that were not detected earlier. These design errors might be as small as a density being wrong on an instrument specification sheet to as large as incorrect tray selection for a distillation column. The assessment of potential design errors can only be made by either a detailed review of design calculations or redoing the calculations.
5) Were there changes in operating conditions that occurred at the same time as the problem began?—These changes in operating conditions may immediately result in the observation of a process problem. However, the most likely scenario is that everything appears normal when the changes are first made. However, at some later point in time, there is a small change in another variable, and the problem begins to be noticed. An example of this might be a reduction in the operating temperature of an exothermic reactor in winter. After the reduction in temperature, all control systems appear to be operating normally. However, temperature control was impossible as the outside temperature and hence cooling water temperature increased with spring conditions. If the process being investigated is integrated with other processes (for example, in a refinery or chemical plant complex), it will be desirable to investigate operating condition changes in these other processes also.
6) Is fluid leakage occurring?—Fluid leakage covers areas such as leakage through heat exchanger tubes, leakage across isolation valves, and leakage through control valves. Leakage can cause reaction rates to be less than desired if the leaking component is an impurity. Leakage can also cause an apparent loss of fractionation efficiency or an apparent loss of pumping or

compression efficiency. The potential for leakage can be determined by a flow sheet review, determination of pressure flow potential using measured pressures, and in some cases detailed calculations of control valve clearances.

7) Has there been either normal or unusual mechanical wear or changes that could impact performance?—Erosion of wear rings or failure of check valves can greatly affect the performance of pumps and compressors. The performance of distillation columns can suffer due to the failure of a single tray segment. Often, the unusual mechanical wear will be caused by a large process upset. Mechanical changes that would affect the process performance might occur. For example, a material of composition change in a mechanical seal might result in decomposition of the seal and contamination of a product.
8) Is the reaction rate as anticipated?—At times, the reaction rate is the problem that the problem solver is attempting to work. However, there are also times when an excessive amount of reaction or lack of reaction is the cause of a problem itself. For example, the alumina desiccant is known to have catalytic properties. The catalytic properties can usually be mitigated by operating techniques. However, if a batch of alumina desiccants has exceptionally high catalytic activity, the normal compensatory operating technique may not be adequate. This may cause problems in the process, which can be traced back to the specific batch of alumina.
9) Are there any adverse reactions occurring?—Adverse reactions are always a potential problem with reactive chemicals. The presence of solids in a distillation column that purified a diolefin such as butadiene or isoprene might be traced to small quantities of oxygen that entered the process from inadequately purged storage vessels and then catalyzed the reaction of the diolefin to form a polymer. Another example might be the utilization of aluminum metal in an analyzer sample system when chlorides and olefins were present. The chlorides would react with the aluminum to form aluminum chloride, which would then cause the olefin to polymerize to an oily material and foul the analyzer. The trays in a distillation column might plug due to the presence of solids formed by the unexpected presence of water.
10) Were there errors made in the construction of the process?—Construction errors are treated as a low priority simply because they almost always involve a unit shutdown or elaborate noninvasive techniques to inspect potential problems. However, these errors do occur. Examples of construction errors are debris left in vessels or piping, improper leveling of distillation trays, beveled orifice meters installed backward, failure to complete the cutting of a "hot tap" so that the flow path is greatly restricted, and installation of an incorrect pump impeller.

This list is meant to serve as a possible checklist that can be used to develop hypotheses. It is not meant to be an all-inclusive list. Certainly, the priorities will change depending on the specific process and status of the process. However, it is believed that this approach can be effective in developing sound hypotheses.

While the thrust of this book is directed toward equipment-related technologies, the approach to developing working hypotheses can be applied to process technologies as well.

6.4 Beauty of a Simplified Approach

The problem solver will often be tempted to invent a very complicated theory to explain observations that he does not fully understand. Normally, this will be counterproductive. A simple theory will often lead to a problem solution that is easier to execute and more effective. The physicist Ernest Rutherford commented over 100 years ago "A theory that you can't explain to a bartender is probably no good."

A real-life example of this concept is found in the history of the building of the Panama Canal. Yellow fever and malaria were serious problems claiming many lives, especially during the French period of canal building. Since the worst epidemics seemed to start during the rainy season, the initial theory was that these diseases were caused by mysterious vapors that came out of the swamps which formed during the rainy season. This complex theory provided essentially no readily apparent problem solution. Prior to the American construction of the canal, Dr. Walter Reed and his coworkers in Cuba had developed a simple competing theory to the "swamp theory." They theorized that yellow fever was spread by the Stegomyia mosquito. In addition, they discovered that this mosquito would only lay eggs in clean water held in an artificial container located in or near a building occupied by humans. With this relatively simple theory, problem solutions became readily apparent. This allowed the yellow fever threat in Panama to be greatly mitigated and was one of the keys for successful completion of the Panama Canal. Similar approaches were developed to mitigate the malaria threat.

6.5 Verification of Proposed Hypotheses

In addition to the technique described above, there are multiple alternative problem-solving techniques that are touted throughout the industrial world. All of them involve developing *possible theoretically sound working hypotheses*. These alternative techniques also put emphasis on maximizing the number of possible hypotheses based on the desire to not overlook any possibility. The techniques are of value for generating possible hypotheses. Unfortunately, these techniques often result in the assumption that a problem solution that seems logical is also technically correct. This is not always so. The process to verify a proposed working hypothesis is a procedure where the hypothesis is carefully examined using the best available techniques. This verification process can rarely be done without calculations. For example, the hypothesis of an operator that the restriction to the flow of a liquid in a 4-in. line is due to a short section of a 2-in. pipe may or may not be technically correct. It can only be assessed as a theoretically sound working hypothesis by calculations. The input of the operator is very valuable since this detail might have been overlooked. However, his conclusion is likely erroneous.

Regardless of how proposed problem hypotheses are developed, they will require verification. It is imperative that the problem solver use his training and calculations to eliminate the hypotheses that are not theoretically correct. These required calculations are part of Step 3 (develop a *theoretically sound* working hypothesis that explains the problem) or Step 4 (provide a mechanism to test the hypothesis). Regardless of whether they are included in Step 3 or Step 4, the calculations should be done before proceeding on a plan for a plant test or operating changes.

Many times, the problem solver is limited in his problem-solving ability by the lack of proven industrial calculation techniques. For example, pragmatic utilization of pump curves is not taught in the academic world. However, it is exceptionally important in the industry. To aid the problem solver, Chapters 7 through 12 contain hints and useful industrial calculation techniques and approaches. These sections will not be inclusive of all the possible technology and know-how associated with various types of equipment but are meant to be a summary of valuable techniques and know-how for the process engineer with plant problem-solving responsibilities.

Rather than discussing specific unit operations, the approach in this book is to discuss the formulation of the working hypotheses in four areas as follows:

- Prime movers (pumps and compressors)
- Staged processes (towers)
- Kinetically limited processes (heat transfer, reaction, and drying)
- Unsteady-state processes

It is recognized that most operating companies may have technical manuals that describe how to design particular pieces of equipment. These manuals are usually very voluminous and are aimed at design as opposed to problem solving. As such, the problem solver does not often use them. In Chapters 7 through 12, an attempt is made to reduce the important equipment concepts to those that are required to solve problems. The judgment about what should be included in these concepts is obviously flavored by the author's own experience. An attempt has been made to cover a broad spectrum of process equipment. It is believed that if a specific type of equipment is not mentioned, there will be similar equipment that is described in sufficient detail to allow one to formulate a working hypothesis for the equipment.

6.6 One Riot–One Ranger

The Texas Rangers are the oldest law enforcement group with statewide jurisdiction on the North American continent. In the late 1800s, according to Texas folklore, there was a small Texas town where a large riot was imminent. The town had limited law enforcement resources. They requested that the governor of Texas send a troop of Texas Rangers to quell the riot. When the train arrived, carrying what the town folks thought would be a troop of Rangers, only one person emerged from the train. When the town folks expressed shock, the Ranger's reply was "One Riot-One Ranger."

While it is true that a committee will always have more knowledge of the facts than any single individual, most effective problem solving is done when one person has the responsibility to solve the problem and uses the input from others. This single individual in process plants is almost always a chemical engineer, process engineer, or an individual acting in one of these roles. It is extremely important for him to bridge the chasm between his training and others that may have important knowledge or data that will help solve problems. For example, very few chemical engineering curriculums cover mechanical seals. However, mechanical seals are an integral part of a centrifugal pump. Thus, when working on a centrifugal pump problem, it is likely that knowledge of mechanical seals will be required. This knowledge can best be obtained by the chemical engineer from discussions with mechanics and/or mechanical engineers.

An example of the need to bridge the chasm between engineering disciplines was a chemical engineer who was asked the question "Is it ok to operate these two identical centrifugal pumps in parallel to obtain increased capacity?" His reply was "I don't know that is an equipment question." In fact, the chemical engineer is the individual who has the necessary knowledge of process conditions to answer the question. He will have to do a more detailed analysis and develop an understanding of centrifugal pumps to finalize the answer to the question. For example, he would have to develop understanding of the following:

- Running pumps in parallel involves determining the exact location on the pump curve (flow rate), determining the capability of the control system to handle the increased pressure drop, and determining the minimum flow that might be encountered during parallel pump operation. These are clearly more than "equipment questions."
- Running pumps in parallel at low flow rates can sometimes result in one of the pumps operating in a "blocked in" condition. If this condition occurs, one of the pumps is pumping the full process flow, while the output from the other pump is very little. This "blocked in" operation could create a serious operation and safety problem. This condition could occur, even though the pumps

are identical, due to different clearances on the wear rings. This clearly requires consultation between various engineering disciplines.
- Although it is a little-known fact, centrifugal pumps often experience a loss of stability, similar to what occurs in centrifugal compressors operating at lower flow than the surge point. If a centrifugal pump is operated at flows below this stability point, serious vibration can occur. The chemical engineer would certainly have to depend on another discipline to determine the stability limit.

The primary point of this discussion is that the chemical engineer as a "problem solver" in a process plant must know or obtain knowledge of the equipment that is involved in the problem he is solving. He will often be required to determine why the piece of mechanical equipment does not have as much capacity as it should rather than just saying "It is not operating at design capacity."

Similar points could be made in the areas of "process chemistry." A chemical engineer serving as a "problem solver" must obtain the necessary knowledge of the process chemistry to formulate theoretically sound working hypotheses. Once the process chemistry is understood, the same five-step procedure can be used to solve process chemistry-related problems, as well as process equipment problems. The time spent with a chemist familiar with the specific process technology chemistry will greatly aid in formulating correct working hypotheses. This need will be obvious in a process where reactors are utilized. However, in almost any process, the possibility for reaction exists. Ignoring the need for completely understanding the process chemistry can lead to formulation of incorrect hypotheses. For example, in a polymer plant, a process engineer developed a hypothesis that the polymer was turning a light shade of pink due to the presence of iron complexes caused by corrosion. A more detailed analysis with a process chemist revealed that one of the additives being used was also a pH indicator and the pink color was due to the slight basicity of the polymer.

In spite of the emphasis on "one ranger one riot" above, the problem solver should never assume or pretend that he has all the knowledge necessary to work the problem. A truly effective problem solver will know and admit his limits. He will then look for help in areas outside of his know-how. However, he will seek this knowledge and apply it to the current problem, as opposed to trying to reassign the problem to someone else.

7

Application to Prime Movers

7.1 Introduction

Prime movers (pumps and compressors) are the workhorses of the process industry. They can be split into two generic categories depending on how they impart energy to the process fluid. These categories are as follows:

- *Kinetic or dynamic systems*: Energy is imparted by accelerating the fluid with an impeller and then decelerating the fluid with an inherent increase in pressure. Examples of this type of equipment are centrifugal pumps and compressors, axial and centrifugal fans, and blowers. Bernoulli's theory can be used to understand how these prime movers operate.
- *Displacement systems*: Energy is imparted by displacing the fluid from a lower pressure to a higher pressure with a pushing or rotating action. Examples of this type of equipment are reciprocating pumps and compressors and rotary blowers and compressors.

In order to formulate theoretically correct working hypotheses for pumps and compressors, knowledge of several relationships and definitions is required. These are summarized in this chapter. In addition, several problems are presented to illustrate the use of these concepts.

7.2 Kinetic Systems

As indicated above, Bernoulli's theory is the key to understanding these systems. While this theory is usually thought of only in terms of the flow of fluids in piping,

Problem Solving Approaches for Maintaining Operational Excellence in Process Plants, Second Edition.
Joseph M. Bonem, Nattapong Pongboot, and Wiroon Tanthapanichakoon.
© 2025 the American Institute of Chemical Engineers, Inc. Published 2025 by John Wiley & Sons, Inc.

it is also valuable when considering the flow of fluids inside a pump or compressor case. Bernoulli's equation is as follows:

$$\Delta P/\rho + \Delta(v^2)/2g_c + \Delta z = -w - \text{lw} \quad (7.1)$$

where

ΔP = the pressure difference between two points
ρ = the fluid density
$\Delta(v^2)$ = the difference in velocity squared between two points
g_c = the gravitational constant
Δz = the difference in elevation between two points
w = the amount of work added by the prime mover
lw = the frictional loss in the piping system

When considering what happens inside the case of a centrifugal pump or compressor, only the first two terms of equation (7.1) are significant. They clearly indicate that the fluid is accelerated by the work of the prime mover to approach the impeller tip velocity. As the fluid enters the outlet piping, it decelerates, and the pressure increases to the discharge pressure. Figure 7.1 shows the typical flow path in a centrifugal prime mover.

Bernoulli's theory as applied to centrifugal prime movers has several implications that are of value to consider: the head ($\Delta P/\rho$) that a pump or compressor develops is directly related to impeller tip speed squared. High-speed impellers are required to develop high heads for either pumps or compressors. An alternative is multiple-staged impellers. The high-speed impellers and multiple-staged impellers can also be combined in a single pump or compressor.

The operating characteristic of a centrifugal system (pump or compressor) is defined by the equipment characteristic "head curve." This head curve is

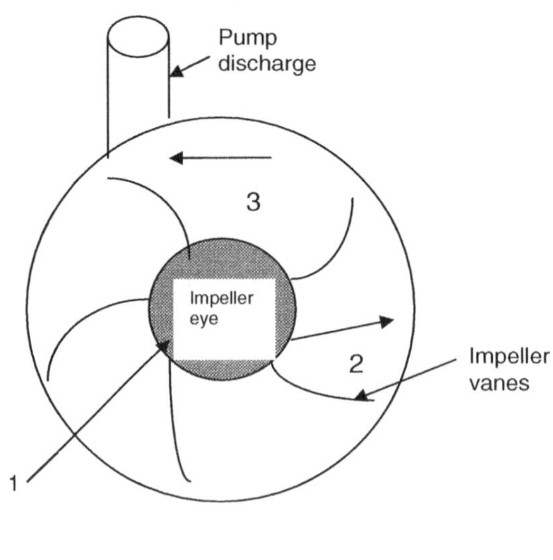

Figure 7.1 Flow path in centrifugal equipment. 1. Fluid enters the eye of the impeller by pressure flow from the pump suction flange. 2. It is accelerated by centrifugal force to a velocity approaching the tip velocity of the impeller. 3. As the fluid approaches the discharge, the internals of the pump casing allow it to slow down with the inherent increase in pressure.

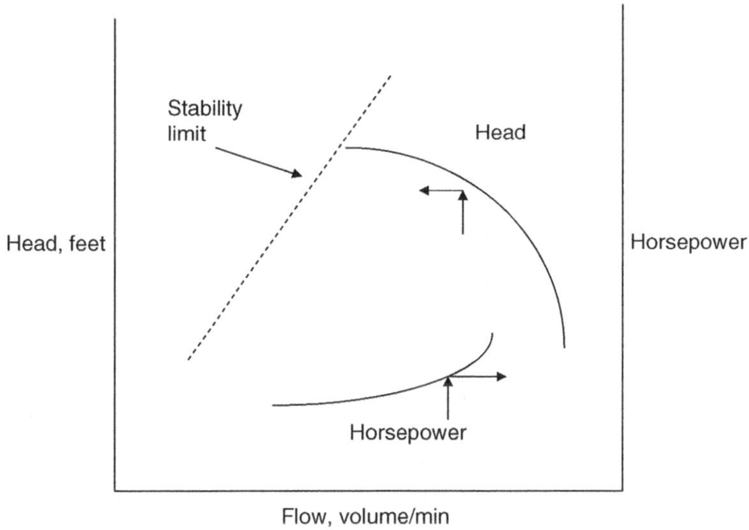

Figure 7.2 Characteristic centrifugal pump or compressor curve.

developed by the equipment supplier and is provided as part of the equipment purchase. An example is shown in Figure 7.2. In this figure, head and horsepower are shown as a function of flow rate. In addition, the stability limit is also shown. Several generalized points can be made about this figure.

Operation to the left of the stability limit will result in flow instabilities as flow surges forward and then backward through the prime mover. The stability is usually well-defined for compressors and blowers. However, for pumps, it is usually 25–40% of the best efficiency point (BEP). If flow reversal occurs, it can be damaging to the equipment. Operation of equipment in parallel can sometimes result in operating in the surging region. This happens because the parallel pieces of equipment have different head curves, or often there are slight differences in the mechanical characteristics of "identical" pieces of equipment.

The BEP is the point on the "head curve," where the hydraulic efficiency is at a maximum. Note that this efficiency does not include motor efficiency. While the head curve is usually developed using water or air, *it is valid for any fluid if the correct units are utilized for flow and head.* These units are defined later.

As shown in Figure 7.2, the horsepower requirements normally peak at the "end of the curve" (maximum flow rate). The driver for the prime mover may or may not be provided with "end of the curve" protection. If the driver is not designed with "end of the curve" protection in mind, operating at this point will normally cause the driver to be overloaded. If the driver is a steam turbine, it will slow down, causing the pump to appear to be operating "off the pump curve." If the driver is an electric motor, it will shut down as the electrical load causes the motor protection device to react and shut down the motor before the motor fails due to overload.

7.3 Pump Calculations

Before considering exact calculation procedures for pumps, a few definitions are required. These definitions are applicable to positive displacement pumps, as well as to centrifugal pumps.

- Cavitation = A condition that occurs if $NPSH_R > NPSH_A$. If this situation occurs, some of the liquid being pumped will vaporize between the pump suction flange and the pump impeller. This will cause the pump to operate off the head curve, and damage may occur to the impeller.
- Off the head curve = This is a condition where the actual operating point, as defined by the flow rate in gallons/min (gpm), and pressure rise, as described by feet of fluid flowing, are below the pump curve, as defined in Figure 7.2.
- $NPSH_R$ = Net Positive Head Required. This is the head in feet required to overcome the pressure loss between the pump suction flange and pump impeller eye. The pump supplier will specify this. A typical $NPSH_R$ versus flow rate curve is shown in Figure 7.3. Note that this curve is usually developed with water, but it is valid for any fluid.
- $NPSH_A$ = Net Positive Head Available. This is the difference in feet of head between the actual pressure at the pump suction flange and the vapor pressure of the liquid being pumped. If the liquid has been stored under a nitrogen, air, or inert gas blanket, some questions may arise regarding the actual vapor pressure of the liquid. The most conservative approach is to assume that the vapor pressure is equivalent to the pressure in the storage vessel.

The equations below can be used in combination with a pump curve supplied by the manufacturer, which is similar to Figure 7.2, to determine essential pump operating characteristics. When using the pump curve for a centrifugal pump, the differential pressure across the pump can be calculated as follows:

$$\Delta P = H^* \rho / 144 \tag{7.2}$$

where

ΔP = the pressure rise across the pump, psi
H = the pump head at the given flow rate, ft
ρ = the pumped fluid density, lb/ft^3

As indicated earlier, pump curves are usually developed using water as the fluid. They are valid for any fluid as long as the density used for calculations is that of the fluid being pumped. Thus, in equation (7.2), the pressure developed by a centrifugal pump can be calculated from the head curve supplied by the pump manufacturer using the head developed at the desired flow rate and the density of the flowing fluid.

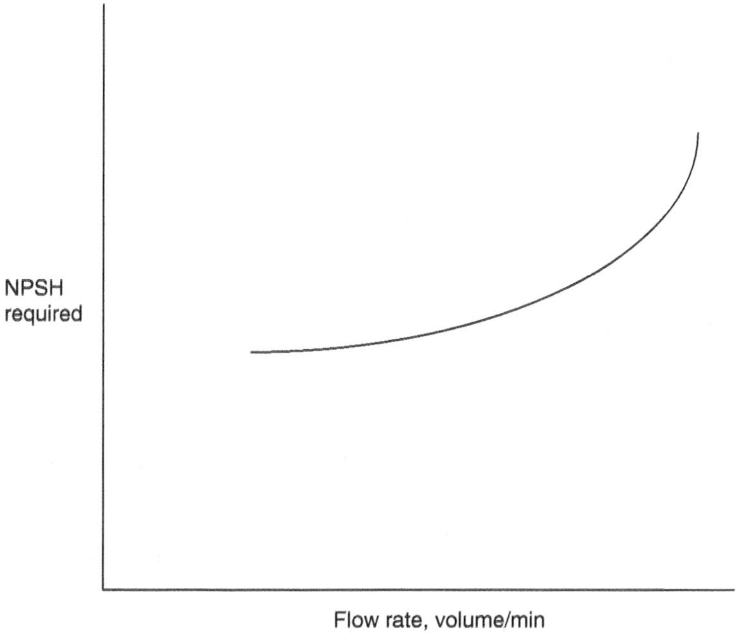

Figure 7.3 Typical NPSH required for the centrifugal pump.

The *hydraulic* horsepower for a pump can be estimated as follows:

$$\text{BHP} = H^{*}F^{*}100/(E^{*}33,000) \tag{7.3}$$

where

BHP = the energy delivered to the fluid, horsepower
F = the flow rate through the pump, lb/min
E = the hydraulic pump efficiency, percent

It should be noted that this calculation only includes the hydraulic efficiency. There will be other efficiencies (mechanical and/or electrical) that must also be considered.

On rare occasions, it is necessary to determine the temperature change across the pump. For example, pumps with very low flow rate, high recirculation rate, and high heads may cause significant heat to be generated and conducted to the fluid recycling to the pump suction line. This higher temperature at the pump suction flange may reduce the NPSH_A. The fluid temperature increase across the pump can be estimated as follows:

$$\Delta T = \text{BHP}^{*}(100 - E)^{*}2545/(100^{*}60^{*}F^{*}C_P) \tag{7.4}$$

where

C_P = the fluid specific heat, BTU/lb-F

This calculation will often indicate that a cooler is required in the fluid recycle line. Even in high head pumps with low flow and no recirculation, heat generation can be a problem. The heat generated by the pump inefficiency will heat up the fluid, which in turn will reject some of the heat to the pump case.

7.4 Centrifugal Compressor Calculations

As indicated earlier, a head curve similar to that shown in Figure 7.2 is valid for both centrifugal compressors as well as for pumps. The compressor flow is normally expressed as *actual cubic ft/min* (ACFM) at the compressor inlet conditions. In addition, for a compressor, the calculation of head is much more complicated than for a pump. This is because gases are compressible, as opposed to noncompressible liquids. The specific volume for a liquid for all practical purposes can be considered a constant.

While there are multiple references to compression calculations in the literature, two of the references used in the preparation of this book were written by Lyman F. Scheel (1968, 1972).

The calculation of head for a compressor depends on the relationship between the pressure and volume of the gas. This relationship is given by the thermodynamic equation:

$$P^* V^k = \text{constant} \tag{7.5}$$

When applying this relationship to compression (centrifugal or reciprocating), the actual value of "k" depends on which of the three thermodynamic assumptions is used. These assumptions are as follows:

1) *Isothermal*: The constant temperature assumption is impossible to achieve in practice.
2) *Adiabatic*: This assumption requires that no heat be added or removed from the system. This path can be approached in a piston compressor where the discharge and suction compressor valves have exceptionally large areas. The stipulation of exceptionally large area valves is necessary since the suction and discharge pressures are measured before and after the compressor valves. These compressor valves are flat ribbon valves that are integral to the compressor. However, if the pressures could be measured inside the valves, no stipulation of exceptionally large area suction and discharge valves would be required. In this idealized compression, $k = C_p/C_v$. C_p/C_v is the ratio of the specific

heats at constant pressure and constant volume. If compression is adiabatic *and* reversible, it is isentropic compression. That is, it occurs at constant entropy. For this special compression, a Mollier diagram determines the compression path. Like isothermal compression, isentropic compression rarely occurs in practice.

3) *Polytropic*: This is an empirical assumption for evaluating the compression path in a compressor. However, it is the normal technique to evaluate compression head. For polytropic compression, the "k" value is replaced by "n," which is obtained either from the compressor manufacturer or from plant test data. This will be discussed later.

The polytropic and adiabatic head can be determined from the same basic equation shown below:

$$H = 1545^{*}T_S{}^{*}Z^{*}(R^{\sigma} - 1)/M\sigma \quad (7.6)$$

where:

H = the polytropic or adiabatic head, ft
T_S = the suction temperature, °R
Z = the average (suction and discharge) compressibility
R = the compression ratio
M = the gas molecular weight
σ = the polytropic or adiabatic compression exponent

There are three different ways to evaluate "σ." They are as follows:

1) The compressor vendor may supply the value based on test data.
2) The compressor vendor may supply either the polytropic or adiabatic compression efficiency. In this case, the compression exponent (σ) can be calculated as follows:

$$\sigma = (k - 1)^{*}100/(k^{*}E) \quad (7.7)$$

where

E = either the adiabatic or polytropic compression efficiency, percent
K = the ratio of specific heats, C_p/C_v

3) The polytropic compression exponent can be calculated from plant data using the following relationship:

$$T_d = T_S{}^{*}R^{\sigma} \quad (7.8)$$

where

T_d and T_s = the absolute discharge and suction temperatures, respectively.

It should be recognized that the relationship shown in equation (7.8) gives the compression exponent at any point in time. If the compressor needs mechanical repairs, this value may be higher than predicted by the compressor vendor. Since the compression exponent calculated from plant data does vary with the mechanical condition of the compressor, this value could be used as a daily monitoring tool for critical compressors. For example, an increase in the compression exponent might indicate that the clearances in a centrifugal compressor had increased to the point that excessive amounts of gas were recirculating to the compressor suction. An increase in the compression exponent for a reciprocating compressor might indicate that the suction and/or discharge valves should be replaced.

Once the head and flow rate are known, the fluid horsepower requirements for the compressor can be easily calculated using equation (7.9):

$$\text{BHP} = F^{*}H^{*}100/(33,000^{*}E) \tag{7.9}$$

where:

F = the flow rate, lb/min
H = the polytropic or adiabatic head, ft
E = either the adiabatic or polytropic compression efficiency, percent

A word of caution is of value at this point. The compression efficiency (adiabatic or polytropic) should be consistent with the method used for the compression exponent. This will normally be the polytropic efficiency and the polytropic compression exponent. It should also be noted that equation (7.9) includes no mechanical or electrical efficiency that will be associated with belts, gears, or motors.

7.5 Displacement Systems

The term "Displacement Systems" refers to the prime movers that displace a fixed amount of fluid (liquid or gas) essentially independent of the differential pressure across the pump or compressor. Typical equipment items that fit this category are reciprocating pumps/compressors and rotary pumps/compressors. While it might be argued that this type of equipment is no longer used in modern plants, an examination of different processes will show that this type of equipment does indeed have a place in a modern process plant. Examples of the use of this type of equipment are as follows:

- Reciprocating compressors where the head requirements are high and the volume rate is low to moderate. An example of this is the high-pressure compression step in low-density polyethylene plants.

- Proportioning pumps with high head requirements or where there is a need to control flow by pump adjustments, as opposed to using a flow controller. These might include additive or catalyst pumps.
- Pumps or compressors where it is desired to transfer a fixed rate of material regardless of the discharge pressure. This would not be possible with a typical centrifugal system since the flow rate would decrease as the discharge pressure increases.

There are important concepts in understanding this class of prime movers. It should be realized that with this type of equipment energy is imparted to the fluid by displacement of a *fixed volume* of that fluid. The mass flow rate will depend on the fluid suction conditions and physical dimensions of the equipment. That is, larger pump or compressor cylinders will allow for displacing a larger volume of fluid, but the mass of the fluid moved also depends on the density at *suction conditions*.

The calculations described in the previous section are applicable to this type of system also. However, head will not be dependent on flow rate. There will be no head versus capacity curve. The pump or compressor head will depend on the suction and discharge pressures only. Figure 7.4 shows a typical flow pattern for a reciprocating pump/compressor.

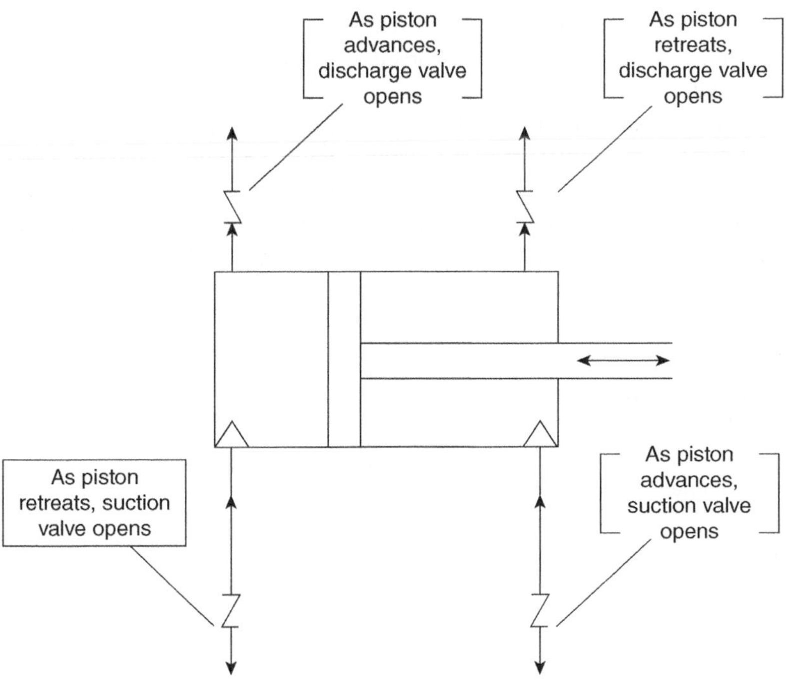

Figure 7.4 Reciprocating flow path in double-acting equipment.

Some definitions will help in evaluating and understanding this type of equipment. They are as follows:

- *Volumetric efficiency*: The actual volume of fluid displaced relative to the volume of the cylinder of a reciprocating pump/compressor or rotating pockets of a rotary pump/compressor. For a liquid, this efficiency approaches 100%. However, for a gas, it is approximately 70%. The differences are due to the compressibility of gases. As the pressure in the cylinder or rotating pocket decreases from discharge pressure to suction pressure, the residual gas in the cylinder/pocket expands to partially fill the cylinder/pocket and reduce the volumetric efficiency of the compressor.
- *Leakage*: This is an additional loss in volumetric efficiency caused by leakage through clearances. An example of this leakage in a reciprocating pump or compressor is the flow of material between the piston and cylinder wall or across the suction and discharge check valves.
- *Horsepower load point*: This is a unique feature of a positive displacement compressor. It is the point on a plot of horsepower versus suction pressure where the required fluid horsepower is at a maximum. On one side of this point, increasing the suction pressure increases the mass flow, which overrides the decrease in the compression ratio and causes an increase in the required horsepower. On the other side of this point, the increasing suction pressure results in a decreased compression ratio, which more than compensates for the increase in the mass flow rate and results in a decreased horsepower. This concept is discussed in more detail later.
- *Clearance*: This is the part of the cylinder in a reciprocating pump/compressor that is not displaced completely by the piston. Figure 7.5 shows a typical sketch of a "clearance pocket." Because of the compressibility of gases, this becomes more important for gases than for liquids. These are often used to reduce the volumetric efficiency of a compressor and hence the required horsepower.

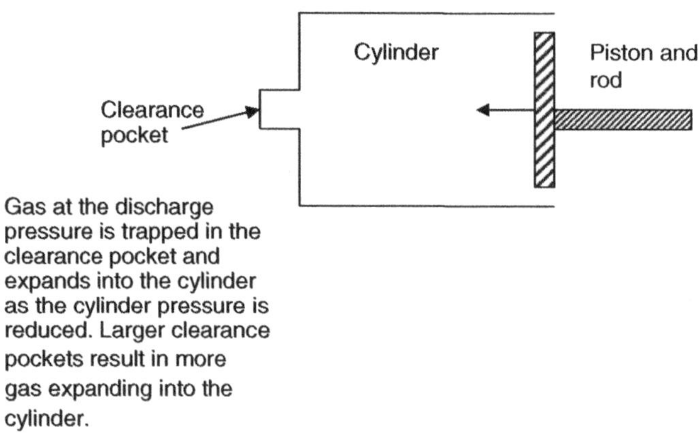

Figure 7.5 Clearance pocket reciprocating compressor.

7.6 Displacement Pump Calculations

Many of the calculations for displacement pumps are identical to those for centrifugal pumps. For example, equations (7.2) and (7.3) can be used to calculate the head/pressure rise relationship and BHP, respectively. The main differences in calculations for the two types of pumps are calculation of the fluid flow rate and $NPSH_A$.

For a single-acting reciprocating pump, the following relationship can be used to estimate the volumetric flow rate.

$$V = \Pi^* D^{2*} L_S^* S^* E_V/(4^* 100) \tag{7.10}$$

where:

V = the pump capacity, ft^3/min
D = the diameter of the pump cylinder, ft
L_S = the length of the pump stroke, ft
S = the pump speed, rpm
E_V = the pump volumetric efficiency, percent

There are several points to consider when examining equation (7.10). The pump volumetric capacity is independent of pressure rise and/or pump head, except as the discharge pressure decreases the volumetric efficiency. This is different than the relationship between flow and head for a centrifugal pump.

Equation (7.10) is for a single-acting pump, that is, one that displaces the cylinder volume one time for each stroke. The pump that is shown in Figure 7.4 is a double-acting pump. It displaces the cylinder two times for each complete stroke. The capacity of a double-acting pump will be somewhat less than two times that of a single-acting pump of the same dimensions. The actual volume displaced on the return stroke of the piston will be slightly less due to the volume occupied by the piston and piston rod. Equation (7.10) can be modified for more complicated pumps.

The pump design (reciprocating, plunger, or rotary) and the mechanical condition of the check valves and internal pump clearances primarily determine the volumetric efficiency. In addition, for high differential pressure pumps, liquid compressibility must also be considered. If the liquid is compressible, the liquid trapped in the clearances will expand when the pressure decreases from discharge pressure to suction pressure. This will cause the volumetric efficiency to decrease. There will also be more leakage across the check valves with high-pressure pumps.

The determination of the amount of $NPSH_A$ is more complicated for reciprocating-type pumps than for centrifugal or rotary pumps. For any type of pump, pumping a fluid at its boiling point, the $NPSH_A$ is simply the difference in elevation head between the liquid level in the suction drum and the pump suction less the frictional pressure drop. This concept is valid for reciprocating pumps also, except the frictional pressure drop must be determined at the actual flow rate rather than the average flow rate. In addition, since the fluid in the pump suction line must be accelerated from zero velocity to the maximum velocity, there is a pressure loss due to this energy requirement. Fortunately, these two pressure losses occur at different times so that their values are not directly additive. For example, the maximum acceleration head occurs at zero flow rate. Since the flow rate is 0, this is the point of minimum frictional loss. Conversely, the maximum frictional head loss occurs at the maximum rate when there is no acceleration required. For most low-viscosity fluids, the acceleration head dominates. It can be estimated as follows:

1) Calculate the maximum flow rate in the suction pipe.

$$V_P = K^* D^{2*} L_S{}^* S / \left(60^* D_P^2\right) \tag{7.11}$$

where

V_P = the peak velocity in suction pipe, ft/second
D = the diameter of the pump cylinder, ft
L_S = the length of the pump stroke, ft
S = the pump speed, rpm
D_P = the diameter of the suction pipe, ft
K = a factor that depends on the pump design. The "K" value for a double-acting pump is approximately Π/2. "K" values for other style of pumps can be obtained from the pump vendors.

2) Calculate the frictional pressure drop, $(h_F)_{MAX}$, in the suction line using conventional techniques based on the peak velocity calculated using equation (7.11).
3) Calculate the acceleration head required to accelerate the liquid in the suction line from zero velocity to peak velocity. This can be determined as follows:

$$(h_A)_{MAX} = 1.35^* L_P{}^* S^* V_P / 307 \tag{7.12}$$

where:

$(h_A)_{MAX}$ = the maximum acceleration head loss, ft
L_P = the *actual* suction pipe length, ft
S = the pump speed, rpm
V_P = the peak velocity in the suction pipe from equation (7.11), ft/second

4) The most conservative approach is to determine the suction piping loss by combining the two types of head loss as follows:

$$(h_T)^2_{MAX} = (h_F)^2_{MAX} + (h_A)^2_{MAX} \tag{7.13}$$

where

$(h_T)_{MAX}$ = the maximum suction piping head loss to be used to determine the $NPSH_A$

5) However, as indicated earlier, these two head losses almost always occur at separate times. Thus, the use of equation (7.13) may give overly conservative predicted head losses. The most reasonable estimate of maximum head loss in the suction piping is the larger of the two losses.

While equation (7.12) appears to be empirical, it can be easily derived from first principles of engineering by realizing that the entire mass in the suction line must be accelerated to the maximum suction line velocity as the suction stroke of the pump begins. The value of 1.35 is an experience-based empirical factor. It is added to allow for the non-sinusoidal motion of the piston and the increased acceleration loss due to elbows and pipe fittings.

Additional information about estimating the $NPSH_A$ for reciprocating pumps can be found in a magazine article entitled "Reciprocating Pumps" by Terry L. Henshaw, *Chemical Engineering*, September 21, 1981, page 105.

Another consideration with displacement pumps is their "on and off" nature. Rather than continuous flow that a centrifugal pump provides, the flow will vary from 0 to a maximum rate. In some specialized applications, it may be important to assure that there is always some minimal amount of flow from a displacement type of pump. An example of this might be the need to maintain a continuous flow of the catalyst to a polymerization reactor. Stopping the flow of the catalyst to the reactor even for a few seconds might result in plugging of the catalyst injection nozzle. Obtaining this continuous flow can be accomplished in a reciprocating pump system by one of two methods. A more complicated duplex or triplex pump can be utilized. These pumps allow for almost continuous flow of the material being pumped. Another alternative is to install a flow surge bottle in the discharge piping of a simplex reciprocating pump. This vessel, which is equipped with an internal bladder and pressured with nitrogen, can be designed so that the discharge flow to the process approaches an average flow rather than a peak flow, followed by a period of zero flow.

7.7 Calculations for Positive Displacement Compressors

As indicated earlier, the head calculations for both types of compressors are identical. Equations (7.6) through (7.9) can be used to evaluate both head and horsepower requirements for a positive displacement compressor. The actual suction and discharge pressure for a reciprocating compressor may be slightly different than measured since the actual pressure is measured in the suction and discharge piping rather than directly at the compressor cylinder during the suction and discharge strokes. In between the compressor cylinder and where the suction and discharge pressures are measured are the ribbon valves. The ribbon valves are pipe diameter-size flat devices containing multiple strips of metal. These serve as check valves to isolate the compressor cylinder from the suction and discharge pressures at the appropriate points in the compressor cycle. For example, when the piston advances, causing a buildup of pressure in the cylinder, the suction ribbon valve closes, keeping gas from flowing back into the suction piping. If the compressor suction and discharge valves are in good mechanical condition, this pressure difference will not be significant. However, if the valves are partially restricted, the pressure at the compressor will be different than measured. Leakage can also occur in these valves. As indicated earlier, equation (7.8) can be used to monitor the status of these valves for critical compressors.

The flow rate from a positive displacement compressor depends on both the volume of the cylinder or rotating pocket and the volumetric efficiency. As indicated earlier, the volumetric efficiency of a compressor is much lower than that of a pump. This is because gases are very compressible relative to liquids. Volumetric efficiency for a positive displacement compressor can be estimated as follows:

$$E_V = 100^*(1 - L - C^*(R^{1/k} - 1)) \tag{7.14}$$

where

E_V = the volumetric efficiency, percent
C = the residual gas remaining in the reciprocating compressor clearance pocket or rotating pocket after discharge, fraction of displacement volume
L = the leakage of gas around the piston or rotating element, fraction of displacement volume. It is roughly 2% of the compression ratio expressed as a fraction
R = the compression ratio

The horsepower load point is a unique feature of a positive displacement compressor. As shown in Figure 7.6, it is the suction pressure at which the BHP reaches a maximum.

To the right of the horsepower load point, the compression ratio and head are decreasing, but the mass flow is increasing. To the left of the load point, the compression ratio and head are increasing, while the mass flow is decreasing. Thus, for a positive displacement compressor, driver overload can occur when the suction pressure is either increasing or decreasing. A single-stage reciprocating compressor, with a constant discharge pressure, constant volumetric efficiency,

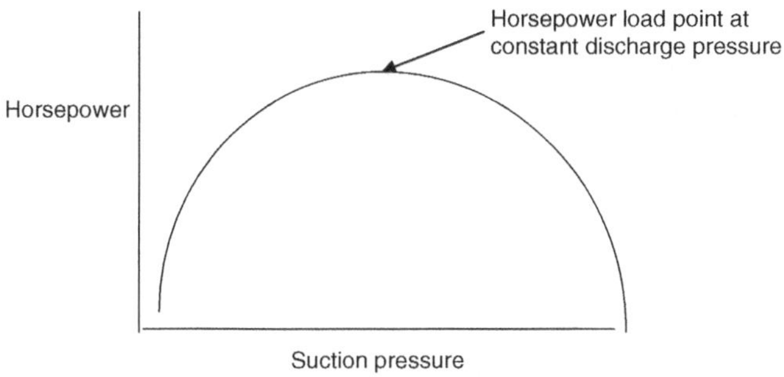

Figure 7.6 Positive displacement compressor load point.

and an approximate clearance pocket volume of 10%, will reach a maximum horsepower loading at an approximate compression ratio of

$$R = 1.1^*(n + 1) \tag{7.15}$$

where

R = the compression ratio
n = the constant used to define the polytropic compression exponent = $(n\ 1)/n$

Multiple-stage compressors are more complicated, but the concept that decreasing the compression ratio does not always decrease the horsepower requirements is still valid.

7.8 Problem-Solving Considerations for Both Systems

7.8.1 Compressors

For critical compressors, the suction and discharge temperatures should be used to monitor compressor performance. As indicated previously, equation (7.8) can be used on a daily basis to monitor the performance of a critical centrifugal or positive displacement compressor. If the compression exponent (α) begins to increase, it is an indication that the compressor is becoming less efficient. Another approach that uses the same inputs is to calculate the "leakage" from the discharge pressure to the suction pressure. An increase in this leakage occurs due to mechanical wear caused by increased clearances in the compressor. This leakage can be calculated as follows:

$$L = \left(\left(T_D/(P_D/P_S)^{\sigma}\right) - T_S\right)/(T_D + \Delta T_H - T_S) \tag{7.16}$$

where

L = the discharge to suction leakage, fraction of volumetric flow
T_D = the discharge temperature, °R
T_S = the suction temperature, °R
ΔT_H = the Joule Thompson cooling effect, °R
P_D = the discharge pressure, psia
P_S = the suction pressure, psia
σ = the compression exponent

The Joule Thompson cooling effect is the level of temperature change that would be expected as the pressure is lowered from discharge pressure to suction pressure.

A similar approach can be utilized for monitoring steam turbines that are often used as drivers on large pumps and compressors. The efficiency of a steam turbine can be determined by comparing the actual change in enthalpy with that predicted assuming an isentropic (constant entropy) expansion. That is:

$$E_T = 100^*(H_1 - H_O)/(H_1 - H_E) \quad (7.17)$$

where

E_T = the turbine efficiency, percent
H_1 = the inlet steam enthalpy, BTU/lb.
H_O = the outlet steam enthalpy, BTU/lb.
H_E = the outlet steam enthalpy with isentropic expansion, BTU/lb.

For critical turbines, the efficiency can be calculated and monitored on a daily basis. This will allow spotting mechanical problems before they become so severe that an immediate repair is required.

For centrifugal compressors, the compressor performance curve supplied by the compressor manufacturer can be used to analyze problems. To do this accurately will require careful planning and consideration. The following items should be included in any planned problem-solving activity associated with compressors:

- Make sure that all field instruments have been calibrated before taking any data.
- Determine the kinetic head as described in equation (7.6).
- Calculate the ACFM as accurately as possible. If necessary, adjust the metered flow rate for differences in pressure, temperature, and molecular weight between the meter specification sheet and actual conditions.
- Make sure that the gas composition is known since it can affect variables such as molecular weight, calculated head, temperature difference between the suction and discharge, and flow rate in lb/hr.

If problem-solving involves a plant test on a centrifugal compressor, beware of increasing the speed at constant volumetric flow. Since the surge point normally increases with speed, the compressor could go into surge if the volumetric flow rate is maintained constant when the speed is increased.

Similar guidelines are also appropriate for positive displacement compressors, except that there will not be a head versus flow curve as for a centrifugal compressor. It will be important to have instruments calibrated and to know the gas composition so that the actual capacity can be compared to the predicted capacity. The calculated compression exponent can also be compared to that supplied by the compressor manufacturer.

As indicated earlier, restrictions between the pressure measurement points (discharge or suction) can cause the actual pressures in the compressor cycle to be different from the measured pressures. This could cause the compression exponent calculated from inlet and outlet temperatures to be higher than anticipated. In a reciprocating compressor, the most likely cause of this difference in pressures is ribbon valves that are partially plugged. These valves can also cause a loss of compression capacity if they are partially plugged or if they are leaking. This problem can be monitored and detected by using equation (7.8). Chronic valve malfunction is often due to liquid or solid entrainment.

7.8.2 Pumps

One of the most common problems causing poor pump performance is an inadequate available NPSH. While there may be pump manufacturers that claim that their pumps have minimal required NPSH, it should be recognized that all pumps have significant NPSH requirements. The problem-solver should look with suspicion on any claims that the NPSH requirements are less than those shown in Table 7.1.

In resolving a centrifugal pump problem, the centrifugal pump performance curve supplied by the pump manufacturer should be used to analyze the problem. If the pump is operating as predicted by the performance curve, then the problem

Table 7.1 Minimum NPSH requirements.

Pump Type	NPSH Required, ft
Centrifugal	6
Centrifugal with booster	1–2
Positive displacement	4

is related to high flow rates or piping limitations. If the pump is not operating on the performance curve, then one of the following five things may be occurring:

1) The pump clearances may have worn so that a large amount of liquid is recirculating from the discharge to the suction. This will cause the pump to actually be pumping more fluid than shown by flow meters, which will result in a lower than anticipated discharge pressure.
2) The $NPSH_A$ is not sufficient. If there is not sufficient NPSH available, some of the liquid will vaporize in the pump inlet, causing the actual amount of liquid being pumped to decrease.
3) A vortex is being formed in the suction side storage tank due to the high velocity in the suction line. The vortex causes the vapor to be sucked into the pump suction line. This will cause the pump performance to be below that predicted even if the NPSH is sufficient. This problem can be remedied with the installation of a vortex breaker or larger suction line.
4) The pump impeller is the wrong size. Essentially, all centrifugal pumps can be fitted with impellers of different sizes. Often during routine maintenance, a new impeller of the wrong size is installed in the pump. It should also be noted that most pump curves show multiple impeller sizes. Thus, before the problem-solver can accurately consider a pump problem, he needs to know the impeller size that is installed in the pump under consideration.
5) The pump is being driven by a steam turbine, and it is not running at the design speed. While this is obviously a steam turbine or steam supply problem, it is often presented to the problem-solver as a pump problem.

As indicated for compressors, all instruments should be calibrated and adequately compensated for non-design conditions before beginning to collect any data. The problem-solver should also be concerned about pumps operating too far from their design point. If a pump is operating at very low flow rates, the resulting instability may cause damage to the pump internals. In addition, high flow can result in excessive $NPSH_R$ or excessive horsepower requirements.

Positive displacement pump problems often are either related to inadequate available NPSH or leaking check valves. The calculation of available NPSH for a reciprocating pump was discussed earlier. Leaking check valves in reciprocating pumps result in a similar loss in capacity, as described for reciprocating compressors.

7.9 Example Problem 7.1

A centrifugal compressor (C-100L) handling a gas stream that is mostly nitrogen does not "seem to have" the required process capacity according to the operations

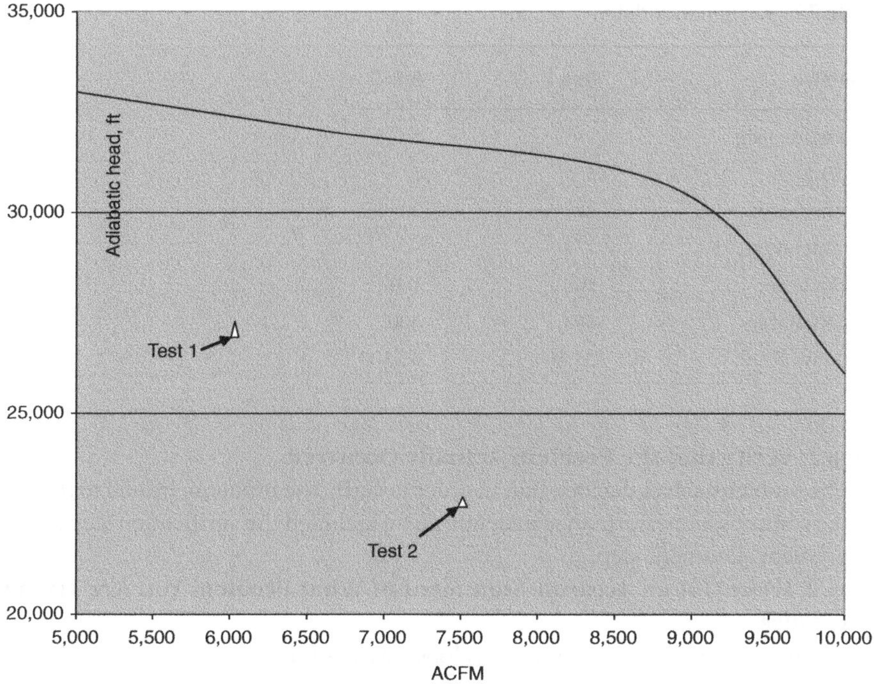

Figure 7.7 Compressor performance curve.

personnel. The compressor is driven by a steam turbine. The problem-solver associated with this plant was asked to determine the cause of the problem and to recommend actions to get the compressor back up to capacity as soon as possible.

Given:

- The compressor curve (Figure 7.7) is available from the compressor vendor.
- Gas molecular weight = 28.
- Gas compressibility (z) = 1.
- Gas specific heat ratio ($k = C_p/C_v$) = 1.4.
- Plant conditions as shown in Table 7.2. All instruments were calibrated prior to taking any of the data.

In analyzing the problem, the problem-solver used the five-step procedure described earlier. He recognized in the words of the operations personnel that they believed the compressor was not performing as it should, but they had not compared the actual compressor performance to the predicted performance.

Table 7.2 Compressor data.

Variable	Run 1	Run 2
Pressure, psig		
Suction	5	5
Discharge	28	24
Temperature, °F		
Suction	100	100
Discharge	308	300
Flow, lb/hr	33,100	41,300

Step 1: Verify that the Problem Actually Occurred

The problem-solver decided that in order to verify the problem, he had to review the compressor performance in detail. So he included the verification step in the problem statement step.

Step 2: *Write* Out an *Accurate* Statement of What Problem You Are Trying to *Solve*

The problem statement that he developed was as follows:

> "Plant personnel indicate that the performance of C-100L is not as good as they remember from the past. The performance of this compressor is causing a production limit which does not seem to be easily overcome. Determine if the compressor is performing as predicted by the compressor manufacturer's performance curve. If it is not performing as predicted, determine what the cause of the problem is and recommend remedial actions."

Note that the above statement emphasizes determining if a problem exists compared to the statement provided by operations personnel, which says determine what the problem is. Problem-solving Steps 3–5 are not initiated until the alleged poor compressor performance can be verified.

To determine whether C-100L was performing as predicted by the performance curve, the problem-solver proceeded and developed Table 7.3.

A reasonable conclusion is that C-100L is performing well below the performance curve at the higher flow rate and is somewhat below the performance curve at the lower rate. In addition to the poor operation relative to the performance curve, the polytropic efficiency is lower at the higher rates. The actual operating points along with the predicted operating curve are shown in Figure 7.7.

Table 7.3 Compressor calculation steps.

1) He calculated **σ** using equation (7.8)		
$T_d = T_s * R^\sigma$		
	Run 1	Run 2
Compression ratio	2.17	1.96
Compression exponent (**σ**)	0.408	0.452
2) He calculated the polytropic head using equation (7.6)		
$H = 1545 * T_S * Z * (R^\sigma - 1)/M\sigma$		
Polytropic head, ft	28,110	24,400
3) He calculated the ACFM (actual cubic ft per minute) at C-100L suction		
Gas density, lb/ft^3	0.0919	0.0919
ACFM	6,000	7,490
4) He looked up the predicted polytropic head from Figure 7.7		
Polytropic head, ft	32,000	31,500
5) He compared the predicted versus actual polytropic heads and expressed these values as a percent		
(Actual – Curve) * 100/Curve	−12	−23
6) He calculated the polytropic efficiency using equation (7.7) and knowing the compression exponent from Step 1 and the specific heat ratio of 1.4		
$\sigma = (k - 1) * 100/(k * E)$		
$(k-1)/k$	0.286	0.286
Polytropic efficiency	70	63
7) He calculated fluid horsepower required using equation (7.9)		
$\text{BHP} = F * H * 100/(33{,}000 * E)$		
Horsepower	672	805

If sufficient data are available, it may be of value to consider when the problem began. The next step is as follows:

Step 3: *Develop* a Theoretically Sound Working Hypothesis that Explains as Many Specifications of the Problem as Possible

Referring back to the list of questions given in Chapter 6, the problem-solver developed the questions and answers shown in Table 7.4. Note that in formulating answers to the questions, the operator's memory of past performance being better than current performance is not taken as a fact. If the operator's memory was taken as factual, the possibility of design or construction errors could be eliminated. Memory that is not backed up by hard data is very elusive.

Table 7.4 Questions/comments for Problem 7.1.

Question	Comment
Are all operating directives and procedures being followed?	All appeared to be correct and being followed
Are all instruments correct?	The instruments had allegedly been calibrated
Are laboratory results correct?	A gas analysis confirmed the molecular weight of the gas
Were there any errors made in the original design?	The piping might not have been designed for the high rates. However, this would not explain subpar operation at the lower rates
Were there changes in operating conditions?	No
Is fluid leakage occurring?	Internal leakage might explain the problem. In addition, the presence of solids might cause a restriction in the suction piping
Has there been mechanical wear that would explain the problem?	Mechanical wear of internals might explain the problem
Is the reaction rate as anticipated?	Not applicable
Are there adverse reactions occurring?	Not applicable
Were there errors made in the construction?	Restrictions in the suction or discharge piping associated with construction might explain the problem

Based on the answers to the questions in Table 7.4, some possible explanations for the problem were proposed as follows:

1) The suction or discharge piping between the pressure gages and the compressor is either restricted or too small.
2) There is a restriction in the compressor suction inlet or discharge outlet piping.
3) The steam turbine driving the compressor is slowing as the compressor horsepower increases.
4) There may be large amounts of compressor internal leakage allowing gases to flow from the discharge to suction side.

Hypothesis 3 was eliminated since the compressor polytropic efficiency decreases with increasing flow rate. An inadequate steam turbine would not cause a decrease in polytropic efficiency. Hypothesis 4 was treated as a lower-priority possibility, even though it would explain a reduced polytropic efficiency. However, it is unlikely to explain why the efficiency decreases as the flow rate increases. The problem-solver decided to test both hypotheses 1 and 2 since they were both theoretically sound working hypotheses. Since he was testing both hypotheses, he postponed formalizing the final working hypothesis until he had done additional work.

Step 4: *Provide* a Mechanism to Test the Hypothesis

Several methods were developed to test these hypotheses. These tests and the results were as follows:

1) The problem-solver calculated the pressure drop in the compressor suction and discharge lines between the pressure instruments and compressor inlet and outlet flanges. These calculations indicated that the calculated pressure drop was minimal in both lines. Thus, he felt like hypothesis 1 could be eliminated.
2) After the pressure drop calculations were made, he arranged for X-rays of the suction and discharge piping to be taken. Based on these X-rays, there appeared to be a significant buildup of solids and/or debris on the bottom of the suction piping before and after the suction pressure instrument.
3) The historical data were reviewed to determine the relationship of the compressor efficiency with time. This relationship is shown in Figure 7.8. It is obvious from this figure that there appears to be a single event in time where the efficiency decreased dramatically. A review of operating data indicated an upset in the suction knockout drum, which might have caused large amounts of solids to be entrained.

Based on this information, the problem-solver developed the following working hypothesis:

> "The loss of capacity of C-100L appears to be due to the accumulation of solids in the suction piping. This accumulation of solids is after the suction pressure gauge. The reduced suction pressure is causing the compressor to appear to operate well below the performance curve at the high rates and 12% below the compressor curve at the lower rates. This reduced suction pressure is also causing an apparent loss in compressor efficiency that is more noticeable at high rates. The loss in performance is more noticeable at high rates because the pressure drop caused by the restriction is greater. The presence of solids in the piping may be related to an upset that occurred in the compressor suction drum."

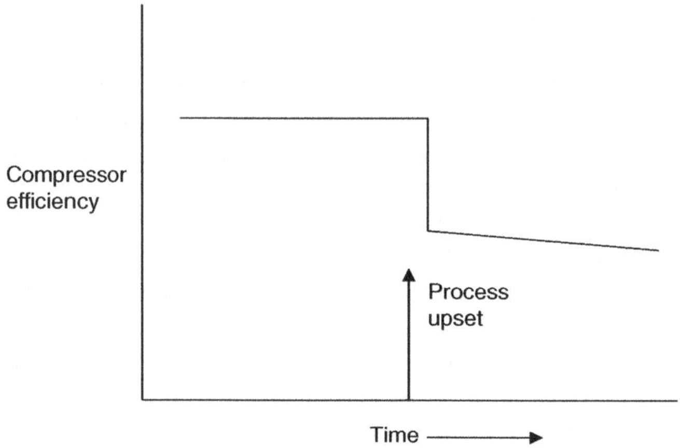

Figure 7.8 Compressor efficiency versus time.

Step 5: *Recommend* Remedial Action to Eliminate the Problem without Creating Another Problem

Since the X-rays indicated the likely presence of solids as the most reasonable cause of the loss of compressor capacity, plans were formulated to shut down the system and clean out the suction piping. While cleaning out the suction piping did eliminate the current compressor limitation, additional problem-solving effort was required. The problem-solver next reviewed the upset that appeared to cause solids to be entrained and developed means to prevent this from happening again.

7.10 Lessons Learned

Problems should always be investigated in as quantitative fashion as possible. For example, in this problem, the first thing that the problem-solver did after developing the problem statement was to determine where the compressor was operating relative to an absolute criterion, such as the head-flow curve provided by the compressor manufacturer. It would have been possible to launch into a hypothesis development period before comparing the actual operation to the predicted head curve. However, it would have led to giving equal validity to erroneous hypotheses and valid hypotheses. For example, it was only when the predicted and actual performances were compared at high and low rates that hypotheses 3 and 4 (that there was internal leakage and that the steam turbine was slowing down) could be eliminated.

While hypotheses 1 and 2 were equally valid and could well have been pursued in parallel, it is almost always faster to make calculations to confirm a hypothesis rather than to arrange for elaborate testing. The calculations of pressure drop in the suction piping indicated that the additional cost and time associated with X-raying the suction piping was indeed of value.

This problem also illustrates the flexibility of the five-step approach. The problem-solver did not feel that he had sufficient data to propose a hypothesis until he had done some additional work (Step 4), including calculations and X-rays. The five-step approach should not be used to fit all problem-solving activities into a single exact method.

7.11 Example Problem 7.2

A process that had been idled for 12 months due to a decrease in demand was now required to operate at full rates. A centrifugal pump (P-25) in the process operated as anticipated until the design flow rate was approached. At that point, it no longer operated as predicted by the pump curve. As the unit was returned to full capacity, the level in the accumulator drum upstream of the pump had been decreased to provide for more surge capacity at the higher rates. The pump curve is shown in Figure 7.9. The pump was handling a liquid hydrocarbon at its vapor pressure.

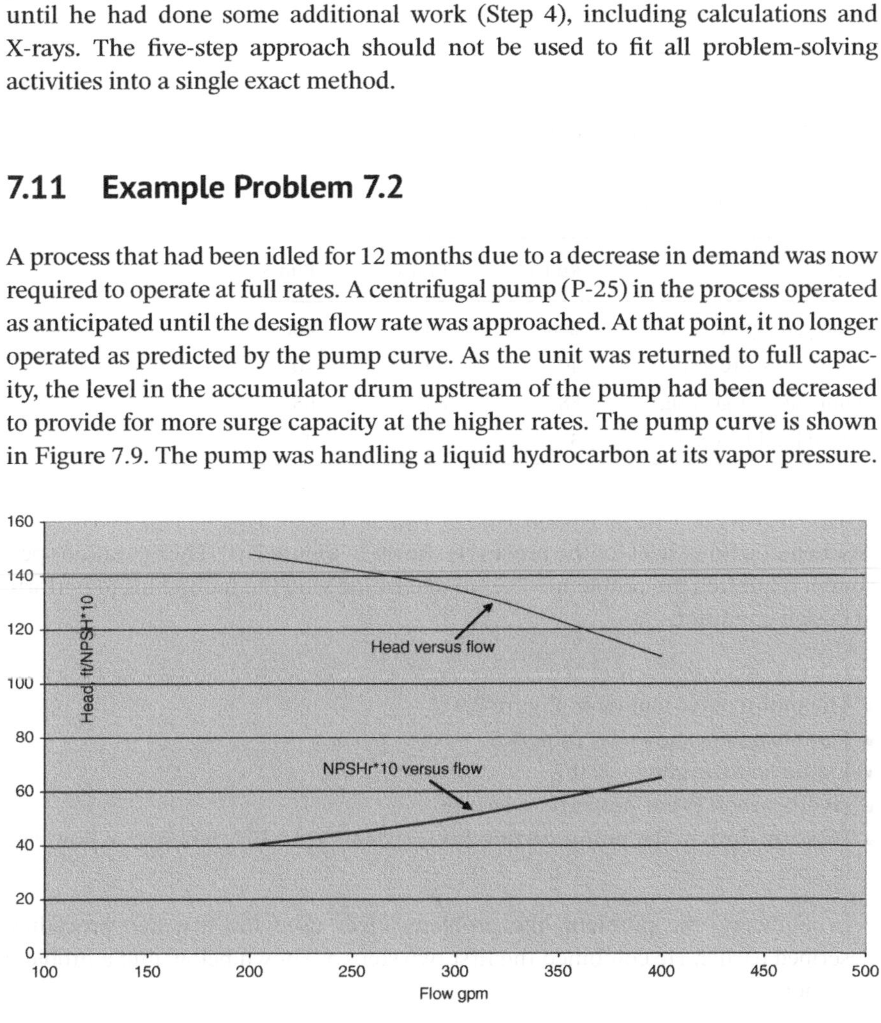

Figure 7.9 Pump curve for Problem 7.2.

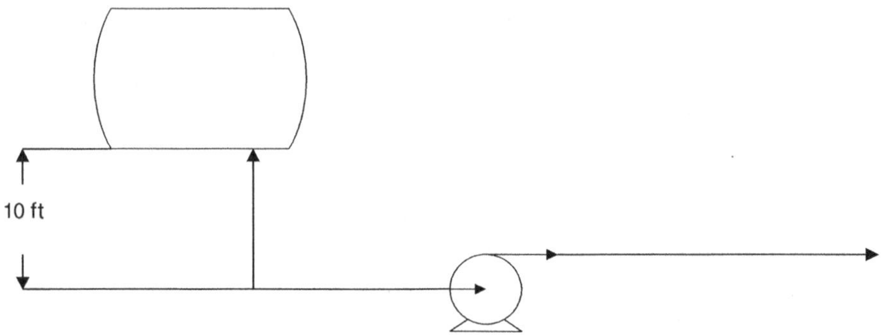

Figure 7.10 Schematic flow for Problem 7.2.

Table 7.5 Plant test data.

Test	Run 1	Run 2	Run 3
Flow rate, gpm	200	300	400
Drum pressure, psig	40	40	40
Pump discharge, psig	88	83	65
Drum liquid level, ft	3	3	4

A schematic flow sheet for the process is shown in Figure 7.10. The operations personnel requested the problem-solver to determine why the pump fails to perform at design conditions.

Given:

- The pump curve shown in Figure 7.9
- Plant test data shown in Table 7.5
- Liquid specific gravity = 0.7
- Hydrocarbon vapor pressure = 40 psig
- Pressure drop in the pump suction line = $0.0000528 * F^{1.8}$, where F = flow rate in gpm.

In analyzing the problem, the problem-solver used the five-step procedure described earlier. He combined the first two steps as shown below into a problem statement.

Step 1: Verify that the Problem Actually Occurred
Step 2: *Write* Out an *Accurate* Statement of what Problem You Are Trying to *Solve*

Table 7.6 Performance calculations.

	Run 1	Run 2	Run 3
1) He calculated the pressure at the pump suction using the relationship given between flow and pressure drop			
Suction line pressure drop, psi	0.732	1.52	2.55
Elevation head, psi	3.94	3.94	4.25
Suction pressure, psig	43.2	42.4	41.7
2) He calculated the pump differential head knowing that the specific gravity relative to water was 0.7			
Discharge pressure, psig	88	83	65
Differential pressure, psig	44.8	40.6	23.3
Differential head, ft	148	134	77
3) He calculated the pump differential head knowing that the specific gravity relative to water was 0.7			
Curve differential head, ft	148	135	110
Error, percent	0	0.7	30

The problem statement that he developed was as follows:

"P-25 has been reported to be operating off the pump curve as the design flow rate is approached. It is probable that this problem was only noticed as the design rates were required after a 12-month period of low-rate operation. The level in the accumulator was decreased when the unit was returned to full rates. Since this pump is creating a major production limitation, work to investigate the problem is planned. The first step will be to confirm that P-25 operates on the pump manufacturer's pump curve until the design rate is approached. If the pump does not operate on the curve at all rates, determine the cause of the inadequate performance." To confirm whether P-25 is operating on the vendor pump curve, the problem solver did the calculations shown in Table 7.6. The problem solver concluded that P-25 was operating close to the pump curve until the flow rate reached 400 gpm. The next step was

Step 3: *Develop* a Theoretically Sound Working Hypothesis that Explains as Many Specifications of the Problem as Possible

Again, the questions given in Chapter 6 were helpful to the problem-solver in formulating possible hypotheses. These questions and appropriate comments developed by the problem-solver for this example problem are shown in Table 7.7.

Table 7.7 Questions/comments for Problem 7.2.

Question	Comment
Are all operating directives and procedures being followed?	All appeared to be correct and being followed. The operating directive for the level in the drum had recently been reduced
Are all instruments correct?	The instruments had allegedly been calibrated
Are laboratory results correct?	Not applicable
Were there any errors made in the original design?	The piping might not have been designed for the high rates or the increasing $NPSH_R$ as the flow increased might not have been considered
Were there changes in operating conditions?	No, except for the increased flow and reduced level
Is fluid leakage occurring?	Internal leakage might explain the problem
Has there been mechanical wear that would explain the problem?	Deterioration of wear rings might explain the problem
Is the reaction rate as anticipated?	Not applicable
Are there adverse reactions occurring?	Not applicable
Were there errors made in the construction?	Not applicable. There had been no recent construction

Based on this series of questions, two possible hypotheses were developed as follows:

1) The suction piping is too small or is restricted, causing insufficient NPSH at the pump inlet flange.
2) Deterioration of internal pump wear rings might be causing internal leakage so that the pump is recirculating large quantities of fluid. This will cause the pump to appear to be operating off the performance curve.

Following the concept of doing calculations prior to making recommendations to do mechanical work, the calculations shown in Table 7.8 were done. Since at the highest rate (400 gpm) the $NPSH_R$ (7 ft including the safety factor) is greater than the $NPSH_A$ (5.6 ft), the problem-solver developed the following working hypothesis:

> "The failure of P-25 to operate on the pump curve at high rates is due to the fact that the NPSH is not sufficient."

Table 7.8 Pump calculations for Problem 7.2.

	Run 1	Run 2	Run 3
Vapor pressure of the liquid, psig	40	40	40
Suction pressure at the pump from Table 7.6, psig	43.2	42.4	41.7
NPSH available, psi	3.2	2.4	1.7
NPSH available, ft	10.5	7.9	5.6
NPSH required, ft			
From curve	4	5	6.5
Safety factor	0.5	0.5	0.5
Total	4.5	5.5	7

Step 4: *Provide* a Mechanism to Test the Hypothesis

The problem-solver developed two mechanisms to test the hypothesis. The suction line could be replaced to reduce the pressure drop in the suction line. A larger suction line that reduced the pressure drop to less than 1 psi would increase the $NPSH_A$ to a value above that required by the pump. Another possible mechanism to test the hypothesis would be to increase the level in the drum. If the level was increased by at least 1.5 ft, the $NPSH_A$ should be adequate.

Step 5: *Recommend* Remedial Action to Eliminate the Problem without Creating Another Problem

The remedial action depends on which of the hypothesis testing mechanisms is used. If the suction line is replaced and the problem is eliminated, this mechanism becomes the remedial action. If increasing the drum level is a successful test, this may or may not be a remedial action. Increasing the drum level reduces the amount of surge volume that can be used. A careful study would be required to determine if this action creates other problems.

7.12 Lessons Learned

While the available data did not allow the problem-solver to eliminate the hypothesis that there was a mechanical problem with the pump that allowed it to recirculate large amounts of fluid, he could by calculations show that the hypothesis of inadequate available NPSH was a correct hypothesis. Thus, if the pump had no mechanical problems at all, it still would not perform as predicted by the pump curve. Therefore, the problem-solver elected to develop solutions to the lack of available NPSH.

Often in a process unit, operating directives that are changed and seem to work fine at reduced rates are not adequate when the rates are increased to design levels. In addition, designs that work at less than design rates may not work at full rates. In the specific example problem, the reduction in the drum level seems to be the cause of the lack of available NPSH. However, additional study would be required before increasing the drum level would be considered an acceptable solution.

7.13 Example Problem 7.3

(This problem is provided to illustrate a calculation technique) A reciprocating compressor (C-100A) was operating in a mode that results in an overload condition on the electric motor. The compressor was equipped with adjustable clearance pockets so that the clearance as a fraction of displacement could be varied. What clearance pocket setting should be used to obtain the maximum flow rate without overloading the electric motor?

Given:

• Clearance pocket setting, percent of displacement	4 or 6 or 10
• Suction pressure, psig	15
• Discharge pressure, psig	85
• Compressibility factor (z)	1
• Molecular weight	42
• Polytropic compression exponent	0.32
• Suction temperature, °F	100
• Ratio of specific heats	1.4
• Polytropic efficiency	90
• Mechanical efficiency	95
• Piston displacement, ft^3/min	300
• Motor rating, HP	55

The following procedure shown in Table 7.9 can be used to analyze this problem.

7.14 Example Problem 7.4

The compressor described in Problem 7.3 began losing capacity. The flowmeter showed about 35 lb/min. Estimates of motor loading based on ampere readings indicated that the compressor was requiring significantly less than 50 BHP. All

Table 7.9 Calculation procedure.

1. Calculate the compression ratio:

$$R = 99.7/29.7 = 3.36 \tag{7.18}$$

- Calculate the polytropic head using equation (7.6):

$$H = 1545^{*}T_S{}^{*}Z^{*}(R^{\sigma} - 1)/M\sigma \tag{7.19}$$

$$H = (1545^{*}(100 + 460) * 1^{*}(3.36^{0.32} - 1))/(42^{*}0.32) = 30,500\text{ feet}$$

- Calculate the volumetric efficiency for each clearance using equation (7.14):

$$E_V = 100^{*}(1 - L - C^{*}(R^{1/k} - 1)) \tag{7.20}$$

The leakage factor (L) can be evaluated as 2% of the compression ratio as follows:

$$L = 2^{*}R/100 = 0.07 \tag{7.21}$$

Clearance, percent	4	6	10
Clearance factor (C)	0.04	0.06	0.10
Leakage factor (L)	0.07	0.07	0.07
$R^{1/k} - 1$	1.38	1.38	1.38
Volumetric efficiency, percent	87	85	79

- Calculate the mass flow rate and BHP. Use equation (7.9) to calculate the BHP:

$$\text{BHP} = F^{*}H^{*}100/(33,000^{*}E)$$

Volumetric flow rate, ft^3/min	261	255	237
From piston displacement and volumetric efficiency			
Gas density, lb/ft^3	0.208	0.208	0.208
Mass flow rate, lb/min	54.3	53	49.3
BHP	58.7	57.3	53.3

Note the BHP includes both mechanical and polytropic efficiency.
Conclusion: The 10% pocket setting must be used to avoid overloading the motor.

operating conditions were identical to those specified in Problem 7.3. In addition, the pocket clearance was set at 10%. When the problem-solver was asked to consider this problem, he used the five-step approach discussed earlier.

Step 1: Verify that the Problem Actually Occurred

Verification that the problem was actually occurring was relatively simple since both the flowmeter and the ampere reading indicated that the compressor output was less than its rated capacity.

Step 2: *Write* Out an *Accurate* Statement of What Problem You Are Trying to *Solve*

The problem-solver wrote out the following problem statement:

> "C-100A seems to be operating well below the design capacity. The flow meter shows 35 lbs/min where the flow should be close to 50 lbs/min. In addition, the horsepower loading is well below the anticipated load of 50 BHP as determined by an ampere reading. The compressor unloading pockets are set so that the clearance is about 10% which is normal. It is not known when the decrease in capacity actually occurred, but it seems to have been a gradual decrease. Determine what has caused the loss of capacity on C-100A (the reciprocating compressor), why it occurred, and what can be done to eliminate the problem."

Step 3: *Develop* a Theoretically Sound Working Hypothesis that Explains as Many Specifications of the Problem as Possible

The questions given in Chapter 6 were used to help formulate possible hypotheses. These questions and appropriate comments developed by the problem-solver for this example problem are shown in Table 7.10.

Table 7.10 Questions/comments for Problem 7.4.

Question	Comment
Are all operating directives and procedures being followed?	All appeared to be correct and being followed
Are all instruments correct?	All instruments except the flowmeter had been calibrated
Are laboratory results correct?	Analysis of gas being compressed had not been obtained for several months
Were there any errors made in the original design?	Not applicable since the compressor capacity loss was a recent occurrence
Were there changes in operating conditions?	No
Is fluid leakage occurring?	Internal leakage might explain the problem
Has there been mechanical wear that would explain the problem?	Deterioration of compressor valves might explain the problem. In addition, the belt on the electric motor was beginning to look worn
Is the reaction rate as anticipated?	Not applicable
Are there adverse reactions occurring?	Not applicable
Were there errors made in the construction?	Not applicable

Several possible hypotheses were developed by the problem-solver as follows:

1) The valves in the compressor are bad, causing loss in capacity.
2) The flowmeter is erroneously low.
3) The worn belt drive between the electric motor and compressor is slipping, causing the compressor to operate at a much lower speed than design.
4) The molecular weight of the gas being compressed is different than that of the design.
5) The actual suction pressure is much lower than shown by the instrumentation.

The problem-solver developed a table to attempt to sort out the various hypotheses by comparing the theoretical impact of the hypotheses to the actual observations. This is shown in Table 7.11.

Since "slipping belts" (hypothesis 3) will almost always be heard even above the noise in a compressor house, the problem-solver proposed the residual theoretically sound working hypothesis as follows:

> "The poor performance of C-100A is due to leakage through the suction and/or discharge valves."

Table 7.11 Hypothesis comparison.

	Would Hypothesis Explain Observations of:	
Hypothesis (number)	**Low Ampere Reading**	**Low Flow Rate**
Bad valves (1)	Yes	Yes
Flowmeter (2)	No	Yes
Slipping belt (3)	Yes	Yes
Molecular weight (4)	No[a]	?
Low suction pressure (5)	?[b]	Yes

a) For a reciprocating compressor, the horsepower load is essentially independent of molecular weight. This is true because head is indirectly related to molecular weight (see equation (7.6)), and the mass rate is directly related to the molecular weight. For example, if the molecular weight of a gas being compressed doubles, the polytropic head will be reduced by a factor of 2 and the amount of gas being compressed will increase by a factor of 2. This assumes that the suction and discharge pressures are constant, and the only change is the molecular weight of the gas.

b) Whether or not a low suction pressure can explain the decrease in horsepower requirements depends on which side of the load point the compressor is operating at normal conditions. See Figure 7.6 for a typical load versus suction pressure curve. For this specific compressor, it is likely that decreasing the suction pressure will have minimal impact on the horsepower requirements.

Step 4: Provide a Mechanism to Test the Hypothesis

The actual mechanism for testing the hypothesis involved determination of "optimum technical depth" required as described in Chapter 4. Some points that were considered in determining the "optimum technical depth" were as follows:

- How urgent was the problem? Was the loss of capacity causing a loss in plant production or could gases be diverted to another location?
- Was a spare compressor available? If a spare compressor were available, the required degree of confidence that the solution was correct would be less than if no spare was available.
- Would a compressor shutdown require an entire unit shutdown? If a unit shutdown were required, the required degree of confidence would be much higher.
- Could inlet and outlet temperatures be easily measured and were historical data available? If these temperatures could be easily obtained and compared to historical data, it would increase the degree of confidence in the solution.

The two possibilities were that the current and historical inlet and outlet temperatures were easily available. In this case, the compression exponent could be calculated using equation (7.8) as shown below:

$$T_{\mathrm{d}} = T_{\mathrm{s}}{}^{*}R^{\sigma} \tag{7.8}$$

Then, archived data could be used to calculate the historical compression exponent (σ). The current data could be compared to historical data, and if a change were obvious, it could be concluded that this was a valid test of the hypothesis. If no historical data are available, equation (7.7) can be used to calculate the theoretical compression exponent from the vendor's performance curve.

$$\sigma = (k-1) * 100/(k^{*}E) \tag{7.7}$$

This theoretical compression exponent can then be compared to the actual exponent based on plant operating data. If compressor inlet and outlet temperatures are not readily available, it may be possible to use infrared techniques discussed in Chapter 13 to approximate compressor temperatures.

In this particular problem, a spare compressor was not available. An analysis of the inlet and outlet temperatures indicated that the deterioration of performance had been a gradual decline. This analysis is shown in Table 7.12.

Step 5: Recommend Remedial Action to Eliminate the Problem without Creating Another Problem

Thus unfortunately, it was necessary to recommend a compressor shutdown to replace the worn valves. Since there was no spare compressor and the compressor was an integral part of the unit, it was necessary to shut down the entire unit. As seems to be the rule, the shutdown occurred at a bad time from a business perspective.

Table 7.12 Analysis of the compression exponent for Problem 7.4.

		Days after Last Valve Replacement			
Variable	**Design**	**0**	**30**	**60**	**90**
Suction					
Temperature, °F	70	70	70	70	70
Pressure, psig	15	15	15	15	15
Discharge					
Temperature, °F	320	320	331	340	360
Pressure, psig	85	85	85	85	85
Compression ratio	3.36	3.36	3.36	3.36	3.36
Compression exponent	0.32	0.32	0.33	0.34	0.36

7.15 Lessons Learned

Replacement of valves on a reciprocating compressor is a "normal maintenance" item. However, since this compressor was in a critical service, the problem-solver developed a system to allow including the compression exponent in a daily monitoring system. His recommendation to do this allowed optimization of the timing for a compressor downtime. In addition, it allowed him to follow the valve performance closely and to possibly correlate valve performance with process upsets or changes that are causing rapid valve wear.

The importance of understanding technology is also illustrated by this problem. The original hypothesis that a change in the molecular weight of the gas being compressed was eliminated because a reciprocating compressor horsepower is essentially independent of molecular weight. In addition, the possibility that a low suction pressure is causing the problem was eliminated since with this specific reciprocating compressor, suction pressure has minimal effect on horsepower load. These conclusions are not intuitively obvious and can only be developed by understanding compression technology.

7.16 Example Problem 7.5

(This problem is provided to illustrate a calculation technique) Determine the $NPSH_A$ for a reciprocating pump given the following:

Pump Type	Double Acting
Cylinder diameter	1/2 in
Stroke length	4 in
Speed	60 rpm
Suction line diameter	1 in
Suction line length	25 actual feet
Suction drum pressure	40 psig
Liquid elevation	10 ft above pump suction
Fluid vapor pressure	40 psig
Fluid specific gravity	0.65 relative to water
Fluid viscosity	0.2 cP

As indicated in Section 7.6, there are two kinds of pressure drop in the suction line of a reciprocating pump. There is the typical type of frictional loss, but it occurs at maximum flow rate, which is greater than the average flow rate. In addition, the entire volume of fluid in the suction piping must be accelerated from a non-flow condition to the maximum velocity. As indicated in this section, these two frictional losses occur completely out of phase with each other. That is, the frictional loss associated with the maximum flow rate occurs when there is no acceleration. Also, the acceleration loss from zero velocity to full line velocity occurs when there is no frictional loss. Thus, the pressure loss in the pipe is generally taken as the larger of the two calculated values. The calculations for estimating these pressure losses are based on equations (7.11) and (7.12). The calculation procedure is as follows:

1) Calculate the maximum flow rate in the suction pipe:

$$V_P = K^* D^{2*} L^* S / \left(60^* D_P^2\right) \tag{7.22}$$

$$= \pi^* (0.0417^2)^* \, 0.33^* 60 / (2^* 60^* (0.0833^2))$$

$$= 0.13 \text{ fps} \tag{7.22}$$

2) Calculate the frictional pressure drop by conventional means. This calculation is not shown.

$$h_F = 0.0069 \text{ ft}$$

3) Calculate the head required to accelerate the fluid.

$$h_A = 1.35^* L_P{}^* S^* V_P / 307 \tag{7.23}$$

$$= 1.35^* 25^* 60^* 0.13 / 307$$

$$= 0.85 \text{ ft} \tag{7.24}$$

4) Select the larger of the two values to determine the actual NPSH for the system.

In this specific problem, the acceleration head will be the critical pressure loss. Actual NPSH at the pump suction will be 10 ft less 0.85 ft or 9.15 ft. This available NPSH will normally be adequate for most reciprocating pumps.

7.17 Example Problem 7.6

A centrifugal process gas compressor C-5A driven by a steam turbine no longer had sufficient capacity for the service that it was designed for. The speed of the turbine was controlled by a steam control valve in the incoming steam. There had allegedly been no changes in operating conditions, except for a decrease in steam pressure from 200 to 190 psig. This decrease in steam pressure was part of an overall optimization of the plant utility system. In order to increase the steam pressure slightly, operations personnel removed the steam meter orifice that was used to measure the steam flow to the steam turbine. Operations personnel had requested technical help to prove that the steam pressure should be increased back to 200 psig.

The design and current operating conditions are shown in Table 7.13.

The maximum design of the steam turbine included a slight safety factor to allow operations at 10,000 rpm with the steam control valve 90% open. As noted in the table, even with the steam valve 100% open, it is not possible to reach 9000 rpm. The design conditions are those required for the process gas compressor.

The problem-solver approached the problem using the five-step procedure as shown below.

Step 1: Verify that the Problem Actually Occurred

As the first step in verifying that the problem was real, the problem-solver had the appropriate instruments checked. He also used independent sources to confirm the flow and pressure meters. While the compressor output had not been followed on a daily basis, key variables shown in Table 7.13 were available from computer archives. All of these indicated that there was a real problem.

Step 2: *Write* Out an *Accurate* Statement of What Problem You Are Trying to *Solve*

The problem-solver wrote out a description of the problem that he was trying to solve as follows:

> "Currently C-5A does not seem to have the desired capacity, even though capacity was adequate in recent history. The steam turbine is operating at slightly lower speed than the design values and the steam control valve is fully open. There has been a reduction in inlet steam pressure from 200 psig to 190 psig. However, it is not clear that this is the cause of the lack of capacity. Determine why C-5A does not have sufficient capacity and provide recommendations for improving the performance of C-5A."

Table 7.13 Design and current operations.

Variable	Design	Current Operations
Suction pressure, psig	5	5
Discharge pressure, psig	45	35
Suction temperature, °F	100	100
Discharge temperature, °F	301	264
Flow rate, lb/hr	25,000	25,000
Gas composition		
Propylene, mol percent	95	95
Nitrogen, mol percent	5	5
Calculated molecular weight	41.3	41.3
Compressibility	1.0	1.0
Specific heat ratios		
Propylene	1.21	1.21
Nitrogen	1.4	1.4
Mixture	1.2195	1.2195
Turbine speed, rpm		
Maximum	10,000	10,000
Design	9,000	
Actual		8,400
Inlet steam conditions psig	200	190
Temperature, °F	500	485
Outlet steam conditions psig	25	25
Temperature, °F	280	295
Steam flow rate, lb/hr		Out of service
Maximum	16,000	
Design	14,700	
Steam control valve Position, percent	80	100

Step 3: *Develop* a Theoretically Sound Working Hypothesis that Explains as Many Specifications of the Problem as Possible

The questions given in Chapter 6 were helpful in formulating possible hypotheses. These questions and appropriate comments for this example problem are shown in Table 7.14.

Table 7.14 Questions/comments for Problem 7.6.

Question	Comment
Are all operating directives and procedures being followed?	All appeared to be correct and being followed, except for the lower steam pressure
Are all instruments correct?	All instruments except the steam flowmeter had been calibrated
Are laboratory results correct?	Laboratory results indicating that the gas being compressed was 95% propylene and 5% nitrogen were confirmed
Were there any errors made in the original design?	The original design was okay, but there was a question about whether the change in steam pressure had received adequate consideration
Were there changes in operating conditions?	Yes. The inlet steam pressure was reduced
Is fluid leakage occurring?	Internal leakage might explain the problem
Has there been mechanical wear that would explain the problem?	Deterioration of the steam turbine nozzles might explain the problem
Is the reaction rate as anticipated?	Not applicable
Are there adverse reactions occurring?	Not applicable
Were there errors made in the construction?	Not applicable

Several hypotheses were proposed as follows:

- When the steam system was optimized, the process designer did not adequately consider the impact of steam pressure on the steam turbine.
- There is excessive leakage around the wear rings on the process gas compressor. This excessive leakage would cause a lower than design compressor efficiency and a reduced gas rate.
- The steam turbine steam jets may have deteriorated. This would cause the steam turbine to have a lower efficiency and thus extract less horsepower per pound of steam than it was designed to do.

At this point, the problem-solver had developed three theoretically correct working hypotheses. All of the three could be evaluated further by additional calculations. Since two of the hypotheses would require compressor shutdowns to inspect and the compressor was not spared, he elected to pursue additional calculations as part of Step 4.

Step 4: Provide a Mechanism to Test the Hypothesis

The problem-solver made the following calculations to help determine if one of these proposed hypotheses was both possible and supported by calculations.

1) He determined the polytropic compression exponent using equation (7.8) as shown in Table 7.15:

$$T_d = T_s{}^*R^{\sigma} \tag{7.8}$$

Since the polytropic compression exponent is the same for the design and operating conditions, the problem-solver concluded based on equation (7.7) that the compressor efficiency was the same as the design efficiency.

$$\sigma = (k - 1) * 100/(k^*E) \tag{7.7}$$

He calculated the steam turbine efficiency using equation (7.17) and Table 7.16:

$$E_T = 100^*(H_1 - H_O)/(H_1 - H_E) \tag{7.17}$$

It is of value to review how the calculations shown in the table were actually done. The inlet and outlet steam enthalpies were taken from steam enthalpy tables given in multiple publications. These values were used to determine the actual enthalpy change across the turbine. The value shown in the table as “at 100% efficiency” assumes isentropic (constant entropy) steam expansion. The most efficient steam turbine process is one that occurs at constant entropy. The outlet steam conditions are then based on the outlet pressure and maintaining the same entropy as the inlet steam conditions. The outlet enthalpy is then determined from steam charts or steam tables. The isentropic enthalpy change can then be determined. The efficiency is simply the actual enthalpy change divided by the isentropic enthalpy change and multiplied by 100.

As shown in Table 7.16, the current efficiency of the steam turbine appeared to be less than design. Thus, it seemed unlikely that the current capacity loss was

Table 7.15 Calculation of the polytropic compression exponent.

Variable	Design	Operating
Suction pressure, psig	5	5
Discharge pressure, psig	45	35
Suction temperature, °F	100	100
Discharge temperature, °F	301	264
Calculated compression ratio	3.03	2.52
Polytropic compression expon.	0.277	0.278

Table 7.16 Calculation of steam turbine efficiency.

Variable	Design	Operating
Inlet steam conditions psig	200	190
Temperature, °F	500	485
Enthalpy, BTU/lb	1267	1260
Outlet steam conditions psig	25	25
Temperature, °F	280	308
Enthalpy, BTU/lb	1176	1191
Enthalpy change, BTU/lb Actual	91	69
At 100% efficiency	140	138
Steam turbine efficiency, percent	65	50
Calculated steam rate, lb/hr	14,700	15,800

associated with the steam pressure change. Since archived data were available, the problem-solver reviewed the historical data to determine when the loss of efficiency occurred and whether it was a one-time loss or a gradual loss. The problem-solver used the techniques shown above and developed Figure 7.11. This figure clearly showed that the steam turbine problem was not related to the drop in steam pressure, but was a gradual decay in efficiency that began well after the change in steam pressure.

The problem-solver had one more question to answer—that is, would an increase in steam pressure back to 200 psig ameliorate the loss of turbine capacity? To answer this question, he calculated the current horsepower load and compared it to the design horsepower load.

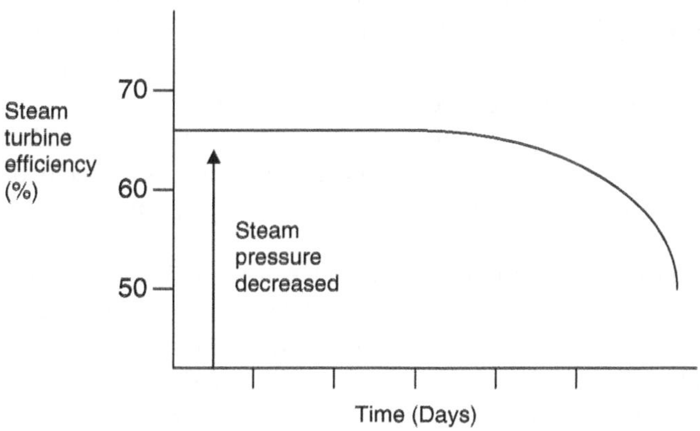

Figure 7.11 Steam turbine efficiency versus time.

Table 7.17 Compressor horsepower calculations.

Variable	Design	Operating
Suction pressure, psig	5	5
Discharge pressure, psig	45	35
Suction temperature, °F	100	100
Discharge temperature, °F	301	264
Flow rate, lb/hr	25,000	25,000
Polytropic efficiency	65	65
Polytropic compression expon.	0.277	0.278
Calculated polytropic head	27,200	22,100
Fluid horsepower	530	430

Equations (7.6) and (7.9) shown below and data given in Table 7.13 were used to calculate the polytropic head and fluid horsepower. These equations were then used to develop Table 7.17.

$$H = 1545^* T_S{}^* Z^* (R^\sigma - 1)/M\sigma \tag{7.6}$$

$$\sigma = (k-1) * 100/(k^* E) \tag{7.7}$$

$$\text{BHP} = F^* H^* 100/(33,000^* E) \tag{7.9}$$

He then considered whether the increase in steam pressure from 190 to 200 psig would cause the compressor to return to normal. To consider this change in steam pressure, the problem-solver first estimated the current steam rate knowing the delivered fluid horsepower (430 BHP) and the enthalpy change across the turbine. He then assumed that steam conditions were returned to the higher pressure, the steam rate increased slightly due to the higher pressure, and that the turbine efficiency remained the same (50%). He then calculated the horsepower that would be delivered to the process as shown in Table 7.18.

Based on the calculations shown in Table 7.18, the problem-solver concluded that raising the steam pressure would not significantly increase the capacity of the process gas compressor. He then developed the following hypothesis:

> "It is believed that the steam turbine steam jets have suffered mechanical damage which has resulted in the gradual deterioration of their efficiency. This would cause the steam turbine to have a lower efficiency and thus extract less horsepower per pound of steam than it was designed to do. Increasing the steam pressure from 190 psig to 200 psig would have minimal impact on the turbine."

Table 7.18 Calculation of BHP delivered to the compressor with increased steam pressure.

	Operating Conditions		
Variable	**Design**	**Current**	**Proposed**
Inlet steam conditions			
psig	200	190	200
Temperature, °F	500	485	500
Enthalpy, BTU/lb	1,267	1,260	1,267
Outlet steam conditions			
psig	25	25	25
Temperature, °F	280	308	TBD
Enthalpy, BTU/lb	1176	1191	TBD
Enthalpy change, BTU/lb			
At 100% efficiency	140	138	140
Steam turbine efficiency	65	50	50
Actual enthalpy change	91	69	70
Steam rate	14,700	15,800	16,300
Horsepower to process	530	430	450

Step 5: Recommend Remedial Action to Eliminate the Problem without Creating Another Problem

Since all indications were that there was some sort of mechanical damage to the internals of the steam turbine and that the turbine efficiency was continuing to deteriorate, the problem solver had no choice but to recommend that the steam turbine be shut down for repairs. In addition, the time relationship indicated that the repairs should be done as soon as possible. Since safety did not appear to be involved, the actual timing of the shutdown was left to the discretion of the management team.

The management team was pleased with the detailed analysis of the problem because

1) The initial apparent cause of the problem (the steam pressure reduction) would have been difficult to reverse since it involved several processes in the plant.
2) The relationship of the turbine efficiency with time that was developed was helpful in determining that the steam turbine should be repaired very quickly.

7.18 Lessons Learned

The value in doing calculations and time-related figures was clearly illustrated by this real-life problem. Before the problem-solver could determine which piece of equipment was the source of the problem, he had to determine the performance of both the compressor and turbine. He determined this by comparing the actual efficiency to the design efficiency. Knowing that the compressor efficiency was essentially the same as the design and that the steam turbine efficiency was less than design allowed him to conclude that the operating problem was related to the steam turbine. In addition, an analysis of the steam turbine supply pressure allowed him to conclude that decreasing the steam pressure was not the cause of the problem and that increasing it back to the original setting would not resolve the performance discrepancy.

If these calculations had not been done, there would be three possible hypotheses that would all be treated as valid. Without the calculations, the most easily identified change would be the reduction in steam pressure. It would also appear to be the easiest route to improving performance. If the problem-solver had not done the discussed calculations, it is likely that increasing the steam pressure would have been chosen as the route to improve performance. Valuable time that could have been spent in doing a better job in assessing the problem would have been spent in getting the steam pressure increased.

7.19 Example Problem 7.7

A new process had recently been put into operation. While the start-up had gone very well, there was a continuing compressor problem. The main recycle gas compressor would mysteriously shut down. The 1500-BHP two-stage reciprocating compressor was driven by an electric motor. This motor was provided with a shutdown device that was triggered if the horsepower load was 3% greater than the motor rating. Thus, the compressor motor would shut down if the load exceeded 1550 BHP.

The compressor was provided with a "first out" indicator that was used to determine which process variable caused the electric motor to shut down. The "first out" indicator included process variables such as oil pressure, high discharge temperatures, high suction temperatures, and low suction pressure. High suction pressure was not included in the "first out" indicator. However, there was a high suction pressure alarm to provide a warning prior to the pressure increasing to the point that the safety valve would release. The indicator panel also had a category called "other reason." Unfortunately, the indicator almost always showed "other reason" for the shutdowns being experienced.

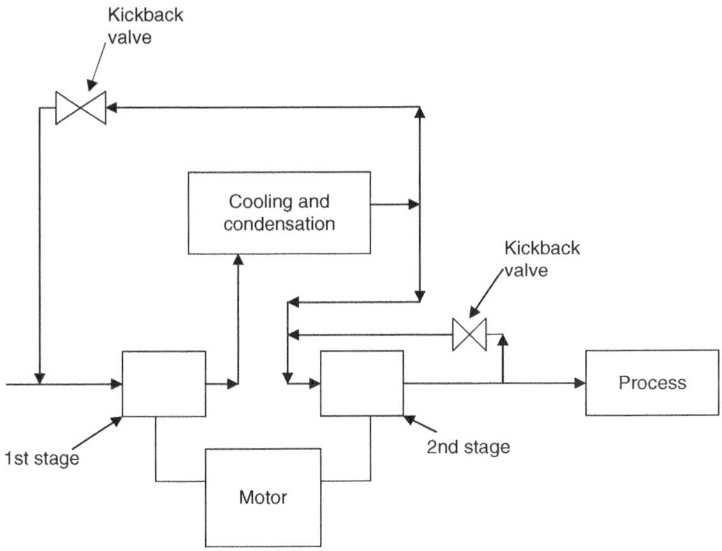

Figure 7.12 Schematic flow for Problem 7.7.

The process gas load was not constant. The gas load varied as the reactor conversion changed. If the reactor conversion decreased, the gas load increased. The process design of the facilities assumed that the conversion would remain constant. Decreases in the conversion were associated with the presence of impurities in the reactor feed or changes in the catalyst flow rate to the reactor.

The compressor was provided with computer-controlled unloaders and recycle capabilities to maintain a constant suction pressure on both stages. If the gas load were such that both the recycle valves were closed and the unloaders were in a position to compress the maximum amount of gas, the suction pressure would increase until the gas density increased to the point that the compressor had adequate capacity. A schematic sketch of the facilities is shown in Figure 7.12.

The key variables for the compressor are given in Table 7.19.

Operations personnel have requested help from the technical organization to determine what was wrong with the "first out" indicator. The problem-solver approached the problem using the five-step procedure as shown below.

Step 1: Verify that the Problem Actually Occurred

There was no doubt that the compressor was shutting down. However, it was not obvious that there was anything wrong with the "first out" indicator. There were times when it showed indications besides "other reason." In addition, simulated signals were used when the system was out of service to confirm that all other

Table 7.19 Compressor operating data for Problem 7.7.

First-stage suction pressure, psig	3
Second-stage suction pressure, psig	60
Interstage pressure drop, psi	2
Discharge pressure, psig	275
Compressibility factor (z)	1
Molecular weight	42
First-stage suction temperature, °F	100
First-stage discharge temperature, °F	283
Second-stage suction temperature, °F	110
Second-stage discharge temperature, °F	280
Interstage condensation, lb/hr	0
Ratio of specific heats	1.21
Polytropic efficiency	90
Mechanical efficiency	95
Piston displacement, ft^3/min	5000
Motor rating, HP	1550 (includes a 3% overload factor)

indicators worked. The problem-solver expanded the original problem scope and stated it as shown in Step 2.

Step 2: *Write* out an *Accurate* Statement of What Problem You Are Trying to *Solve*

> "C-122 has been shutting down for no apparent reason. The 'first out indicator' shows the cause is 'other reason'. At about the same time as the mysterious shutdowns, the suction pressure tends to increase. However, it has been impossible to determine whether this causes the shutdowns or is the result of the shutdowns. Investigate the shutdowns of C-122, the recycle gas compressor, to determine what is causing the unknown shutdowns. There may be a problem with the 'first out indicator' or there may be an unknown reason for the compressor shutdowns. When the cause is determined provide recommendations to eliminate the problem."

Step 3: *Develop* a Theoretically Sound Working Hypothesis that Explains as Many Specifications of the Problem as Possible

Table 7.20 Questions/comments for Problem 7.7.

Question	Comment
Are all operating directives and procedures being followed?	All appeared to be correct and being followed, except there were times when the low-stage suction pressure was not well-controlled
Are all instruments correct?	All pressure instruments had been calibrated
Are laboratory results correct?	Not applicable
Were there any errors made in the original design?	Since this was a new process, this was a consideration. However, the major concerns were the "first out indicator" and the capability of the compressor to handle swings in recycle gas rates with subsequent changes in suction pressure
Were there changes in operating conditions?	No, except for suction pressure swings
Is fluid leakage occurring?	Internal leakage might explain the failure of the compressor to handle the gas flow under some conditions
Has there been mechanical wear that would explain the problem?	Leaking compressor valves might explain the problem
Is the reaction rate as anticipated?	Normally yes. However, decreases in the reaction rate cause increased recycle gas rates
Are there adverse reactions occurring?	Not applicable
Were there errors made in the construction?	Construction errors in the "first out" system might explain the problem

The questions given in Chapter 6 were used to help formulate possible hypotheses. These questions and appropriate comments for this example problem are shown in Table 7.20.

Based on the above table, the problem-solver formulated the following hypotheses:

1) There is a compressor mechanical problem such as bad valves. This problem is not apparent at low flow rates, but becomes apparent at high rates. This mechanical problem causes the compressor to have insufficient capacity at high gas rates. As the suction pressure increases, the compressor shuts down for some unknown reason.
2) The compressor is shutting down for some unknown reason when the suction pressure increases slightly due to normal process variability.

3) There is an intermittent instrumentation failure with the “first out” system. This failure is causing some problems such as low oil pressure to shut down the compressor, but does not show up on the display panel.

Note that there is similarity between the first and second hypotheses. The first hypothesis implies that there is a mechanical condition that can be repaired and the problem will be eliminated. The second hypothesis indicates that the mysterious shutdowns are due to process variability and that there is nothing mechanically wrong with the compressor.

When faced with the need to decide which of these hypotheses to pursue, the problem-solver recognized that it might be difficult to trace an intermittent instrument failure associated with the “first out” system. On the other hand, it would be possible to quickly do some calculations to confirm whether or not either hypotheses 1 or 2 were theoretically correct. He used the following relationships and the basic data shown in Table 7.19 to develop Table 7.21.

$$T_d = T_s{}^* R^{\sigma} \tag{7.8}$$

The problem-solver used equation (7.8) with the suction and discharge temperatures in absolute temperature units and calculated a compression exponent of 0.193. He then used the compression exponent (σ), the specific heat ratio, and equation (7.7) shown below to calculate the polytropic efficiency. The ratio of specific heats is given in Table 7.19.

Table 7.21 Calculations of compressor efficiency.

Variable	Design	Current
Low-stage suction pressure, psig	3	3
Low-stage discharge pressure, psig	62	62
High-stage suction pressure, psig	60	60
High-stage discharge pressure, psig	275	275
Low-stage suction temperature, °F	100	100
Low-stage discharge temperature, °F	278	283
High-stage suction temperature, °F	110	110
High-stage discharge temperature, °F	276	280
Compression exponent (from temperatures)	0.188	0.193
Polytropic efficiency, percent	92	90
Polytropic head, ft	67,200	67,400
Mass flow rate, lb/min	620	620
Total horsepower, BHP	1,440	1,480

$$\sigma = (k - 1) * 100/(k^* E) \tag{7.7}$$

As shown in Table 7.21, the polytropic efficiency is slightly lower than the design efficiency (90 versus 92). This slightly low efficiency is likely within the accuracy of the calculations. The mass flow and calculated total horsepower are also shown. The problem-solver concluded that the hypothesis that the compressor required mechanical repairs was not a valid hypothesis. He then considered the question, "would process upsets such as a sudden increase in the recycle gas rate be sufficient to overload the compressor?" He developed a spreadsheet to allow calculation of the horsepower load at various low-stage suction pressures. The results of these calculations are shown in Table 7.22.

The problem-solver believed that his calculations clearly indicated that the reason for the mysterious compressor shutdowns was associated with a slight increase in process gas rates that caused the compressor suction to increase to the point that the compressor motor was overloaded. This was in spite of intuitive thinking that said the horsepower load should decrease as the compression ratio is decreased. The 1500-BHP motor with a 3% overload factor would likely shut down if the compressor suction pressure exceeded 4.5 psig. He expressed his hypothesis as follows:

Table 7.22 Calculated horsepower load versus suction pressure.

Variable	Case 1	Case 2	Case 3
LS suction pressure, psig	3	4	5
LS discharge pressure, psig	62	62	62
HS suction pressure, psig	60	60	60
HS discharge pressure, psig	275	275	275
LS suction temperature, °F	100	100	100
LS discharge temperature, °F	283	275	267
HS suction temperature, °F	110	110	110
HS discharge temperature, °F	280	280	280
Compression exponent	0.193	0.193	0.193
Polytropic efficiency	90	90	90
Polytropic head	67,400	65,900	64,500
Mass flow rate, lb/min	620	655	690
Total horsepower	1,480	1,523	1,575

"It is believed that the mysterious recycle compressor shutdowns are caused by spikes in the recycle gas rate that cause the motor to overload as the suction pressure is increased."

While it might seem that this hypothesis could be confirmed by a simple analysis of the low-stage suction pressure and compressor amperes versus time, the exact point of shutdown was not obvious. As the compressor suction pressure increased and the compressor shut down, the pressure continued to increase rapidly. It was impossible to determine whether the shutdown occurred when the pressure reached 4, 4.5, 5, or 6 psig.

Step 4: Provide a Mechanism to Test the Hypothesis

The problem-solver considered that his best option to test his hypothesis was to run a test at reduced production rates. A test was run at a low enough production rate that the recycle gas never got high enough to cause the suction pressure to go above 4 psig. During this test, there were no mysterious compressor shutdowns.

Step 5: Recommend Remedial Action to Eliminate the Problem without Creating Another Problem

While this test was successful at preventing the mysterious compressor shutdowns, it was obviously not a permanent solution. Experience with the process over the early start-up period indicated that the recycle gas rate was likely to increase to 15% above the steady-state design rate during decreases in reactor conversion. The problem-solver looked at an increase in gas rate of 15% and concluded that the suction pressure would increase from 3 to 5.7 psig if this occurred. At 5.7 psig, the calculated horsepower load would be 1610 BHP. This would be well above the motor acceptable load with the 3% overload factor. He discussed the situation with the motor manufacturer and found out that operating the motor continuously at 10% overload would be expected to shorten the life of the motor slightly. Since the 10% overload would only occur during times when the reactor conversion dropped, a larger overload switch was installed.

7.20 Lessons Learned

While this problem solution may seem obvious, it should be recognized that the initial assessment was that there was a problem with the "first out" instrumentation system. If this idea had been followed with no thought of other possible hypotheses, there would have been a significant delay in solving the problem since the problem was thought to be an intermittent failure. In addition, while the 3% overload rating may seem too conservative, without the calculations done by the problem-solver, it would not be known whether the 10% overload rating would cover the range of loads to be expected. The calculations also helped steer the problem-solver away from the conclusion that there was a mechanical problem with the compressor.

This problem also illustrated that what might seem to be intuitively correct (lowering the compression ratio by increasing the suction pressure will decrease the horsepower load) is not always true. While it is true that lowering the compression ratio by increasing the suction pressure lowers the polytropic head, it also increases the mass flow rate. Both polytropic head and mass flow rate are important in determining the horsepower load. The exact relationship between suction pressure and horsepower load can only be determined by calculations.

7.21 Example Problem 7.8

An industrial gas plant as shown in Figure 7.13 has a steam reforming furnace to produce hydrogen using methane as the feed with a forced draft blower to supply air to burners.

Despite no apparent increase in the methane feed flow rate, the blower continued to have increased air flow rate to burners. The operation team was wondering why this happened. Air consumption gradually increased until the blower seemed to reach its maximum capacity limit (around 15,000 kg/hr. as shown in Figure 7.14). If no remedial action was taken, the blower would likely need to be

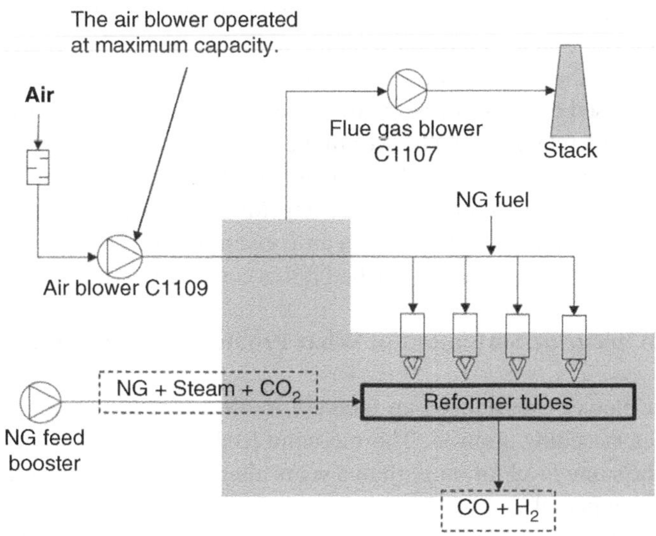

Figure 7.13 Schematic flow.

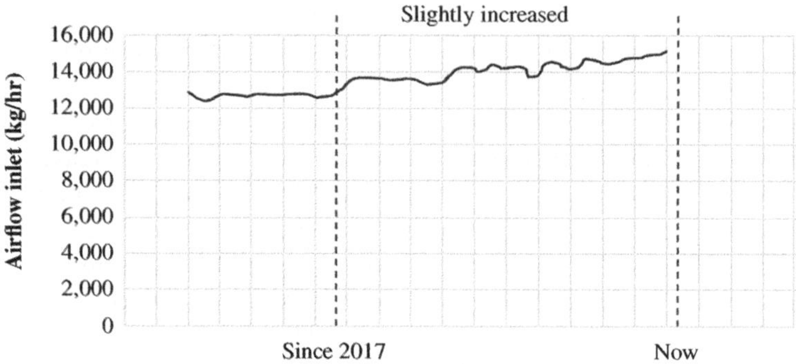

Figure 7.14 Blower flow versus time.

replaced with a new one, causing the plant a large capital investment. The root cause remained mysterious at the time.

The problem-solver was called in to determine what was wrong. The problem had been observed for more than 5 years before the problem-solver was called in. There had been another steam reformer furnace expert who was called in to solve this problem, but he failed to solve the problem despite his knowledge of the steam reformer furnace technology.

The problem-solver approached the problem using the five-step procedure as shown below.

Step 1: Verify that the Problem Actually Occurred

The problem solver had an important problem-solving philosophy to not rely on one source of information to determine a problem. In this case, a flow meter error could have caused the air flow rate reading to mysteriously increase. What else could have been used to cross-check the increased air consumption? Looking at all the parameters related to the blower that could be measured or being monitored apart from the air flow rate, the problem-solver noticed the amperage and voltage reading at the Motor Control Center (MCC) of the plant to also increase, which shows a sign of real increase in air consumption. However, the magnitude of amps increase needs to be verified if it is consistent with air consumption increase or not.

Step 2: *Write* out an *Accurate* Statement of What Problem You Are Trying to *Solve*

"A forced-draft air blower experienced an increase in air consumption as the discharge flow meter reading showed. The methane feed rate to the furnace had remained at the same level for more than 5 years after this strange air consumption increase happened."

Step 3: *Develop* a Theoretically Sound Working Hypothesis that Explains as Many Specifications of the Problem as Possible

Table 7.23 Questions/comments for Problem 7.8.

Question	Comment
Are all operating directives and procedures being followed?	Yes. The operator followed the same operating and control philosophy they had followed for years before the incident started to happen
Are all instruments correct?	The flowmeter might have been in error. Another instrument that could be erroneous is the stack excess oxygen analyzer
Are laboratory results correct?	Not applicable. Methane feed gas composition was confirmed to remain the same and normal
Were there any errors made in the original design?	No. The blower was working fine before the incident
Were there changes in operating conditions?	No. the operator did not manually and intentionally change operating conditions, but air consumption was higher due to excess oxygen control requiring more air into the burners
Is fluid leakage occurring?	Air leakage to atmosphere might explain the problem, but it was difficult to prove during operation due to very low pressure in the pipe as low as 40 mbarg
Has there been mechanical wear that would explain the problem?	Blower efficiency deterioration over time might explain the problem
Is the reaction rate as anticipated?	Not applicable
Are there adverse reactions occurring?	Not applicable
Were there errors made in the construction?	No. The blower was working fine before the incident

The questions given in Chapter 6 were used to help formulate possible hypotheses. These questions and appropriate comments for this example problem are shown in Table 7.23.

Based on this table, the problem-solver formulated the following hypotheses:

1) Flow meter reading showing increased air consumption might have been incorrect.
2) Air leakage to the atmosphere might be happening before entering burners, causing high flow to air manifold but same flow as before to burners.
3) Blower efficiency deterioration over time.
4) The stack excess oxygen analyzer might have been erroneous.

Plant people noted a few hypotheses above that had not been previously identified by the previous problem-solver. Note that the problem had been so chronic for more than 5 years because some of the hypotheses were not identified. This was likely because previously all hypotheses had been identified out of a genius approach instead of a series-of-questions approach. This emphasized the importance of using a series-of-questions approach instead of jumping into any specific root cause without having a holistic view of the potential root causes of the problem.

Step 4: Provide a Mechanism to Test the Hypothesis

Three hypotheses were possible to be tested by calculation using actual plant data. Test hypothesis 1 by BHP calculation from amps and voltage.

$$\text{BHP} = (1.732^* I^* V^*(\text{motor efficiency})^*(\text{motor power factor})/746) \quad (7.25)$$

where

BHP = brake horsepower (hp)
I = electrical current (A)
V = Electrical voltage (V)

Test hypothesis 2 by air piping and manifold hydraulic calculation. The concept of pressure drops and equivalent length was applied as follows.

$$\Delta P = 10^{-5} * \left(\frac{4f_f L_{EQ}}{D}\right) * \left(\frac{\rho v^2}{2}\right) \quad (7.26)$$

$$L_{EQ} = L + \left(\frac{D}{4f_f}\right)\Sigma K = (\text{Typical multiplier of 3}) * (\text{Pipe straight length excluding fittings}) \quad (7.27)$$

where,

ΔP = pressure drops across the pipe
f_f = fanning friction factor
L = straight pipe length section
L_{EQ} = equivalent length
D = inner diameter of the pipe
ρ = mass density of fluid
v = linear velocity of fluid in the pipe
L = actual length of the pipe
K = loss coefficient

The equivalent length (L_{EQ}) in equation (7.27) is normally calculated from counting the number of fittings and elbows and then applying corresponding loss coefficients (K). However, for the preliminary calculation purpose only for root cause assessment, equivalent pipe length calculation may require a simpler-yet-valid approach using a typical rule-of-thumb multiplier of 3. Therefore, we multiplied pipe straight length (20 m) with the multiplier of 3 to obtain equivalent pipe length ($3 \times 20 = 60$ m) as a rough calculation. Table 7.24 shows that if the problem case (15,600 kg/hr) actually had occurred without any piping air leakage to the atmosphere before reaching the burner manifold inlet, air supply pressure would likely have dropped to negative vacuum pressure (−15.34 mbarg), which is *not* possible to put air into burners. However, in actual operation, air could still be continuously supplied to burners without

Table 7.24 Calculation results for normal case and problem case.

Description	Unit	Normal Case	Problem Case
Blower air flow rate	kg/hr	12,500	15,600
Air density	kg/m^3	1.2	1.2
Air viscosity	Pa·s	0.00002	0.00002
Blower discharge pressure, (1)	barg	0.0437	0.041
	mbarg	43.7	41.0
	bara	1.05695	1.05425
Blower discharge temperature	°C	35.0	35.0
	K	308.2	308.2
Pipe inside diameter (approximate)	in	18	18
Pipe straight length	m	20	20
Pipe equivalent length, L_{EQ}	m	$3 \times 20 = 60$	$3 \times 20 = 60$
Darcy friction factor	—	0.0136	0.0134
Piping pressure drop, (2)	bar	0.00644	0.00984
Venturi flow meter pressure drop, (3)	bar	0.010 (flow meter data)	0.0155 (flow meter data)
Air preheat equipment pressure drop, (4)	bar	0.020 (design data)	$(0.020)(15\,600/12500)^2 = 0.031$ (prorated design data)
Burner manifold inlet pressure, (5) = (1)–(2)–(3)–(4)	barg	0.00726	−0.01534
	mbarg	7.26	−15.34
Conclusion		Positive pressure that is possible to supply air to burners	Negative vacuum pressure that is not possible to supply air to burners

any interruption, implying that air supply pressure still remained positive at the burner manifold inlet, as also confirmed by the field pressure gage at that location. Therefore, it is very likely that air leakage occurs in the air supply piping between the air blower discharge and the burner manifold inlet.

Another supporting evidence for hypothesis 2 is that if the methane feed flow rate had remained constant while the air flow rate had continued increasing, excess oxygen level in flue gas should have increased significantly. However, from actual observation, the excess oxygen analyzer did not show any significant increase. Thus, air leakage to atmosphere in the air supply piping (hypothesis 2) was the most likely root cause.

Test hypothesis 3 by blower BHP and efficiency equation. The increase in power should be based on an incremental flow rate at nearly the same efficiency value.

$$\text{BHP} = \frac{(Q)(\text{Static Pressure})}{6356 * (\eta_{\text{overall}})} \tag{7.28}$$

$$\eta_{\text{overall}} = \eta_{\text{fan}}{}^{*}\ \eta_{\text{motor}}{}^{*}\ \eta_{\text{driver}} \tag{7.29}$$

where,

BHP	=	Brake horsepower of blower (hp)
Q	=	Air flow rate (ft^3/min)
ΔP	=	Static pressure increased by the blower (in H_2O)
η_{overall}	=	Overall or total efficiency of the blower
η_{fan}	=	Efficiency of the fan impeller (between 0 to 1)
η_{motoor}	=	Efficiency of the motor (between 0 to 1)
η_{driver}	=	Efficiency of the driver or VFD (between 0 to 1)

For example, in normal case, substitute Airflow = (12,500 kg/hr)/(1.2 kg/m^3) = 10,416.67 m^3/hr. or 6131.02 ft^3/min, Static Pressure increased = 43.7 – (–6.18) mbarg = 49.88 mbarg or 20 in H_2O, fan efficiency = 0.82 and assuming motor and driver efficiency are 1 into equation (7.28). We will obtain BHP = 23.55 hp. or 17.57 kW.

Based on all calculations above, hypothesis 2 remained the most likely root cause. This is because the electrical load delivered to the fluid as calculated from both the electrical consumption in equation (7.25) and the flow rate in equation (7.28) was very close. Therefore, the flow meter must be correct. However, plant people refused to believe the problem-solver that hypothesis 2 was the real root cause. The reason was because plant people had a shop leak test record of the air preheater in the air supply manifold loop and also a record of air piping leak testing with a leak detecting liquid during furnace pre-commissioning.

However, the problem-solver insisted on hypothesis 2 because this system has a very low leak test pressure at <100 mbarg level, thus needing to use a manometer tube instead of a pressure gage for leak test pressure observation. The problem-solver believed that this low-pressure level might be difficult to visually detect piping leaks using a leak detecting liquid because there might not be sufficient pressure to cause visible bubbling on the piping surface.

In order to decisively prove hypothesis 2, during plant shutdown, the plant people conducted a test run to block off all burner individual valves and then ran the blower at shut-off condition with all air blocked in the manifold. As soon as the blower was completely blocked off, the audible noise of air leak from the manifold header end flange appeared. This noise was not audible during normal operation because there was not enough air leakage flow through the manifold header end flange to make an audible noise. The problem-solver got a call from the plant people soon after the blower block-off test and was delighted to know that all of the 5-year-long mystery was finally solved.

Step 5: Recommend Remedial Action to Eliminate the Problem without Creating Another Problem

The remedial action was of course to fix the manifold header end flange and stop the air leakage. Another potential future measure was to use a technology to help identify the leak early such as the ultrasonic imaging leak detector where the leak is detectable even without audible noise because the leak detector would catch the leak in the frequency range that human ears cannot hear.

7.22 Lessons Learned

This problem illustrated that what might seem to be intuitively correct that air leakage should not happen if the equipment and piping have passed leak testing at shop is not always true. Therefore, it is important to not over-believe or blindly believe any data without challenging that the data might be wrong or misleading. This constructive doubting habit leads to more effective problem-solving and reveals something that could not have been identified without doubting first.

Nomenclature

A	air flow
BHP	brake horsepower or the energy delivered to the fluid, hp
C	the residual gas remaining in the positive displacement compressor clearance pocket or rotating pocket after discharge, fraction of displacement volume
C_P	the fluid-specific heat, BTU/lb-°F
D	the diameter of the pump cylinder, ft

D_P or D	the diameter of the suction pipe, ft
E	the hydraulic pump efficiency or adiabatic/polytropic compression efficiency, percent
E_T	the turbine efficiency, percent
E_V	the pump or compressor volumetric efficiency, percent
F	the flow rate, lb/min
f	Fanning friction factor
g_c	the gravitational constant
H	the pump head or adiabatic/polytropic head, ft
H_I	the inlet steam enthalpy, BTU/lb
H_O	the outlet steam enthalpy, BTU/lb
H_E	the outlet steam enthalpy with isentropic expansion, BTU/lb
$(h_A)_{MAX}$	the maximum acceleration head loss for a reciprocating pump, ft
K	a factor for reciprocating pumps that depends on the pump design or loss coefficient
k	the ratio of specific heats, C_p/C_v
L	the leakage of gas around the piston or rotating element, fraction of displacement volume. It is roughly 2% of the compression ratio expressed as a fraction. Also used as length of straight pipe
LEQ	Equivalent length
L_S	the length of the pump stroke, ft
L_P	the *actual* suction pipe length, ft
Lw	the frictional loss in the piping system
L	actual pipe length
L_{EQ}	equivalent line length
M	the gas molecular weight
n	the constant used to define the polytropic compression exponent $\sigma = (n - 1)/n$
P_D	the discharge pressure, psia
P_S	the suction pressure, psia
Q	Air flow rate ft^3/min
R	the compression ratio
S	the pump speed, rpm
T_D	the absolute discharge temperature, °R
T_S	the absolute suction temperature, °R
V	the reciprocating pump capacity, ft^3/min
V_P	the peak velocity in the suction pipe of a reciprocating pump, ft/s
v	linear velocity
w	the amount of work added by the prime mover
Z	the average (suction and discharge) compressibility

P or ΔP	the pressure difference between two points or pressure rise across a pump
ΔT_H	the Joule Thompson cooling effect, °R
$\Delta(v^2)$	the difference in velocity squared between two points
Δz	the difference in elevation between two points
η	efficiency—overall, fan impeller, motor, and driver/VFD
ρ	the pumped fluid density, lb/ft^3
σ	the polytropic or adiabatic compression exponent

8

Application to Plate Processes

8.1 Introduction

Examples of plate processes are fractionation towers, extraction towers, absorption towers, or any process that depends on mechanical design to provide *intimate contact* between two phases, followed by a zone where *phase separation* is achieved. The intimate contacting followed by phase separation allows equilibrium between the two phases to be achieved or approached.

The information provided in this chapter is intended to provide a basis for the problem-solver to successfully complete Step 3 of the problem-solving discipline:

> "Develop a theoretically sound working hypothesis that explains as many specifications of the problem as possible."

Since much of what occurs in a plate process is not visible to the naked eye and may not be easily understood by X-rays, it is important to be able to correctly imagine what is occurring inside the equipment.

Fractionation using sieve trays is discussed in this chapter since it is the most common application. However, the principles are applicable to all fractionation tray designs and any other plate processes.

8.2 Fractionation with Sieve Trays

The purpose of a sieve tray is to provide as close an approach to equilibrium between the liquid and vapor phase as is reasonably possible. To do this, the following three zones are required:

1) A high-intensity vapor–liquid contact zone. In this zone, liquid must be the continuous phase. That is, vapor bubbles must exist as discrete entities surrounded by a liquid-phase regime.

Problem Solving Approaches for Maintaining Operational Excellence in Process Plants, Second Edition.
Joseph M. Bonem, Nattapong Pongboot, and Wiroon Tanthapanichakoon.
© 2025 the American Institute of Chemical Engineers, Inc. Published 2025 by John Wiley & Sons, Inc.

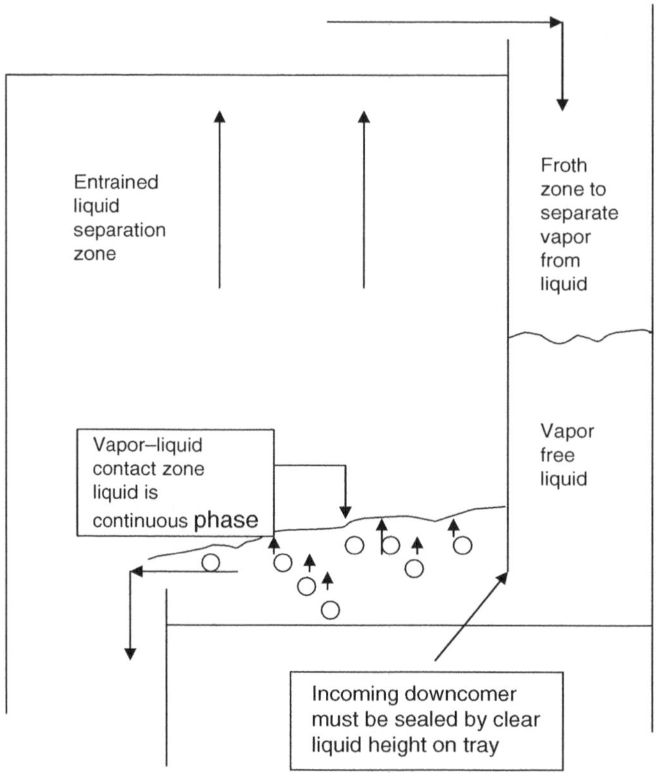

Figure 8.1 Typical fractionation tray.

2) An entrained liquid separation zone. This vapor continuous zone is immediately above the continuous liquid phase. In this zone, the liquid droplets entrained with the rising vapor disengage and return to the liquid phase.
3) A vapor froth separation zone. The liquid leaving each tray contains vapor that has not yet disengaged from the liquid. The downcomer provides time for disengagement to occur.

Figure 8.1 shows a typical sieve tray illustrating these three zones. In addition, it indicates two other important parameters for a fractionating tray design. These are as follows:

1) The inlet downcomer must be sealed. That is, the head of the clear liquid on the tray must be greater than the clearance between the downcomer and the tray deck. This criterion will force liquid to build up in the downcomer so that

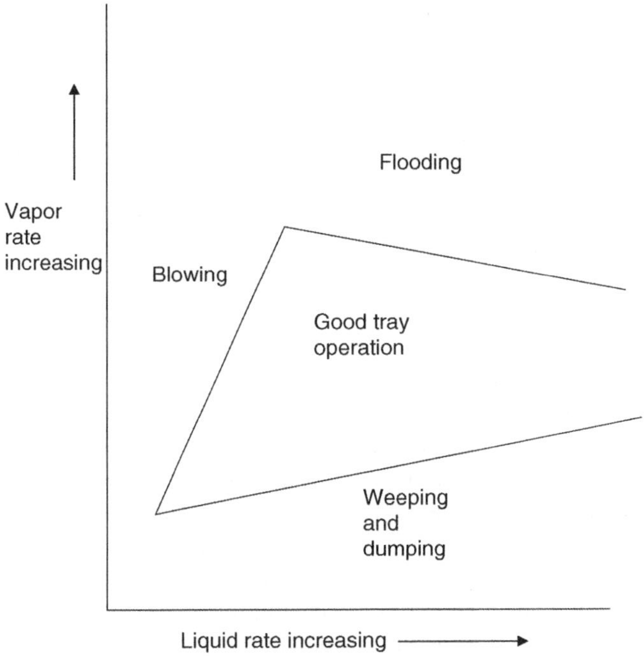

Figure 8.2 Typical tray stability diagram.

the downcomer is sealed. If this criterion is not met, vapor can flow up the downcomer rather than through the holes on the tray above.

2) The hydraulic gradient (the difference in clear liquid height between the inlet downcomer and outlet downcomer) must be minimized. If the hydraulic gradient is too great, the trays will have a tendency to weep liquid on the inlet side of the tray.

Tray stability diagrams provide an analytical means to help visualize what is occurring on a fractionating tray. Figure 8.2 provides an example of a tray stability diagram. The four areas of unacceptable operation are as follows:

1) *Flooding*: This condition is marked by vapor velocities based on the tower cross-sectional area being so high that large amounts of liquid are carried up into the continuous vapor-phase regime. This liquid does not adequately disengage from the vapor and is carried up into the tray above. This condition leads to excessive loading in the tray outlet downcomer as the entrained liquid returns to the tray below via the downcomer. This excessive loading causes the liquid level in the downcomer to build up to an unacceptable level.

Flooding tends to start at one tray and propagate upward through a section of the tower. It may often be detected by a measurement of differential pressure across a section of the tower.

2) *Downcomer filling*: This condition is marked by the downcomer either completely filling or filling to the point where adequate vapor disengagement cannot occur. It can be caused by flooding, as indicated earlier, or excessive liquid rates. It can also be caused by tray or downcomer restrictions. In a similar fashion to flooding, it can propagate up a fractionating tower.
3) *Blowing*: In this condition, vapor velocity through the holes in a fractionating tray is so high that the vapor phase becomes the continuous phase in the high-intensity vapor–liquid contact zone. It is usually caused by a combination of low liquid rates and high vapor rates.
4) *Weeping/dumping*: In this condition, the vapor rates are so low that liquid pours down the holes in the tray. This results in a very low liquid level on the trays. This usually results in the loss of the downcomer seal, causing vapor to flow up the downcomer. This single unsealed downcomer will likely cause a high degree of frothing, which will lead to liquid holdup in the downcomer. This may result in downcomer flooding in the trays above the tray that is weeping and/or dumping. It should be noted that tray weeping/dumping could also be caused by mechanical damage such as a tray segment that has come loose and is hanging down from its support.

As a general rule, the symptoms of "flooding" and "downcomer filling" are a higher-than-normal pressure drop across the tower and loss of fractionation. The "blowing" and "weeping/dumping" symptom is loss of fractionation in the tower. Pressure drop for these conditions may be normal or slightly lower than normal.

The tray stability diagram can be developed by the following calculation procedure:

1) Assume a liquid rate.
2) Vary the assumed vapor rate at this assumed liquid rate until the weeping, blowing, flooding, or downcomer filling limits are encountered.
3) Repeat this calculation for several different liquid rates until the tray stability diagram can be completed.

The calculation techniques for determining the various tray limitations are beyond the scope of this book. Adequate references are available in the open literature or in individual company literature. The key point is that a tray stability diagram provides a means to visualize what is happening in the tower and thus will be helpful in Steps 3 through 5 of the disciplined problem-solving approach.

8.3 Problem-Solving Considerations for Fractionating Towers

Assuming that the fractionating tower was designed correctly, problems will almost always be associated with process changes or mechanical damage. Confirmation that the fractionating tower was designed correctly can be done by developing a tray stability diagram. This tray stability diagram will allow plotting operating conditions on the diagram to confirm that they are in the regime of "good operations." Developing this tray stability diagram is also consistent with the premise of this book—Calculations should be done prior to developing hypotheses that might explain the problems being encountered.

Mechanical damage to trays can create tray performance problems. Some of these possible areas of tray damage or improper tray installations are described in the next few paragraphs.

A tray segment can fail. Most sieve trays are designed to be installed and/or removed through manways that are 18–36 in. in diameter. This approach also facilitates tray inspection and replacement. Thus, a single tray will consist of several segments connected together. If one of these connections should fail, a condition similar to weeping/dumping will occur. As indicated earlier, this can propagate upward, causing a loss of several theoretical trays.

The trays may not be level. While this is almost always an installation problem, it might also occur after an extended period of operation. This delayed manifestation could be due to extreme foundation settling. In addition, problems associated with uneven trays might not show up under all operating conditions. A tray slope of less than 0.2% is generally acceptable, and slopes of up to 0.7% have provided good operations. Highly critical towers (vacuum towers, for example) are special cases.

There may be restrictions in the tray. The presence of solids in a fractioning tower will often lead to plugging of a downcomer or plugging of holes in sieve trays. Solids can be present in the tower due to multiple causes such as

- *Construction debris*: It is not unheard of to discover items such as rags, tools, or even safety equipment left behind in a tower after construction or repairs. Unfortunately, there have been occasions when an inspection of the tower prior to start-up did not find the debris that blocked a downcomer. Following the start-up of the tower, problems were experienced, and a subsequent shutdown and inspection revealed the debris in the downcomer.
- *Entrainment from a drum containing solids*: Very fine particles can often be entrained from a vapor–liquid disengaging drum and carried into a tower. If these particles are not soluble in the liquid on the trays in the tower, they can accumulate in either the downcomer or sieve tray holes.

- *Corrosion of the tower internals*: If the liquid on the trays contains a corrosive compound, the trays may begin to corrode. This could cause the holes in the sieve trays to enlarge. In addition, it is possible that some of the corrosion products could be deposited lower in the tower as the composition of the liquid changed. Because of this possibility, sieve trays are often fabricated from a corrosion-resistant material such as stainless steel.
- *Reaction of trace components inside the tower*: While this would seem like an unlikely event, there have been instances where small quantities of water reacted with a soluble material in the liquid phase on the tower trays and formed an insoluble material. This material would accumulate in the sieve tray holes and the downcomers and create plugging problems in a similar fashion to the item discussed earlier.

The vapor inlet flow pattern may change. For example, the holes in a sieve tray may enlarge or become irregular due to corrosion, or the valves in a valve tray may become loose. Either of these may cause significant disturbances to the vapor inlet flow pattern, resulting in poor tray performance.

Process changes causing poor tray performance may be due to known changes such as an increase in rates or very subtle changes. For example, there may be instrumentation errors that cause excessive vapor or liquid rates. Even at constant liquid and vapor rates, foaming caused by trace quantities of a surface-active ingredient may occur. An even more subtle change might be the tower that appears to be operating perfectly normal until an event such as described earlier. The water present from an exchanger leak reacts with a soluble material to form solids. These solids lead to plugging of the holes in the sieve tray. The first indication of a problem might be the increased differential pressure across the tower.

8.4 Development of Theoretically Sound Working Hypotheses

Once Step 1 (*Verify* that the problem actually occurred) and Step 2 (*Write* out an accurate specification of the problem) have been completed, the following guidelines can be used to develop a working hypothesis.

Calculations can be performed to determine the following:

- *Tray stability diagram*: This will highlight the areas of unacceptable tray operation, as well as indicating the current point of operation. If the current operations are in an unacceptable area, this must be corrected before considering any other calculations.
- *Number of theoretical stages required for the degree of separation being encountered*: Knowing the concentrations at various points in the tower, the number of

theoretical stages required to make this separation can be estimated by computer or manual techniques.

- *Estimated tray efficiency*: The tray efficiency is the theoretical stages divided by the actual stages and expressed as a percentage. It is a technique to allow estimating the number of actual stages required if the number of theoretical stages is known. While this book does not cover the techniques, there are multiple methods available to estimate the tray efficiency.

It should be noted that these calculations will require good plant data and tower/tray design information. Instrumentation should be calibrated before using plant data to perform these calculations to avoid having to redo the calculations.

Assuming that the calculations described above indicate that the tower should be performing better than what the actual data indicate, the list of questions given in Chapter 6 can be used to help formulate a working hypothesis. Once a hypothesis is formulated, it may be of value to develop a plant test to confirm the hypothesis. Measurements of tray temperatures, pressures, and compositions will be helpful in determining what areas of the tower are worthy of future analysis. Several meaningful approaches are shown in Table 8.1. In addition, X-rays can be utilized to examine the suspect area of the tower to either help formulate a hypothesis or provide a mechanism to test the hypothesis (Step 4: *Provide* a

Table 8.1 Evaluation of trays by sampling/data analysis.

Measurement	Typical Values	Meaning of Atypical Values
Pressure drop across trays	Should be measured across as few trays as possible Check measurements with calculations Normal values 0.05–0.20 psi/tray. Lower for vacuum towers	Trays plugged, damaged, or flooded
Temperature change across trays	Should be equivalent to that estimated from fractionation calc. or equivalent to historical values	Tray damage or process changes
Composition of liquid on trays	Should be equivalent to that estimated from fractionation calc.	Numerous
Venting sample bomb of tray liquid	For pressure towers, the vaporizing liquid should cool off the bomb	Liquid phase is not present on the tray
Composition of vapor on trays	Should be equivalent to that estimated from fractionation calc.	Numerous
Venting sample bomb of tray vapor	No temperature change	Tray flooding

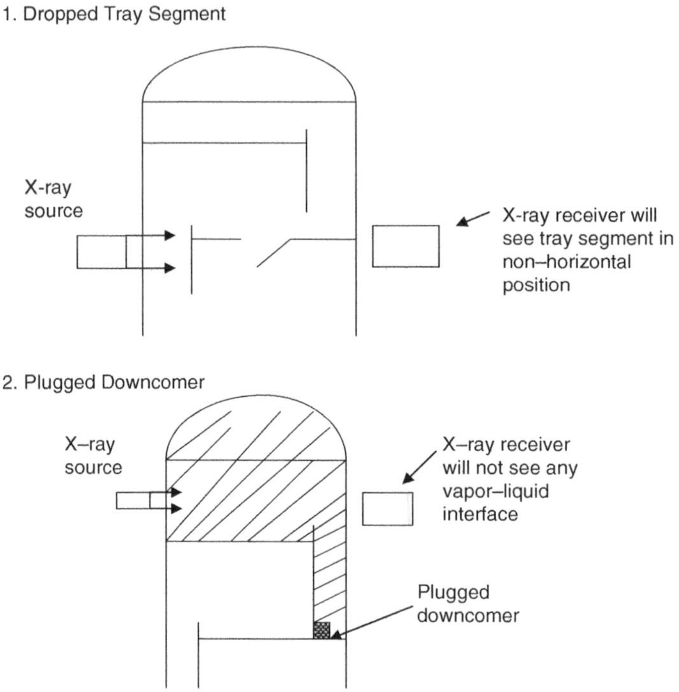

Figure 8.3 Examples of tray problems detectable by X-rays.

mechanism to test the hypothesis). Some examples of the use of X-rays are shown in Figure 8.3.

8.5 Problem-Solving Reboiler Circuits

Reboilers which are often straightforward in the conceptual stage are often a significant part of a plant fractionation problem. The most frequent culprit is the vertical thermosiphon reboiler. A typical flow sheet is shown in Figure 8.4.

Thermosiphon reboiler problems are often due to hydraulics. The operation of this class of reboilers depends on a delicate balance of elevation, fluid densities, and pressure drop. The density of the return line to the tower is a function of the percent vaporization. These thermosiphon reboilers almost never operate with pure vapor in the return line to the tower. Thus, the amount of vaporization (percent vaporization) in the reboiler will determine the density in this tower return line. There must be sufficient head to cause the system to circulate with no application of external work. Some of the possible causes of hydraulic problems are described in the following paragraphs.

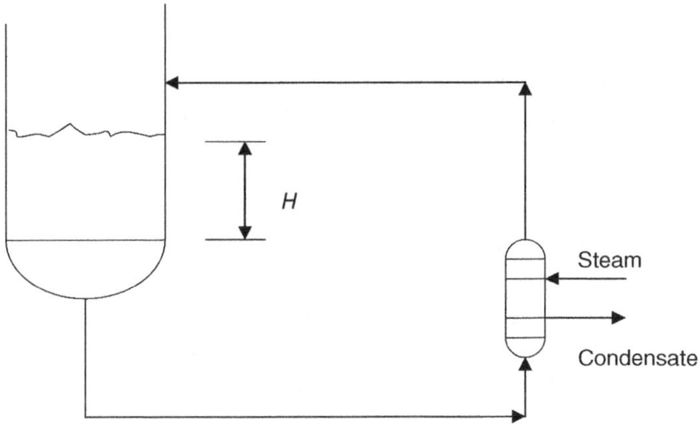

H = liquid height that causes reboiler to circulate
Feed to reboiler is all liquid
Vapor return line contains a mixture of liquid and vapor. Typically this line contains 5 to 25 percent vapor. The amount of vapor in this line which is also referred to as the percent of vaporization is set by the pressure drop in the liquid and vapor return line.

Figure 8.4 Thermosiphon reboiler operating principles.

There may be inadequate head of liquid in the tower bottom. If the liquid level in the bottom of the tower is not high enough, the driving force to cause the process liquid to flow will be inadequate. This may result in a lower-than-design liquid level in the reboiler and/or a lower-than-design circulation rate. The low reboiler level will cause the surface area available for vaporization to decrease, with a likely loss of heat input to the tower. In addition, if the process liquid flow decreases and the heat input remains the same, the percent vaporization will increase. In some applications, this may create reboiler fouling.

There may be restrictions in the reboiler inlet or outlet piping. This will create additional frictional piping loss. The result of this problem will be similar to that discussed in the earlier paragraph. That is, the increased circuit pressure drop may result in a lower-than-design liquid level in the reboiler and/or a lower-than-design circulation rate.

There may be excessive head in the tower bottoms. A high level in the tower bottom may cause cycles in a reboiler operation. This high level leads to a high circulation rate through the reboiler circuit. This will result in a high fluid density (low percent vaporization) in the reboiler outlet. With the higher fluid density, the circulation rate will decrease. The reduced circulation rate will cause the percent vaporization rate to increase, which will increase the circulation rate. The result of this can be a widely cyclic operation of the reboiler circuit.

The tower level may be higher than the reboiler return line. While avoiding this seems obvious, the instrumentation sometimes is designed so that what seems to be a reasonable tower level is actually above the reboiler inlet.

If an inadequate reboiler performance is thought to be associated with a hydraulic problem, the best approach to generating a theoretically sound working hypothesis is to either redo or review the original hydraulic calculations. These calculations might be in error or be based on an incorrect length of equivalent piping. It is possible that the actual piping detail is radically different than that assumed by the process designer. This review should also include confirming that the tower elevation relative to the reboiler is the same as in the original design. If the original calculations appear to be correct and consistent with the actual installation and operating details, then it is likely that there is a piping restriction somewhere in the reboiler piping circuit. The more common problem-solving tool of changing the level in the tower without first considering the hydraulic calculations will not be helpful.

Another possible reboiler problem is associated with inadequate drainage of the steam condensate. This will cause a high level of the condensate in the reboiler, resulting in a portion of the reboiler surface area being covered with the condensate rather than condensing steam. This can also lead to poor stability in the reboiler control circuit. Some of the reasons for poor removal of the condensate are described in the following paragraphs.

The steam traps can be improperly sized or can be malfunctioning. If this occurs, the condensate will build up in the reboiler until the trap opens to discharge the condensate.

Steam pressure modulation control systems often create problems when low-pressure steam is used. If the reboiler is controlled by a control valve in the inlet steam line, it is possible for the pressure on the steam side of the reboiler to be lower than the pressure of the condensate return system. Since the steam pressure in the reboiler is lower than that in the condensate system, the condensate level in the reboiler will increase, covering some of the tube area with the condensate. As the tube area is covered, the steam pressure must increase to compensate for this. As the steam pressure increases, the condensate is drained from the reboiler. This cycle will be repeated, leading to instability in the condensate removal system. This instability in the condensate removal system is often transmitted to the tower.

The temperature driving force needs to be at an optimum level. The temperature difference between the heating medium and process side must not be too great or too small. A large difference (greater than 100 °F) can produce film boiling instead of nucleate boiling. This film boiling causes a vapor film to exist at the tube wall. Under some conditions, this will result in a much lower heat transfer coefficient than design. On the other hand, operating with a close approach between the heating fluid and process fluid can cause the reboiler to surge. That is, the reboiler will cycle between no heat input and greater-than-design heat input. This surging can be caused by depletion of volatile materials in the reboiler. If low-pressure steam is used on a reboiler that heats a mixture of a volatile material (propane) and nonvolatile material (octane), depletion of the volatile material will cause the reboiler to stop condensing any steam until the inventory of the volatile material is replenished. In addition, the same effect can be observed due to depletion of water from a water—low volatility hydrocarbon two-phase mixture. In a similar fashion to that described above, this will result in a rapid change in boiling point and cause the reboiler to stop condensing steam until the water concentration is replenished.

Solving a reboiler problem can require from almost no engineering to a great deal of engineering problem-solving skills. However, regardless of the technical complexity, the principles of problem-solving can still be applied. The questions given in Chapter 6 will still be of value. In addition, simple instruments can be utilized for Step 3 (Develop a *theoretically sound* working hypothesis that explains the problem) and Step 4 (Provide a mechanism to test the hypothesis of the problem-solving approach). Examples of these instruments are as follows:

- An infrared thermometer can be used to measure the temperature of a suspect steam trap.
- A gage glass and pressure gage can be used to measure the condensate level and pressure in a reboiler that is cycling excessively.

More complicated instrumentation might be used as follows:

- A highly sensitive pressure drop instrument might be connected to a high-speed recorder or a process control computer with a rapid scan frequency. This could be used to diagnose reboiler hydraulic problems.
- A device such as a calorimeter that is capable of determining the enthalpy of a process stream could be used to measure the enthalpy of a stream leaving a reboiler. This will allow determination of the percent vaporization that is occurring in the reboiler.

8.6 Example Problem 8.1

A propane/butane fractionation tower (T-1) had ceased to operate like it should. The data shown in Table 8.2 represent good and current operation.

A schematic flow diagram is shown in Figure 8.5. In addition to the data shown above, the chronological information shown in the following paragraphs was available.

Table 8.2 T-1 operating conditions.

	Good Operation	Current Operation
Feed rate, lb/hr	100,000	100,000
Feed composition, wt% C_3	30	40
Feed enthalpy	Saturated liquid	Saturated liquid
Vapor boil-up, lb/hr	79,500	124,000
Reflux rate, lb/hr	31,300	65,000
Reflux enthalpy	Saturated liquid	Saturated liquid
Distillate comp., wt% C_3	95	80
Control tray comp., wt% C_3	60	60
Bottoms, comp., wt% C_3	2	2

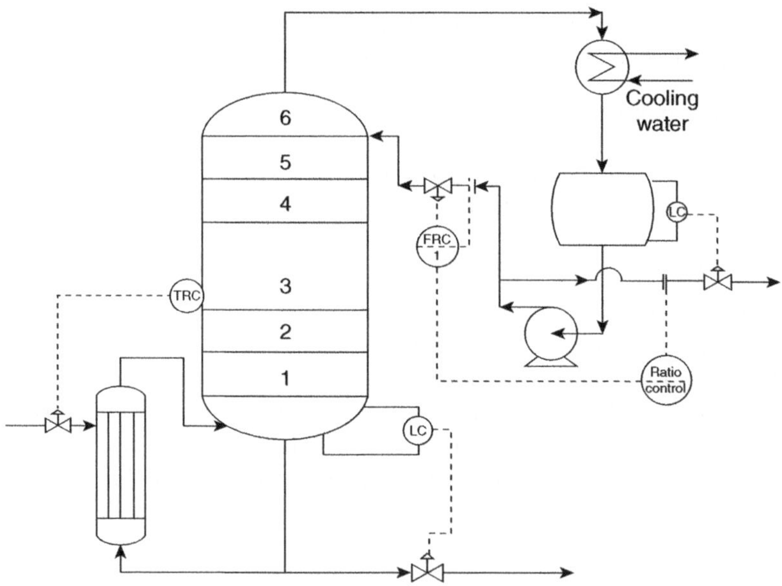

Figure 8.5 Schematic flow for Problem 8.1.

A routine repair and inspection downtime on the tower occurred sometime after the "good operation" data were recorded. During this downtime, the existing trays, which showed evidence of corrosion, were removed, and new trays with the identical design were installed. While this seemed to go well, the normal mechanical supervisor was off sick during the tray installation procedure. No obvious change in performance was noticed immediately. However, it was several weeks before full rates were achieved due to limited product demand.

The laboratory began using a different gas chromatograph (GC) than was used during the "good operation" period. This GC was thoroughly checked out in a series of cross-checks with the old GC and other laboratory GCs. One of the laboratory technicians that had analyzed this sample for 10 years was reporting that he finds a film of heavy hydrocarbons or oils in the GC after each analytical run. All flowmeters and temperature instruments had been confirmed to be accurate.

When the problem-solver was assigned the problem, he began by using the five-step procedure discussed earlier. He combined Steps 1 and 2 as follows.

Step 1: Verify that the Problem Actually Occurred

Since all indications were that there was a real problem, the problem-solver simply combined the verification with the problem statement given in Step 2.

Step 2: *Write* Out an *Accurate* Statement of What Problem You Are Trying to *Solve*

The problem statement that he developed was as follows:

> "Currently T-1, the plant Propane/Butane Splitter, is operating poorly based on historical standards. The current plant data indicates that the rectification section is not performing well even though the reflux rate is well above that required for good performance. There were several changes made recently. These changes consisted of new trays, a new GC, and a feed composition change. All of these occurred prior to observation of the loss of fractionation. The loss of fractionation was not noticed until the product demand increased so that the tower began operating at design rates. Determine the reason for the poor performance of the rectification section of T-1."

Step 3: *Develop* a Theoretically Sound Working Hypothesis that Explains as Many Specifications of the Problem as Possible

When the chronological history was reviewed, it was tempting to believe that the trays were installed wrong. While this was a strong possibility, it did not represent a theoretically sound working hypothesis without adequate calculations and additional analysis. The questions in Chapter 6 were utilized along with a tray stability diagram to formulate hypotheses. (see Table 8.3)

While this series of questions did not provide an exact diagnosis of the problem, it was apparent that the tower was operating well beyond the region of experience and operating directives. The first question that the problem-solver sought to answer was "Do the current operating rates of the tower variables by themselves explain the poor tray performance?" Two types of calculations were made using the data to allow answering this question.

Table 8.3 Questions/comments for Problem 8.1.

Question	Comment
Are all operating directives and operating procedures being followed?	No. Operating conditions appeared to be well outside the normal targets
Are all instruments correct?	All instruments had been calibrated
Are laboratory results correct?	The new GC had been thoroughly calibrated
Were there any errors made in the original design?	Not applicable since no design changes were made
Were there changes in the operating conditions?	Yes. In addition to the known changes, it is possible that whatever caused the percent C_3 in the feed to increase might have also caused an extraneous component to be introduced into the tower
Is fluid leakage occurring?	Damaged trays might cause internal tray leakage, which might explain the problem
Has there been mechanical wear that might explain the problem?	No
Is the reaction rate as anticipated?	Not applicable
Are there adverse reactions occurring?	Not applicable
Were there errors made in the construction?	Maybe. The tray installation might not have been correct

A tray stability diagram was developed. This is shown in Figure 8.6. This diagram indicates that the top part of the tower is being overloaded and probably is operating in a flooded condition.

The internal tower vapor rate was calculated by material balance and heat balance. These calculated rates were then compared. If flooding was occurring, the vapor rate as determined by material balance would be higher than that determined by an overall heat balance. This is because as the tower floods, liquid would be carried over with vapor from the top tray into the reflux condenser. The total vapor (including entrained liquid) going into the condenser and accumulator would be greater than that estimated from a heat balance on the tower. Regardless of what control scheme was being used, this excess liquid must be pumped back to the tower and would show on the reflux flowmeter.

The problem-solver calculated the vapor and liquid loading in the top of the tower by material balance and heat balance as follows:

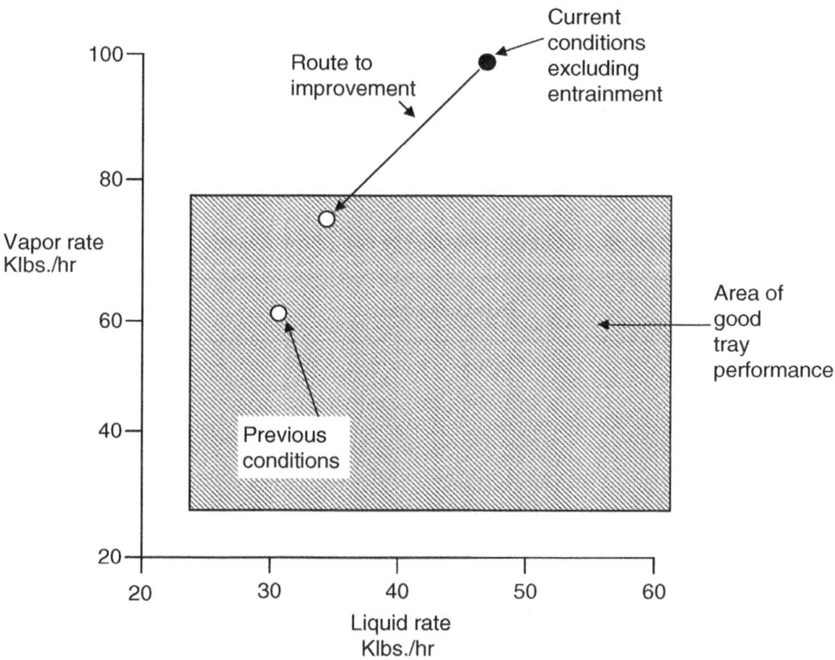

Figure 8.6 Tray stability diagram for Problem 8.1.

1) Q!~He calculated the distillate rate by material balance:

$$F = B + D \tag{8.1}$$

$$F^* X_F = B^* X_B + D^* X_D \tag{8.2}$$

where

F = feed rate, lb/hr
B = bottoms rate, lb/hr
D = distillate rate, lb/hr
X_F = feed concentration, wt%
X_B = bottoms concentration, wt%
X_D = distillate concentration, wt%

For the two cases given in Table 8.2, the calculated results are shown in Table 8.4.

2) He estimated the vapor and liquid loading in the top of the tower. Since the reflux is not subcooled, the liquid rate (L) in the top of the tower is simply the

Table 8.4 Estimated material balance rates.

	Past Operation	Current Operation
Distillate (*D*), lb/hr	30,100	48,700
Bottoms (*B*), lb/hr	69,900	51,300

Table 8.5 Estimated vapor and liquid rates in the top of the tower.

	Past Operation	Current Operation
Liquid (*L*), lb/hr	31,300	65,000
Vapor (*V*), lb/hr	61,400	113,700

reflux. In addition, the vapor rate (V) in the top of the tower can be estimated as follows and shown in Table 8.5.

$$V = L + D \tag{8.3}$$

Using the principle of "equal molal overflow," he calculated the top vapor rate (V) based on the heat input to the reboiler. The "equal molal overflow" principle states that for systems with minimal nonideality, the vapor rate throughout the tower expressed in mols/hr is equal, as long as no other vapor or heat input is introduced. The tower feed is a saturated liquid; therefore, there is no other vapor generated when the feed is added to the tower. In addition, there is no other heat input to the tower. Thus, the equation below was used to estimate the vapor rate at the top of the tower from the heat input to the bottom of the tower. The results of this calculation for past and current operations are shown in Table 8.6.

$$V_T = V_B{}^* M_T / M_B \tag{8.4}$$

where

V_T = vapor rate at the top of the tower, lb/hr
V_B = vapor rate at the bottom of the tower, lb/hr
M_T = molecular weight of vapor top of the tower
M_B = molecular weight of vapor bottom of tower

The problem-solver compared the top vapor rate as estimated from the overhead material balance and from the bottoms vapor rate and developed Table 8.7.

The problem-solver now had two independent calculations that indicated that the tower was flooding. The tray stability diagram indicated that the

Table 8.6 Top vapor rate based on heat input to the tower bottom.

	Past Operation	Current Operation
Vapor rate bottom, lb/hr	79,500	124,000
Molecular wt, top	44.54	46.23
Molecular wt, btm	57.63	57.63
Vapor rate top, lb/hr	61,400	99,500

Table 8.7 Vapor rates calculated from material balances and heat balances.

Calculation Method	Past Operation	Current Operation
Top material balance, lb/hr	61,400	113,700
Bottom heat balance, lb/hr	61,400	99,500
Comments	Good check	Poor check

tower was operating in the flooding regime. In addition, the comparison of the heat balances and material balances indicated that there was more material leaving the top of the tower than could be accounted for by a heat balance. The fact that previously this approach gave a good comparison was proof that there was a significant change in the operation of the tower. The problem-solver then developed the following hypothesis.

> "The poor performance of T-1 is due to flooding of the rectification section that is being caused by the excessive vapor and liquid loading. This excessive vapor and liquid loading might be caused by one of the following:

- *Operator error*: The reboiler steam rate or tower reflux rate was set too high by operator error.
- *Tray installation error*: There may have been an error in the tray installation which resulted in poor fractionation and caused the control system or operator to increase the tray loadings to attempt to compensate for this. This caused the tower to operate in a flooded regime.
- *Foaming*: The presence of a surface-active material might have caused the tower to originally function poorly which then caused a manual or automatic intervention to compensate for this uncovered problem. This intervention caused the tower to begin to operate in a flooded regime.
- *Other*: There may be other as yet undiscovered explanations for the changes which moved the tower operation into the flooding regime."

Note that in this problem hypothesis, the problem-solver expressed the need to explore what may have caused the excessive tower loadings rather than to just assume that it was due to an operator or control system error. It should also be noted that while the increased C3 concentration in the feed to the tower is unlikely to be the root cause of the problem, it did indicate that the source of the C3/C4 liquid might have changed. This new source might have resulted in a different trace impurity or an increase in the concentration of an existing trace impurity that would cause foaming in the tower. The possible trace impurities in a C3/C4 liquid stream could include materials used for de-icing exchangers (methanol), materials used for removing water (glycol), or materials used for neutralizing acidic compounds (amines).

Step 4: Provide a Mechanism to Test the Hypothesis

The problem-solver decided to test the hypothesis by reducing the tray loadings to safe levels, as indicated by the tray stability diagram. If this test returned the operation to "past operation" levels, then the mechanism to test the hypothesis becomes a partial solution.

He also considered what would happen if the operation did not return to the conditions and results previously experienced when the tray loadings were reduced. If that contingency occurred, he believed that additional tests or investigation would be required. These additional tests would include one or more of the following:

- Perform more detailed testing using some of techniques shown in Table 8.1.
- Conduct some sort of foaming test. This would probably require some type of pressure-rated laboratory equipment with a sight glass.
- Take X-rays of the appropriate parts of the tower to determine if tray damage is obvious.
- Determine if there has been a change in the source of the C3/C4 liquid and if there was any information on the trace impurities that might be present in this source.

If the reduction of the tray loadings does not solve the problem, it is likely that an expedient solution of reducing the feed rate to the tower may be required to ensure producing an overhead product that meets the specifications. However, this is not the solution to the problem, but only a "stop-gap" approach to allow making on-specification product.

Step 5: Recommend Remedial Action to Eliminate the Problem without Creating Another Problem

The actual recommendation for remedial action will depend on the results of Step 4. If reducing the tray loadings does not solve the problem, then the other tests listed in Step 4 must be executed and carefully analyzed to determine the required remedial action.

If reducing the tray loadings does solve the problem, it should be recognized that the basic questions outlined in Step 3 still remain. It is unlikely that improper tray installation would be a transient problem. It is also unlikely that an operator purposefully set the ratio controller outside the range of standard operations without a reason. Thus, the most logical possibility is that there was a transient condition that was due to the presence of a trace impurity or due to some other external event. When the operator encountered this external transient event, he tried to respond to the unmeasured and unknown disturbance by increasing the ratio controller, which caused the tower to begin flooding. Since this transient condition might well reoccur, the tower should be monitored daily to determine when the problem reoccurs. It will be of value to be proactive and plan what analyses or actions will be taken when daily monitoring indicates that the transient condition is returning.

The test of reducing the reflux rate so that the operations were in the good tray operating region was successful in returning the operations to normal rates and purities. A review of the possibility that the sources of the C3/C4 liquid had changed indicated that the source had changed, but since the problem only occurred as the production rate increased, it was impossible to connect this change with the change in tower operation.

As indicated earlier, in this circumstance, it is necessary to monitor the tower closely and develop a contingency plan to allow the response to the likely return of the transient condition. The problem-solver took the following actions:

- He developed a fractionation index, which allowed him to follow the operation of the tower on an hour-by-hour basis.
- He developed a list of samples that were to be obtained the next time that the fractionation index dropped below a specified value.
- He worked with the laboratory to develop new analytical techniques for the bomb samples to allow determining if there were surfactants present in any of these samples. The analytical techniques and GCs used for volatile hydrocarbons would likely not be sensitive enough for very low concentrations of surfactants.
- He developed a foaming test that would be used in case of a significant decrease in the fractionation index.
- He collected the specified samples and used the new analytical techniques to determine the possible presence and concentrations of surfactants in the base case with good operations. He also conducted the foaming test on these samples.

8.7 Lessons Learned

While this problem is a fictitious example, it has many elements of real-life problem-solving. The multiple events occurring at the same time that make isolating a single root cause difficult is a real-life element. The presence and subsequent disappearance of a trace impurity that leads to an unsuspected problem is also a real-life event. There are lessons to be learned from this semi-fictitious example.

There was great value in doing calculations to prove that the tower was flooding. The tray loading diagram was a useful tool for problem-solving and also in selling the problem solution to the management.

It should be recognized that if daily monitoring had been utilized that the problem would have been spotted immediately, as opposed to having to wait until the demand increased to the point of requiring the tower to operate at full capacity. While this may seem like an isolated occurrence, many industrial problems lie dormant until it is necessary to increase rates. Also note that it is never too late to begin a daily monitoring system.

This problem also illustrates that there will be occasions when the exact root cause of the problem cannot be determined. In these instances, the problem-solver should develop a system that will be effective in collecting data when the next occurrence of the event happens.

While it could be argued that all problems have an obvious root cause, it should be realized that multiple events that may seem to be related to the problem do occur. Thus, the multiple events (mechanical supervisor being sick, new GC, possible change in source of C3/C4 feed, and oil in the GC) for this problem are often typical of industrial problem-solving. The problem-solver must not discard any set of data or observations, but rather must incorporate them in his problem statement or problem analysis.

Nomenclature

B	bottoms rate, lb/hr
D	distillate rate, lb/hr
F	feed rate, lb/hr
M_{B}	molecular weight of the vapor bottom of the tower
M_{T}	molecular weight of the vapor top of the tower
V_{B}	vapor rate at the bottom of the tower, lb/hr
V_{T}	vapor rate at the top of the tower, lb/hr
X_{B}	bottoms concentration, wt%
X_{D}	distillate concentration, wt%
X_{F}	feed concentration, wt%

9

Application to Kinetically Limited Processes

9.1 Introduction

A kinetically limited process is one that does not go to completion because it is limited by the speed of mass or heat transfer. Completion can be defined as equilibrium or a state that would be reached if an infinite amount of time is available. Examples of this process are heat exchange, reaction, and diffusion-limited operations.

The purpose of this chapter is to illustrate how the five-step approach to problem-solving can be used to solve problems in this type of process. The chapter shows how a generalized approach can be used to develop theoretically sound working hypotheses for any kinetically limited process. The emphasis is on utilizing this approach in an industrial setting. Thus, the more theoretical approach of using multiple constants has been replaced with an empirical "lumped parameter" approach. A specific example of this that most process engineers are familiar with is the use of an overall heat transfer coefficient rather than individual film coefficients. The overall heat transfer coefficient can be determined from vendor specification sheets or from plant test data when the exchanger is known to be clean.

9.2 Kinetically Limited Models

Any kinetically limited process can be described by the generalized equation (9.1):

$$R = C^{*}\mathrm{DF} \tag{9.1}$$

where

R = the rate of change with time of the variable under study
C = a constant referred to as the "lumped parameter constant"
DF = the driving force or incentive for mass/heat transfer to occur

Problem Solving Approaches for Maintaining Operational Excellence in Process Plants, Second Edition.
Joseph M. Bonem, Nattapong Pongboot, and Wiroon Tanthapanichakoon.
© 2025 the American Institute of Chemical Engineers, Inc. Published 2025 by John Wiley & Sons, Inc.

For a kinetically limited process, the lumped parameter constant "C" can be used for problem-solving in multiple ways. The value of "C" can be determined on an hourly or daily basis and monitored as part of a daily monitoring system. Changes in "C" will be trigger points to start active problem-solving. In addition, based on an estimated value of "C," studies can be conducted to estimate what impact changes in the driving force will have on the rate. This may allow process conditions to be modified to compensate for changes in "C."

As indicated earlier, the heat transfer relationship shown below is the best-known example of equation (9.1).

$$Q = U^*A^* \ln \Delta T \tag{9.2}$$

where

Q = R = rate of heat transfer
U^*A = C = the lumped parameter or for heat transfer, the heat transfer coefficient multiplied by the area
$\ln\Delta T$ = DF = the driving force or for heat transfer, the log mean temperature difference

A more complicated form of equation (9.1) is encountered for reaction and diffusion-limited drying. The driving force in these cases is often related to the concentration of a molecule or the difference between the concentration and equilibrium concentration. For example, for diffusion-limited drying, equation (9.1) becomes

$$dX/dt = K^*(X - X_e) \tag{9.3}$$

where

dX/dt = the rate of removal of a volatile from a polymer
K = the lumped parameter constant which is somewhat related to diffusion. Since it deals with mass transfer, it is also referenced in this chapter as a mass transfer coefficient
$X - X_e$ = the driving force
X = the actual concentration of a volatile in a polymer
X_e = the equilibrium concentration of a volatile in the polymer at the same point in time

For a simple reaction between two components where the reaction rate is first order, equation (9.1) becomes

$$dJ1/dt = C_R{}^*J1^*J2 \tag{9.4}$$

where

$dJ1/dt$ = the rate of disappearance of component $J1$
C_R = the lumped parameter constant (reaction rate constant)
$J1$ = the concentration of component $J1$
$J2$ = the concentration of component $J2$

Relationships for other kinetically limited processes can be developed starting with equation (9.1). The key to using this equation for developing theoretically sound working hypotheses is the correct selection of the driving force and use of the "lumped parameter" constant.

The need to select the correct driving force can be illustrated by equation (9.3). The correct driving force is the difference between the actual and equilibrium concentration. If the equilibrium concentration is ignored, the constant determined from plant operating data will be lower than actual. Another example where the use of the incorrect driving force will give improper constants is estimation of heat transfer constants in a fuel-fired furnace. Heat transfer from the flames and refractory surfaces to the metal tubes (primarily radiant heat) depends on the driving force for radiant heat transfer rather than convective heat transfer. Where the driving force for convective heat transfer is simply the log temperature difference, the driving force for radiant heat transfer is expressed as shown below:

$$\mathrm{DF_R} = T_\mathrm{g}^4 - T_\mathrm{m}^4 \tag{9.5}$$

where

$\mathrm{DF_R}$ = driving force for radiant heat transfer
T_g = absolute temperature of gas
T_m = absolute temperature of tube metal

Lumped constant parameters that are developed from plant data will be risky to use for large extrapolations. A lumped parameter constant that can be developed from fundamentals will always have a broader range of applicability. These constants that are developed from fundamentals can be used regardless of how close the operating conditions approach those under which the constant was determined.

However, some constants can be determined from fundamentals only by such an elaborate procedure that it becomes impractical for plant problem-solving activities. For example, in diffusion-related drying (equation (9.3)), the lumped parameter constant or mass transfer coefficient depends on the diffusion rate constant through the polymer, the effective length of the flow path through the polymer particle, the actual particle surface area, and the mass transfer coefficients from the polymer surface to the bulk of the gas. Most polymer particles are irregular shaped and have a large number of internal pores. This makes determining variables such as particle surface area and flow path through the particle difficult, if not impossible to determine. If only one of these individual variables is not available and cannot be developed or determined, then utilizing a lumped parameter empirical constant (overall mass transfer coefficient) provides the only means to evaluate diffusion-related drying. This overall mass transfer coefficient can be determined from plant data, from pilot plant tests, or from bench-scale tests.

9.3 Limitations to the Lumped Parameter Approach

The lumped parameter approach has sometimes been referred to as a "black box" approach. This implies that the person using this approach does not understand the details of what is occurring inside the "black box." For example, the use of an overall heat transfer coefficient does not take into account the individual film side coefficients, metal resistance, and fouling factors. However, in most process plant problem-solving activities, the significant observation is the change in the overall heat transfer coefficient. Whether the decrease of the heat transfer coefficient is due to the tube or shell side film coefficient decrease can usually be determined based on experience. For example, the decrease in the overall heat transfer coefficient for a fractionating tower overhead condenser that utilizes cooling tower water is almost always due to fouling on the cooling water side. Similar logic would apply to the use of the lumped parameter approach for problem solving other kinetically limited processes.

The limitation of this approach is more severe when designing new facilities based on a lumped parameter coefficient determined from an operating plant. If a more fundamental analysis is not done, the lumped parameter approach can lead to design errors. An example of this is the design of a specialized vertical condenser for a new plant based on a condenser in identical service in an existing plant. The condensation in this vertical condenser took place on the shell side, and cooling water was utilized in the tubes. The new plant had slightly higher capacity. To evaluate the overall heat transfer coefficient from pure fundamentals would have required an elaborate model that involved the condensate thickness at various points in the vertical exchanger, as well as the traditional resistances such as water side coefficient, condensing side coefficient, and metal resistance. Rather than doing this highly theoretical analysis, the design team chose to utilize the overall heat transfer coefficient experienced in the existing plant. However, based on the desire to minimize the overall height of the structure in the new plant, the height-to-diameter (H/D) ratio of the vertical condenser was reduced. In order to compensate for this reduced H/D ratio, the tube diameter and exchanger shell diameter were increased. No consideration was given to the effect of these increased diameters on the overall heat transfer coefficient.

When the unit was started up, it was noticed that the heat transfer coefficient was 20% below anticipated values. This underscores the need for geometric similarity when basing a design on empirically developed constants.

In spite of the limitations, when considering plant-related problem-solving, the utilization of an empirically derived lumped parameter constant, along with a theoretically correct driving force, is almost always adequate because of the three reasons given below:

1) Daily monitoring of the process requires only an empirical constant.
2) Time is always critical. Laboratory or elaborate investigations to develop fundamental data are rarely appropriate.
3) If it is desirable to make operating changes, the new conditions are generally not a large extrapolation from current conditions. Thus, there is minimal danger of extrapolation.

9.4 Guidelines for Utilization of this Approach for Plant Problem Solving

Utilizing the "lumped parameter constant" approach for monitoring and solving plant problems is a powerful tool. However, like all tools, it must be used in an appropriate fashion. Some guidelines for using this tool are as follows:

1) Develop a meaningful driving force that is theoretically correct. *It should be noted that there is often a difference between a driving force that appears logical and one that is theoretically correct.* Logic cannot be substituted for the utilization of sound engineering fundamental knowledge. Without a meaningful driving force, the empirically developed lumped parameter constant will not be valid over any range of data.
2) Monitor this lumped parameter constant on a daily basis and over as wide a variety of conditions as possible. If the constant varies, look for correlations between the constant and *independent* variables that make theoretical sense. For example, a correlation between the lumped parameter constant in a diffusion-limited drying process and gas rates makes good theoretical sense. The increased gas rate should increase the rate of mass transfer by creating more turbulence. If a lumped parameter constant varies with no process changes, it is an indication that an extraneous factor has caused a process deviation. If this extraneous factor can be eliminated, then the constant should return to normal. If this factor cannot be discovered or eliminated, then process changes may be required.
3) If it appears desirable to recommend a change in the driving force to increase the kinetically limited rate, minimize the length of the extrapolation that uses an empirically developed lumped parameter constant.

9.5 Example Problem 9.1

A polymer plant was experiencing problems in stripping the residual volatile from the polymer product. A counterflow agitated dryer with pure nitrogen sweep gas was utilized to strip the volatile. All operating conditions appeared to be normal. In addition, all instruments were checked and appeared to be accurate. A timeline indicated that the problem seemed to begin when a new catalyst was introduced into the process for a plant test. Even though using the new catalyst was considered a test run, it was mandatory that the plant be switched to this new catalyst as soon as possible. One of the key advantages of the new catalyst was the higher bulk density product that could be produced with the catalyst. This had been well-demonstrated in pilot plant studies. No drying studies had been conducted in the pilot plant. Drying capability was thought to be associated with sweep gas rate, residence time, and temperatures, which were not changed when the plant was switched to the new catalyst.

The problem-solver was asked to determine what could be done to reduce the volatile levels to the previous concentrations while continuing to operate with the new catalyst. Operating conditions are given as follows in Table 9.1.

In addition to the plant data shown, laboratory results indicated that the equilibrium relationships for polymers produced with the old and new catalysts were identical and could be expressed as shown below:

$$X_E = 295,000^*(P/\mathrm{VP})^{1.5} \tag{9.6}$$

where

X_E = the equilibrium concentration of volatile in the polymer
P = the partial pressure of the volatile in the vapor phase
VP = the vapor pressure of the volatile at the dryer temperature

Table 9.1 Operating conditions with the old and new catalyst.

	Old Catalyst	**New Catalyst**
Polymer rate, lb/hr	20,000	20,000
Polymer density, lb/ft^3	25	28
Pure nitrogen sweep gas rate, lb/hr	1,000	1,000
Dryer temperature (isothermal), °F	200	200
Hexane vapor pressure, psia	28.7	28.7
Dryer pressure, psia	15	15
Inlet hexane content, ppm	1,000	1,000
Outlet hexane content, ppm	25	75

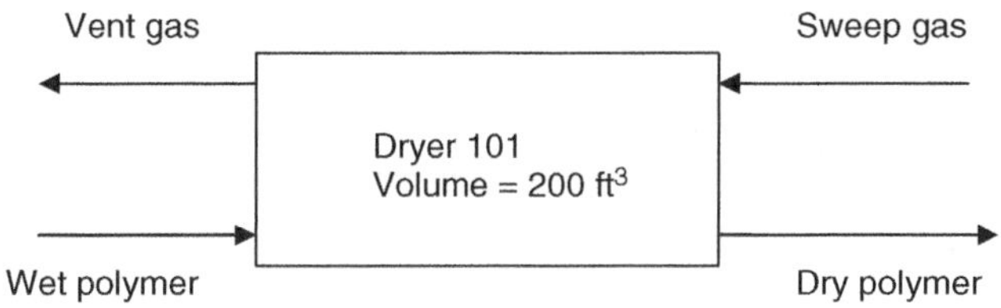

Figure 9.1 Schematic sketch of the dryer.

A schematic sketch of the dryer is shown in Figure 9.1.

The problem-solver began working on the problem by using the five-step problem-solving approach. There was minimal need to verify that the problem had actually occurred. However, he did not skip this step.

Step 1: Verify that the Problem Actually Occurred

The problem-solver confirmed that all of the meters were correct and that the hexane content was indeed three times the normal levels.

Step 2: *Write* Out an *Accurate* Statement of What Problem You Are Trying to *Solve*

He wrote out the following problem statement:

> "The polymer leaving Dryer-101 contains excessive amounts of volatile. This condition started soon after the introduction of a new catalyst in the polymerization reactors. All operating conditions are normal and the instrumentation is correct. It is desirable to convert the plant to using this catalyst as soon as possible. Determine what operating condition changes or what new equipment is required to achieve 25 ppm hexane in the final product at the normal polymer production rate."

Step 3: *Develop* a Theoretically Sound Working Hypothesis that Explains as Many Specifications of the Problem as Possible

Referring back to the list of questions given in Chapter 6 was helpful in formulating possible hypotheses. The problem-solver reviewed these questions and formulated appropriate comments for this example problem as shown in Table 9.2.

As might be expected, these questions pointed strongly to the idea that the problem began when the new catalyst was introduced to the process. They also provided some possible explanation for why the new catalyst would cause a drying problem to occur. It was theorized that the new catalyst created a less porous polymer particle. The lower porosity would cause the diffusion of the volatile through the particle to be slower. This less porous particle would also have a higher particle density, which would explain the higher bulk density. It should be noted that if this list of questions was not utilized, the change in the catalyst

Table 9.2 Questions/comments for Problem 9.1.

Question	Comment
Are all operating directives and procedures being followed?	All appeared to be correct and being followed
Are all instruments correct?	The instruments had allegedly been calibrated
Are laboratory results correct?	Yes
Were there any errors made in the original design?	Not applicable
Were there changes in operating conditions?	New catalyst was started
Is fluid leakage occurring?	Not applicable
Has there been mechanical wear that would explain the problem?	Not applicable
Is the reaction rate as anticipated?	The new catalyst might cause the polymer particles to be less porous. This would cause a lower than anticipated "mass transfer coefficient"
Are there adverse reactions occurring?	See above
Were there errors made in the construction?	Not applicable

would have been pinpointed as the cause, but the explanation of the formation of a less porous polymer particle might have been missed.
As indicated in equation (9.3), the relationship between the rate of drying and driving force for a diffusion-limited drying process can be expressed as follows:

$$dX/dt = K^*(X - X_e) \tag{9.3}$$

Since the problem statement indicates that all operating conditions were normal and laboratory results indicated that the equilibrium relationship for both polymers was the same, there must be a difference in the "K" values between the two polymers.

While a working hypothesis that the "K" value has changed is a perfectly valid hypothesis, it does not help determine what the next step should be because it does not specify the magnitude of the change. If the magnitude of the change is not known, it will be impossible to determine if operational changes or addition of new equipment will be required to allow operation with the new catalyst. Thus, the theoretically sound working hypothesis must include an estimate of the magnitude of the change in the mass transfer coefficient ("K" value).

The value of "K" can be estimated by simple integration if X_e (the equilibrium concentration of volatile in the polymer) is equal to 0. In this special case, integration yields the following:

$$X_f/X_o = e^{-Kt} \tag{9.7}$$

where

X_f = the concentration of the volatile in the outlet polymer
X_o = the concentration of the volatile in the inlet polymer
K = the rate constant or mass transfer coefficient, 1/minutes
t = the amount of time in the dryer, minutes

For the general case where X_e is not equal to 0, numerical integration must be used to determine the unknown variable of "K." This numerical integration can be used to determine the mass transfer coefficient (K) or the outlet concentration of the volatile in the polymer (X_f). This numerical integration can be developed by visualizing a dryer segment as shown in Figure 9.2.

The equilibrium concentration in Figure 9.2 can be determined by a Henry's Law type relationship as shown below:

$$X_e = C_E{}^*Y \tag{9.8}$$

where

C_E = a constant determined from experimental data or application of theoretical principles
Y = the vapor phase composition at any point in the dryer

For the example problem, the equilibrium relationship shown in equation (9.6) can be utilized in place of that shown in equation (9.8) to determine the equilibrium concentration based on the vapor phase concentration.

The integration of equation (9.3) when X_e is not equal to 0 yields the following:

$$\int dX/(X - X_e) = Kt \tag{9.9}$$

Figure 9.2 Dryer calculation segment.

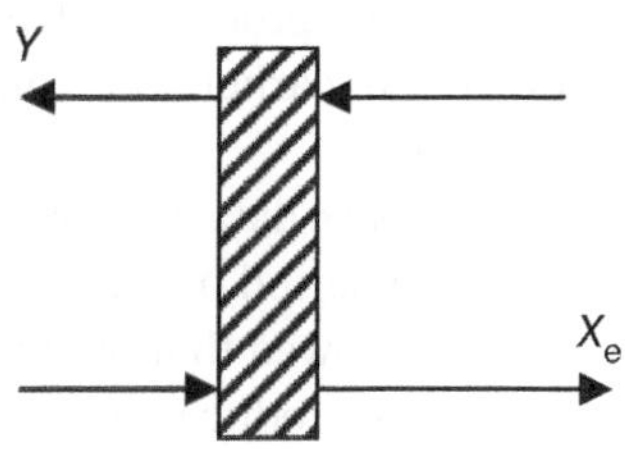

For this problem, it is desired to obtain the "K" value for each catalyst system. To determine the "K" value from this relationship using numerical integration requires the following steps using the data given above and a spreadsheet:

1) Use values of the polymer and gas rates, along with the inlet and outlet hexane concentrations in the polymer, to determine the concentration of hexane in the outlet gas. Note that the data indicated that pure nitrogen was used as the sweep gas.
2) Calculate the total residence time in the dryer assuming no "back mixing." This can be calculated by dividing the polymer holdup in the dryer by the polymer rate.
3) Split the dryer into several segments (100–500) and calculate the *dx* for each segment. The change in the volatile concentration (*dx*) will simply be the total change in volatile concentration divided by the number of dryer segments. The time increment (*dt*) can be determined in a similar fashion.
4) From the relationship given in equation (9.6) and the outlet vapor composition, calculate the X_e of the polymer leaving the first segment.
5) Since this segment represents only an exceptionally small part of the dryer, the actual volatile concentration (X) can be considered constant throughout this segment. For this small segment:

   $$(Kdt)_i = dX/(X - X_e) \qquad (9.10)$$

   In equation (9.10), the term $(Kdt)_i$ is simply the product of the lumped parameter drying constant (mass transfer coefficient) and the residence time in the small segment. The lumped parameter drying constant is assumed to be the same in all segments.
6) Using *dx*, the gas and polymer rates and the gas concentration leaving the first segment, calculate the gas concentration entering the segment by a material balance around the segment. This will be the gas concentration leaving the next segment.
7) Continue this process until the same calculations have been made for each segment. Then, "K" can be calculated as follows:

   $$K = \xi(Kt)_i/t \qquad (9.11)$$

When the problem-solver performed these calculations for the two cases described earlier, the following values of "K" were obtained as shown in Table 9.3.

Extensive data were available from computer archives so that the value of "K" over the period of interest could be evaluated. When this was done, Figure 9.3 clearly showed that the mass transfer coefficient decreased when the new catalyst was introduced into the system.

Table 9.3 Results of numerical integration.

Catalyst	*K* Value, 1/minute
Old	0.25
New	0.16

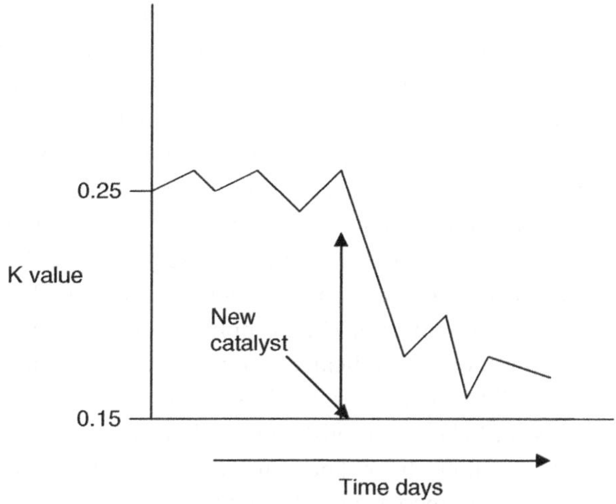

Figure 9.3 *K* value versus time.

A theoretically sound working hypothesis that fits the data would be

> "The recent loss of drying capability is associated with a decrease in the mass transfer coefficient (K) that occurred when the change to the new catalyst occurred. It is believed that this change is due to the lower porosity of the polymer particle produced with this catalyst. Evidence that supports this is the 35 percent lower 'K' value and the higher bulk density."

Step 4: Provide a Mechanism to Test the Hypothesis

At this point, the problem-solver gave consideration to the question of "optimum technical depth." Several alternatives were available to test the hypothesis. The optimum alternative depended on the cost of the solution, the confidence level required, and the cost of continuing poor drying performance. In this polymer process, the customer specifications and/or needs were considered more important than operating at full capacity. Thus, in this case, the cost of the poor drying performance was the lost profits associated with operating at lower than design production rates. Alternatives for testing the hypothesis that were considered were as follows:

- Conducting a test run using the old catalyst to confirm that the old catalyst really had a "K" value higher than the new catalyst.
- Using the "K" value for the new catalyst to estimate what changes in operating conditions would be required to reduce the hexane concentration from 75 to 25 ppm
- Modifying the new catalyst so that it produced a polymer similar to the old catalyst.

In order to determine if operating conditions could be modified to reduce the hexane to 25 ppm, the problem-solver made calculations using the approach described earlier. However, in these calculations, the "K" value was known to be 0.16 1/minutes. The value to be estimated was the outlet volatile concentration. The same spreadsheet and approach were used. The problem-solver changed the operating variables as desired and then varied the outlet volatile concentration in the polymer until the calculations based on the spreadsheet indicated a "K" value of 0.16 1/minutes. Based on the calculations, he concluded that increasing the sweep gas by a factor of 10 (1000–10,000 lb/hr) would reduce the hexane content from 75 to 65 ppm This was obviously not a valid approach. The calculations also indicated that if the residence time in the dryer was increased from 17 to 24 minutes, the hexane concentration would be reduced to 25 ppm This could be accomplished in a plant test by reducing the polymer production rate to 14,000 lb/hr. The volumetric holdup in the dryer was constant, so the only way to increase residence time was to reduce the production rate. The calculations also indicated that the approach of increasing the dryer temperature was not a fruitful route to improved performance. They showed that if the dryer temperature was increased from 200 to 225 °F, at the same residence time and sweep gas rate, the hexane concentration would decrease from 75 to 65 ppm.

From these calculations, it was obvious that the only valid test of plant operating conditions was to reduce the polymer production rate from 20,000 to 14,000 lb/hr. The plant test of operating at reduced rates was the preferred test compared to a plant test of the old catalyst since it would be quicker. Introducing the old catalyst into the reactors would require significant residence time to completely displace the polymer produced with the new catalyst from the entire system prior to the dryer. In addition, continuing to use the new catalyst was mandatory. So the plant test of returning to the old catalyst just confirms that the new catalyst is responsible for the problem. A plant test of increasing the dryer residence time by reducing the production to 14,000 lb/hr was successful at reducing the hexane content from 75 to 25 ppm The "K" value did not change as the residence time was increased.

Step 5: Recommend Remedial Action to Eliminate the Problem without Creating Another Problem

Since the plant test of operating at reduced rates was successful and a study of changing the catalyst to produce a more porous structure indicated that other desirable catalyst attributes would be lost if the porosity was changed, it was decided to add an additional dryer. This additional dryer would provide the necessary residence time. Since the dryer addition would take several months, it was decided to return to using the old catalyst in the interim. When use of the old catalyst was resumed, the mass transfer coefficient increased from 0.16 1/minutes to the previous value of 0.25 1/minutes.

9.6 Lessons Learned

If daily monitoring of the "K" value was being done, it would have been possible to determine immediately that there was likely a drying problem associated with the new catalyst. As often happens in a plant test, there is only a minimal potential problem analysis conducted prior to the test. This often leads to a panicky approach to problem-solving. Chapter 14 discusses approaches for conducting successful plant tests.

If drying problems had been anticipated prior to the plant test of the new catalyst, it would have been possible to use laboratory techniques to determine the magnitude of the difference in mass transfer coefficients. An apparatus such as a thermal gravimetric analyzer (TGA) could have been used to determine the mass transfer coefficient for the polymer produced with both catalyst systems. While the exact absolute values of the mass transfer coefficient determined in this fashion may not have been the same as that in the plant, the relative values would have been very accurate. That is, the TGA would have likely predicted that the mass transfer coefficient (K) would have been 35% lower with the new catalyst.

The model of the dryer that was developed is a good example of a simple, but a valuable tool for doing process analysis work. The driving force of the actual concentration less the equilibrium concentration ($X - X_e$) is theoretically correct and easy to determine. Experimental relationships such as equation (9.6) may not be available. However, there are calculation methods available to approximate the relationship between the vapor phase concentration of the volatile and the equilibrium concentration of the volatile in the solid. One of these techniques is discussed in Sections 9.7. The mass transfer coefficient can be determined from plant data. It is subject to error if there is a significant change in the polymer morphology. For example, when the catalyst was changed, the mass transfer coefficient changed. However, it did not change when the production rate was reduced.

The calculations that were done to estimate the impact of changing plant operating conditions were performed very quickly once the spreadsheet described above was completed. The more expensive alternative of running multiple plant trials to test any *reasonable* hypothesis would have taken much more time and been much more expensive.

9.7 Technique for Estimating Polymer–Volatile Equilibrium

The problem-solver will often not have available to him simulations that allow predictions of polymer–volatile equilibrium. Faced with this dilemma, he often will select the heaviest component in the computer database for simulating a polymer. This may be adequate for polymer–volatile systems that contain volatile concentrations greater than 10–20 vol.%. However, below this level, the polymer begins to act to reduce the volatility of the volatile and the partial pressure of the volatile will be less than anticipated when the polymer is simulated using the heaviest component in the database. The converse will also be true. That is, at volatile concentrations below 10–20 vol.%, the amount of the volatile in the polymer will be greater than that calculated based on the partial pressure of the volatile and simulating polymer as the heaviest component in the database.

There are techniques for dealing with this nonideality. Perhaps the best-known relationship dealing with the equilibrium between a polymer and volatile is the Flory–Huggins relationship. The use and limitations of this relationship are described in *Polymer Chemistry an Introduction* by Seymour and Carraher (2003).

The Flory–Huggins relationship predicts the equilibrium of a volatile in the *noncrystalline* polymer phase. Equation (9.12) shows the original relationship. The modification developed by the author is shown in equation (9.13).

The original Flory–Huggins relationship:

$$\mathrm{Ln}(P/\mathrm{VP}) = \mathrm{Ln}V1 + (1 - M1/M2)^{*}V2 + U^{*}V2^{2} \tag{9.12}$$

where

P = the volatile partial pressure
VP = the vapor pressure of the volatile
$V1$ = the volume fraction of the volatile in the polymer
$V2$ = the volume fraction of the polymer
$M1$ = the molar volume of the volatile, cc/g-mol
$M2$ = the molar volume of the polymer, cc/g-mol
U = the interaction parameter between the volatile and the polymer

Manipulation and assumptions ($V2 = 1$ and $M1/M2 = 0$) by the author yields

$$\mathrm{Ln}(P/\mathrm{VP}) = \mathrm{Ln}(D2^{*}X1/(D1^{*}1000000)) + 1 + \mathrm{U} \tag{9.13}$$

where

$D1$ = the density of the volatile
$D2$ = the density of the polymer
$X1$ = the weight of the volatile in the polymer, ppm

The interaction parameter can be evaluated knowing the "solubility parameters":

$$U = M1^*(S1 - S2)^2/(R^*T) + Z \tag{9.14}$$

where

$S1$ = the volatile solubility parameter $(\text{cal/cm}^3)^{0.5}$
$S2$ = the polymer solubility parameter $(\text{cal/cm}^3)^{0.5}$
Z = the lattice constant. In theory = 0.35 ± 0.1

The values of solubility parameters are given in various handbooks. The units of solubility parameters $(\text{cal/cm}^3)^{0.5}$ are due to the definition of solubility parameters as the square root of the cohesive energy density. It is beyond the scope of this book to cover the derivation of these thermodynamic terms.

It should be emphasized that the concentrations of the volatile in the polymer are that which exist in the noncrystalline region. The standard assumption is that there is no volatile in the crystalline region.

While this book cannot delve into all aspects of polymer–volatile equilibrium, the approach described above is meant to show that there are techniques available to estimate the equilibrium as opposed to simulating a polymer as the heaviest component in the database. While there are computer simulations that allow prediction of a polymer–volatile equilibrium, the above analysis will be of value in understanding the basis for these simulations or in doing the simulation by manual calculation.

9.8 Example Problem 9.2

Several of the tube metal temperature indicators in an operating furnace were indicating considerably hotter than anticipated when compared to historical values. Operations personnel were not concerned since this furnace was operating at normal heat duties; normal process fluid temperatures and the furnace had never shown any tendency to foul. They concluded that the tube metal temperature indicators must have failed due to the harsh furnace environment. The technical problem-solver did not believe that *all* of the tube metal indicators could have failed. A schematic diagram of this problem is shown in Figure 9.4.

The current operating data and the historical operating data are shown in Table 9.4. There were no other data available in the computer archives.

Even though this problem was discovered by the technical problem-solver rather than being delivered to him by a request from operations management, it is still a valid problem that can be approached using the five-step procedure. Because of the belief of operations personnel that there was not really any problem,

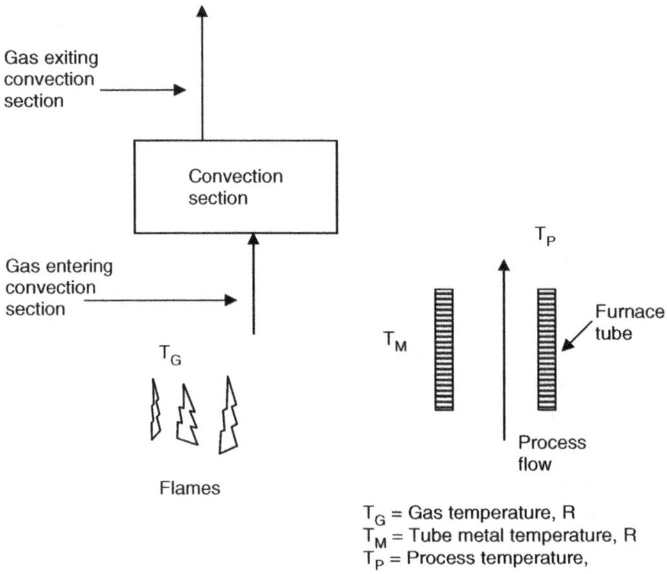

Figure 9.4 Furnace schematic.

Table 9.4 Furnace data.

	Historical	**Current**
Tube metal temperature, °F	600–650	750–900
Gas temperatures		
Entering the convection section, °F	1200	1300
Leaving the convection section, °F	650	700
Excess oxygen, percent	15	15
Heat absorbed, MBTU/hr	80	80
Circulation to furnace k-l/hr	700	700

Step 1 had to be approached in a different fashion than in other problems described previously.

Step 1: Verify that the Problem Actually Occurred

Step 1 required that careful examination be given to the operation management's contention that there really was no problem with the indicated high tube metal temperatures. It was important that the problem-solver convince himself, as well as the operations personnel, that there was a problem. If operations personnel were going to be expected to cooperate on any plant tests, they needed to be convinced that there was a problem. One of the first things that the problem-solver had to do was understand how the tube metal temperature indicators were designed and installed. These thermocouples are different than normal ones, in that they must measure the temperature of the furnace tube without being impacted by the radiant heat from the furnace flames. The problem-solver obtained information from both instrument technicians and instrument engineers to make sure that he understood the technology of tube metal temperature indicators. Once he understood this technology, he recognized that while tube metal temperature indicators do fail, they almost always fail in such a fashion that the failure is obvious. That is, they read ridiculously high or low. Failed tube metal temperature indicators will normally have "upscale burnout." That is, they will read ridiculously high. He also recognized that if fouling had occurred on the inside of the tube, there would be other indications of fouling besides elevated tube metal temperatures. The problem-solver decided to investigate whether there were other indications of fouling. He used the basic equation shown below to describe the radiant heat transfer that is occurring in the furnace:

$$Q = C_1{}^* \left(T_g^4 - T_m^4\right) = C_2{}^*(T_m - T_P) \tag{9.15}$$

where

Q = the heat transferred to the process. This is equal to the heat transferred from the gas flame by radiation. In addition, it must also be equal to the heat transferred through the tube to the process fluid.

T_g = absolute temperature of the gas in the radiant section of the furnace

T_m = absolute temperature of the tube metal

T_P = absolute temperature of the process fluid in the tubes

C_1 = a constant that is related to the type of flame. It should not change unless the type of fuel changes.

C_2 = the product of the heat transfer coefficient and the area of the tubes. It will only change if the heat transfer coefficient changes.

He noted that if the tube metal temperature (T_m) increases at constant heat duty (Q), it must be due to a decrease in the heat transfer coefficient since the area of the tubes will not change. This decrease in the heat transfer coefficient could be caused by a change in the fluid circulation rate or tube fouling. Since the process flow rate to the furnace is constant, the decrease in the heat transfer coefficient is likely due to tube fouling. In order to achieve the same heat duty at the higher tube metal temperature, the gas temperature (T_g) must also increase. Thus, the increased gas temperature shown in Table 9.4 was confirmation that the tube metal temperature had increased. A secondary confirmation was the higher than normal gas temperature leaving the convection section, which is also shown in Table 9.4. The higher temperature gas entering the convection section from the radiant section would not be cooled to as low a temperature as historical data would indicate since the heat transfer area in the convection section was fixed. Using this logic, the problem-solver convinced the operations management that there *might* be a fouling problem in the furnace tubes.

Step 2: *Write* Out an *Accurate* Statement of What Problem You Are Trying to *Solve*

An accurate statement of the furnace problem is as follows:

> "Tube metal temperatures on the process furnace are indicating that the tubes are partially fouled with deposits. Since a furnace shutdown to inspect and possibly replace the tubes will require a plant shutdown, a detailed investigation was undertaken to confirm that the tube metal temperatures are higher than normal. This investigation indicated that the tubes are very likely fouled. There is no archived data to indicate when the fouling began or if it was a gradual fouling or a one time event. While there is strong evidence that the furnace tubes are partially fouled, it is desirable to provide operations management with further evidence that the furnace is fouling and determine what steps should be taken to avoid a tube failure with a subsequent furnace fire. In addition, it will be necessary to determine why the tubes in this furnace are fouling when based on operation personnel's memory this has never happened in the past."

Step 3: *Develop* a Theoretically Sound Working Hypothesis that Explains as Many Specifications of the Problem as Possible

In this example problem, Step 3 is different from that in previous problems. The thrust is to convince operations that a problem exists with the furnace. As discussed earlier, the logic path that was developed was partially successful in convincing operations management that there really was a fouling problem. In addition to this, consideration was given to alternative means to measure the tube metal temperatures. One possibility was to use infrared temperature measurements through a furnace peephole. Besides confirming the tube metal temperature measurements, this would also allow determining where the hottest points were. The measurements based on tube metal temperature indicators are only single points in a few tubes. A much hotter point might exist in another location. An infrared scan was conducted on the

Table 9.5 Questions/comments for Problem 9-2.

Question	Comment
Are all operating directives and procedures being followed?	All appeared to be correct and being followed
Are all instruments correct?	The instruments had allegedly been calibrated
Are laboratory results correct?	Not applicable
Were there any errors made in the original design?	Not applicable
Were there changes in operating conditions?	No
Is fluid leakage occurring?	Not applicable
Has there been mechanical wear that would explain the problem?	Fouling of tubes
Is the reaction rate as anticipated?	Not applicable
Are there adverse reactions occurring?	Fouling is caused by the reaction to produce coke. This reaction might be catalyzed by the material present in the process. The fouling rate seems much higher than in previous experience
Were there errors made in the construction?	Not applicable. It had been 36 months since mechanical work was done on the furnace

furnace tubes. The thermocouples were verified, and even hotter temperatures were discovered at spots where no tube metal indicators existed.

The furnace history was considered by asking such questions as

When did the fouling begin?

Did it coincide with an upset of a magnitude never before experienced?

Has the furnace been operating for a longer continuous period than previous experience?

In addition to these questions, the questions from Chapter 6 were used to help develop a working hypothesis (see Table 9.5).

The following was developed as a working hypothesis:

> "The higher-than-normal tube metal temperatures in the furnace appear to be correct based on both gas temperatures and an examination of the instrument specifications which indicates that the thermocouples are 'up-scale burnout' (if they fail they will read ridiculously high). In addition, an infrared scan indicates that these thermocouple measurements

are correct and that there are even higher temperatures on other tubes. A review of the furnace history indicates that the current run (36 months without water washing the tubes) is nearly twice the length of previous periods of operation. Historically the furnace tubes were water washed at every 18-month furnace inspection downtime. However, they were not water washed at the last downtime because it was expedient to get the furnace back in service. It is believed that this water washing removes the fouling material before it becomes thick enough to impact the furnace operation. The fouling is believed to be initiated by process catalyst residue that is entrained during process upsets."

Step 4: Provide a Mechanism to Test the Hypothesis
In this case because of the risk of a tube failure, the only mechanism was to shut down the furnace and water wash and inspect the tubes. The downtime was timed to coincide with a period of reduced product demand so that the lost revenue was minimized. An inspection of the tubes indicated that many of them were approaching the point of incipient failure. If the tube metal temperatures had been ignored as proposed by the operating personnel, it is highly likely that a furnace fire would have occurred. This not only would have caused serious damage to the furnace, but likely would have occurred at a time when a full production rate was required.

Step 5: *Recommend* Remedial Action to Eliminate the Problem without Creating Another Problem
Three remedial actions were recommended. The first was to carefully monitor the tube metal temperatures during the interim period between the discovery of the problem and the furnace shutdown. This allowed for a controlled plant shutdown to replace furnace tubes rather than an emergency shutdown during a period of high product demand. The second was to water wash the furnace tubes every 18 months. The third recommendation was to monitor the C2 value as determined from equation (9.15) on a daily basis. This would provide information on how the fouling occurred. It could have been a one-time event that caused a sudden decrease in the value or a slow continuous decrease.

9.9 Lessons Learned

This actual real-life problem illustrates the value of a careful technical analysis of all operating data as opposed to only working on problems that operations or mechanical personnel consider to be important. The initial reaction of the operations personnel to ignore the high tube metal temperature indicators was carefully considered as part of Step 1. In addition, the actions of the problem-solver to understand the technology of tube metal thermocouples were consistent with the concept of knowing the technology before trying to solve problems.

This problem also illustrates the fact that historical data are of great value in solving problems. However, *conclusions* (the furnace tubes never foul) based on historical data or memories that are not well supported are counter-productive. In this case, all of the current data indicated that the furnace tubes were indeed fouled. Thus, it was apparent that the conclusion based on memories and/or historical data that indicated the tubes could not be fouled was in error. It is also likely that the tubes had been fouling in the past, but because the C_2 value was not being monitored, the fouling went undetected. The fouling was likely removed every 18 months when the tubes were water-washed.

If the problem-solver had not been aggressive to the point of creating tension, it is likely that a furnace tube failure with a subsequent furnace fire would have occurred.

9.10 Example Problem 9.3

A plant that produced and shipped a polyolefin polymer in railroad hopper cars was contacted by a TV station with a video of one of their hopper cars on fire. These hopper cars were loaded at the manufacturing plant by discharging pellet storage bins directly into the closed hopper cars. While the hopper cars were closed to protect the product from the elements (rain, sunshine, wind, debris ...), there were vents in the front and back of the hopper cars. Thus, the vapor space in the hopper car was essentially air at the ambient conditions plus any hydrocarbon that evolved from the polymer.

When the public relations contact appeared on the TV station to discuss the hopper car fire, he indicated that the particular polymer was not flammable and that could not have possibly caught on fire without some external source such as sabotage or excessive heat generated by a mechanical failure or malfunctioning of the hopper car equipment.

Unfortunately, shortly after this occasion, several other hopper cars arrived at different customers with blackened vapor vents and some charred polymer indicating that there had been a fire of a limited magnitude during transportation. The material that was being shipped in the hopper cars was in the form of small pellets that had been extruded after being produced in the polymerization section. The polymerization section stripped the volatile and unreacted monomer from the polymer at a temperature of 220 °F and a dryer residence time of 30 minutes. In addition, during the extrusion operation, the polymer was heated in the extruder to 550 °F before being pelletized and cooled in a water bath. It seemed very unlikely that there could be sufficient residual monomer or volatile that could create a fire. In addition, operations personnel knew that when the hopper cars were loaded, that they could see a great deal of static electricity being discharged inside the hopper car. They reasoned that if there was an explosive atmosphere inside the hopper car, that it would be ignited during the loading operation. They believed that the maximum concentration of hydrocarbons in the vapor space would occur during loading. They believed that during transit, the vents would create a sweep of air through the vapor space, reducing the concentration of hydrocarbons.

However, because of the multiple indications of fires in the hopper cars transporting the polymer, the operations personnel requested that a technical evaluation be made of what was causing the hopper car fires.

The problem-solver used the five-step procedure to methodically develop a problem solution as described in the following paragraphs.

Step 1: Verify that the Problem Actually Occurred

There was no doubt that something unusual had happened. The problem-solver decided that he needed to know what product had been loaded into these hopper cars that experienced the major fire and evidence of minor fires. When he investigated the loading records, he found that all of the unusual incidents had occurred when a specific product (experimental product X3) was being produced and loaded into the hopper cars. However, not all hopper cars loaded with this product showed evidence of flash fires. He recognized that this product was one that had to be produced at reduced rates to ensure that the volatiles (unreacted monomer and volatile) were adequately stripped. He knew that rates had recently been increased by modifications to the operating procedures and directives.

Step 2: *Write* Out an *Accurate* Statement of What Problem You Are Trying to *Solve*

The problem solver wrote out the following problem statement:

> "There are indications that flash fires are occurring in some of the railroad hopper cars loaded with experimental product X3. The indication of these flash fires was noticed after beginning to use the recently instituted operating procedures and directives. While actual damage was minimal, there was a great deal of customer dissatisfaction. Determine the cause of the flash fires. This analysis should include the observation that not all hopper cars loaded with this product showed signs of a fire. Once the cause has been determined, provide recommendations for eliminating the flash fires. In addition, any recommendations should provide for shipping a product that meets the Department of Transportation (DOT) regulations."

Table 9.6 Questions/comments for Problem 9.3.

Question	Comment
Are all operating directives and procedures correct and being followed?	All were being followed. However, there had been recent changes to the directives and procedures. Some of these appeared highly questionable
Are all instruments correct?	The instruments used to monitor the stripping operation were calibrated weekly
Are laboratory results correct?	The volatiles' results were not routinely measured. It was believed that maintaining adequate temperature was all that was required to maintain volatiles control
Were there any errors made in the original design?	Not applicable
Were there changes in operating conditions?	Yes. See above
Is fluid leakage occurring?	Not applicable
Has there been mechanical wear that would explain the problem?	Not applicable
Is the reaction rate as anticipated?	Not applicable
Are there adverse reactions occurring?	Unknown changes in the catalyst might create a polymer with a less porous structure that would make volatile removal more difficult. This could be detected by bulk density measurements. In addition, the polymer might be decomposing in the extruder, forming volatile materials
Were there errors made in the construction?	Not applicable

Step 3: *Develop* a Theoretically Sound Working Hypothesis that Explains as Many Specifications of the Problem as Possible

In addition to the observations described above, the questions from Chapter 6 were used to help develop a working hypothesis. These questions are shown in Table 9.6.

After reviewing the data and the questions, three possible hypotheses were developed as follows:

1) The new operating directives and procedures were causing excess quantities of the unreacted monomer or volatile to be left in the polymer, which then accumulated in the hopper car and formed an explosive mixture. This mixture could then be ignited by an undefined ignition source.

2) There were changes in the catalyst, which now resulted in a less porous polymer particle. Thus, the temperature and residence time that were previously used successfully in a plant test to demonstrate the new procedures for X3 were no longer adequate. Again, if this happened, it was theorized that the excessive quantities of volatile and unreacted monomer accumulated in the hopper car and formed an explosive mixture.
3) There was a change in the adequacy of stabilization so that the polymer decomposed in the extruder. This decomposition resulted in the formation of volatile materials, which then did not vent out of the extruder because of the short residence time. They were trapped in the pellet and were carried into the hopper car where they evolved from the pellet and accumulated and formed an explosive mixture.

It was clear after considering these three hypotheses that all of them had a common thread. The common thread was twofold. In the first place, the hydrocarbons (volatile and unreacted monomer) had to build up to the point of forming an explosive mixture. The second aspect of this hypothesis was that the explosive mixture was ignited by an ignition source. This ignition source could have been something dramatic like a lightning strike or something as commonplace as static electricity. The movement of pellets in the hopper car while it was in transit could well generate sufficient static electricity to create an ignition source. This would be particularly true if the relative humidity of the air in the vapor space was very low. The problem-solver used the traditional process engineering safety assumption—that is, when dealing with an explosive mixture, an ignition source will always be found if the mixture is in the explosive range for a sufficient amount of time. Thus, he only considered which hypothesis should be investigated for how an explosive mixture formed.

All of the three hypotheses required that the gases accumulate in the vapor space to form an explosive mixture. The first question that the problem-solver had to answer was "How much hydrocarbon had to be left in the polymer for an explosive mixture to form in the vapor space of the hopper car?"

It was necessary to develop some basic data before the problem-solver could begin to assess this question. The data that he developed to help assess this problem are shown in Table 9.7.

In order to test the hydrocarbon accumulation hypothesis, the following calculations were done to determine how much hydrocarbon would be left in the polymer entering the hopper car for an explosive mixture to accumulate in the hopper car vapor space. These calculations assumed that the polymer had been in the hopper car for such an extended period of time that equilibrium between the vapor space and the polymer was reached. The calculations also ignore the

Table 9.7 Data for evaluation.

Item	Value
Hopper car capacity, lbs. of polymer	180,000
Hopper car fill volume, percent	70
Bulk density of the polymer, lbs./ft^3	25
Skeletal[a)] density of the polymer, lbs./ft^3	56
Vapor pressure of the monomer at 100 °F, psia	220
Molecular weight of the monomer	42
Explosive range of the monomer[b)], vol.%	
Lower	2
Upper	11

a) Skeletal density is the density of the polymer if the particle had no voids.
b) Since essentially the entire hydrocarbon was the unreacted monomer, no consideration was given to other volatiles. In addition, it should be noted that for gases the volume percent and mol percent are identical.

purging of the vapor space that might occur as the hopper car traveled to the customer. As such, they represent the worst-case scenario. However, the case might occur if the hopper car sat stationary on the railroad track for an extended period of time with a minimal amount of wind. It was assumed that the equilibrium relationship given in equation (9.6) shown below was applicable for this polymer.

$$X_E = 295,000^*(P/\mathrm{VP})^{1.5} \tag{9.6}$$

$$F = V + S \tag{9.16}$$

where

F = total monomer in the hopper car, lb.
V = monomer in vapor space, lb.
S = monomer remaining in the polymer at 100 °F, lb.

$$\begin{aligned} T &= \mathrm{Ca}^*100/(\mathrm{BD}^*\mathrm{FV}) \\ &= 180,000^*100/(25^*70) = 10,285\ \mathrm{ft}^3 \end{aligned} \tag{9.17}$$

$$\begin{aligned} \mathrm{VT} &= T - \mathrm{Ca}/\mathrm{SD} \\ &= 10,285 - 180,000/56 = 7070\ \mathrm{ft}^3 \end{aligned} \tag{9.18}$$

where

T = total volume of the hopper car, ft^3
Ca = polymer capacity of the hopper car, lb.
BD = bulk density of the polymer, lb/ft^3
FV = volume of the hopper car filled with the polymer, percent
VT = vapor volume of the hopper car, ft^3
SD = skeletal density of the polymer, lb/ft^3

Note that VT includes the volume associated with the pores in the polymer.

$$P = \text{LEL}^* \Pi/100$$
$$= 2^* 14.7/100 = 0.294 \text{ psia} \tag{9.19}$$

where

LEL = the lower explosive limit, volume or mol percent
Π = the total pressure, psia

$$D = \text{MWM}^* 520/(379^*(460+100))$$
$$42^* 520/(379^* 560) = 0.1029 \ \text{lb/ft}^3 \tag{9.20}$$

where

D = density of monomer gas, lb/ft^3
MWM = molecular weight of the monomer

$$V = \text{LEL}^* D^* \text{VT}/100$$
$$= 2^* 0.1029^* 7070/100 = 14.55 \text{ lb} \tag{9.21}$$

$$X_E = 295{,}000^*(P/\text{VP})^{1.5}$$
$$= 295{,}000^*(0.294/220)^{1.5} = 14.4 \text{ ppm} \tag{9.6}$$

$$F = V + S$$
$$= 14.55 + 14.4^* 180{,}000/1{,}000{,}000 = 17.14 \ \text{lb}$$
$$= 95 \ \text{ppm of monomer in the incoming polymer} \tag{9.22}$$

Thus, if the polymer going to the hopper contained more than 95 ppm of the monomer, the vapor space could well be at the LEL.

The problem-solver investigated the current operating directives and procedures and their bases. When he did this investigation, he found that the approach to setting the new directives and procedures had been strictly empirical with no consideration to the theory and time elements of diffusion. A test run had been conducted by establishing the new conditions and rates in the polymerization and extrusion sections. When the conditions were well established, samples downstream of the extrusion and pelletizing operations were collected in an open-mouth container. A conventional gas explosivity analyzer was inserted into the container immediately after the sample was collected, and the percent LEL was measured. The explosivity analyzer gives a reading of the percent of the LEL (lower explosive limit). For example, if the vapor space was at the LEL, the analyzer would read 100%. Thus, percent LEL is a measure of the approach to the lower explosion limit.

In this test of X3, it was believed that measuring the vapor space immediately would give the highest value since the monomer in the vapor phase of the open container would not have time to diffuse into the surrounding air. A single measurement was made, and since the percent LEL was less than 100%, it was concluded that the new conditions and rates would produce a product that would satisfy the DOT regulations. These regulations stated that the vapor space of products shipped in hopper cars should be below the LEL. Obviously, if the vapor space in the hopper car was below the LEL, it would not ignite.

In addition, the problem-solver reviewed the bulk density measurements. The experimental polymer (X3) actually had a slightly lower bulk density than the normal polyolefin products. Thus, it appeared that the product would be more porous than the conventional product. He also reviewed the current stabilizer and found that it was the same that had been in use for several years. This cursory review did not mean that he had ruled out hypotheses 2 and 3 conclusively. But it did indicate that these were unlikely to be the simplest root cause. Following the concept of taking the simplest route whether it is the root cause or a calculation technique, the problem-solver developed the following theoretically correct working hypothesis.

> "It is believed that the hopper car fires and product charring that occurred when shipping X3 are associated with the residual monomer that remained with the polymer when the new rates and operating conditions were used. While the vapor space above the pellets was not in the explosive range immediately after they were loaded into the hopper car, the evolution of vapors from the polymer over time allowed an explosive concentration to build up in the hopper car vapor space. While it is uncertain what the source of ignition was, it is known that static electricity is almost always present due to the movement of the polymer particles. Whether this explosive mixture ignited and to what degree a fire occurred depended on three factors:

1) The movement of the hopper car. If the hopper car was continuously in motion, there was a high probability that some of the vapors would be replaced by air and the vapor phase concentration might be reduced to a level below the lower explosive limit.

2) The humidity of the ambient air. If the humidity was high in the hopper car, it is unlikely that a static discharge would occur.
3) The temperature. If the temperature of the polymer in the hopper car was elevated (due to the ambient temperature), then there is a higher probability of a major fire occurring rather than a flash fire that chars some of the polymer on the top of the hopper car.

   In order to eliminate this problem, the production rates of X3 should be reduced to the previous levels. In addition, the new operating directives and procedures should be abandoned."

Step 4: Provide a Mechanism to Test the Hypothesis

Two mechanisms were provided to test the hypothesis. On the next X3 run, the rates were reduced to the previous levels. The percent LEL was measured in the vapor space of each hopper car after 24 hours and prior to shipping. In each case, the percent LEL was well below 100%. There were no reports of charred polymer or fires, even though the customers had been alerted to watch for such events. In addition, during the run, samples were taken of the product leaving the pelletizing section in the same fashion as discussed earlier. The percent LEL was determined as a function of time, and Figure 9.5 was used to illustrate the buildup of the percent LEL as a function of time. Thus, it was concluded that reducing the production rates was a successful test of the hypothesis.

Step 5: *Recommend* Remedial Action to Eliminate the Problem without Creating Another Problem

The recommendation to operate at reduced rates could only be considered an interim recommendation. The problem-solver used some of the techniques

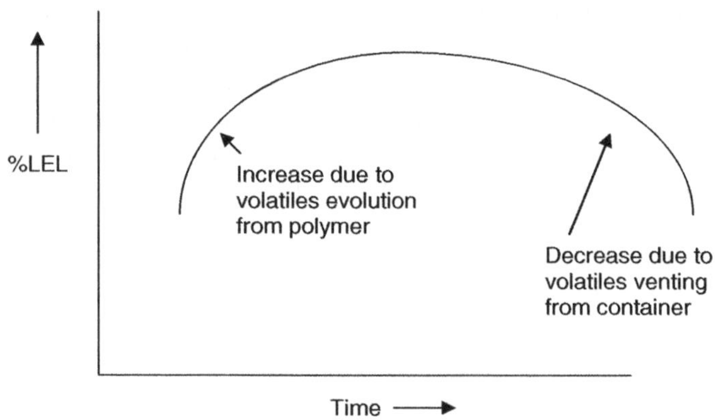

Figure 9.5 Percent LEL versus time.

described in Example Problem 9.1 and concluded that the only way to operate at full rates while producing this polymer was to add additional dryer residence time. He recommended that a study be initiated to determine whether the economics of producing X3 could justify either operating at reduced rates or adding a larger dryer.

9.11 Lessons Learned

There were pressures to increase the production rates of X3. It was obvious in hindsight that these pressures caused a blatantly flawed empirical test to be developed. The conclusion that rates could be increased was based completely on this flawed test with no theoretical calculations. This empirical test overlooked the obvious—that vapor evolution from the polymer that occurred over time might cause a maximum in the relationship between percent LEL and time. This maximum could well be above the LEL. The risk of this approach could have been discovered if theoretical calculations had been done to determine what the maximum equilibrium concentration of the monomer in the vapor phase of the sample container would be. This calculation would require that the concentration of the monomer in the polymer leaving the pelletizing section be known. The concentration of the monomer in the vapor phase could then be calculated using the techniques described earlier and with the assumption that none of the monomer dispersed into the atmosphere from the container.

There are times when theoretical calculations are more accurate than obtaining laboratory results. In this case, if the concentration of the monomer in the polymer could be analyzed and the equilibrium relationship was available (equation (9.6), for example), then the *maximum* equilibrium vapor phase concentration and hence the percent LEL could be determined easily. This would be more accurate than the maximum percent LEL determined by the explosivity meter. This is because the explosivity meter only measures the percent LEL at a point in time, which is likely not to be the maximum concentration.

The failure to do a detailed theoretical analysis of the proposal to increase rates by modified operating directives and procedures was costly. It gave the company adverse publicity and adverse relationships with their customers. Their customers had never seen charred polyolefin products from the company or any other company that they purchased the material from. The approach also cost problem-solving time. Instead of starting to understand the monomer removal limitation when X3 was first discovered to have a limitation, the empirical approach caused the start of problem-solving activities to be delayed.

In addition, problem-solving activities were diverted to understanding the hopper car fires, as opposed to eliminating the drying limitation.

The approach of asking customers to be on the watch for the charred polymer might seem to be "just asking for trouble." However, it was mandatory to confirm that the plant test of operating at reduced rates was truly successful. The failure to get this information would have likely resulted in a failed test since enough data would not be present to prove or disprove the hypothesis.

Nomenclature

BD	the bulk density of polymer, lb/ft^3
C	a constant referred to as the "lumped parameter constant." If it is related to heat transfer, the constant is generally taken as $U*A$. If the reaction is involved, it is simply set equal to "C_R" in this chapter. If drying or stripping is involved, the value "K" is used in this chapter
C_R	the lumped parameter constant for reaction (reaction rate constant)
C_E	a constant that relates equilibrium of the volatile in the polymer to the vapor phase composition. It can be determined from experimental data or approximated by application of theoretical relationships
C_1	a constant that is related to the type of furnace flame. It should not change unless the type of fuel changes
C_2	the product of the heat transfer coefficient and the area of the tubes. It will only change if the heat transfer coefficient changes
Ca	the polymer capacity of the hopper car, lb
D	density of monomer gas
$D1$	the density of the volatile
$D2$	the density of the polymer
DF	the driving force or incentive for mass or heat transfer to occur
DF_R	the driving force for radiant heat transfer
dX/dt	the rate of disappearance of component X by reaction or stripping
F	the total monomer in the hopper car, lb
FV	the volume of the hopper car filled with the polymer, percent
$J1$	the concentration of component 1
$J2$	the concentration of component 2
K	the lumped parameter constant for drying or stripping. It is somewhat related to diffusion. Since it deals with mass transfer, it is also referenced in this chapter as a mass transfer coefficient
LEL	the lower explosive limit, volume or mol percent
$\ln\Delta T$	the log mean temperature difference or the driving force for heat transfer
MWM	the molecular weight of monomer
$M1$	the molar volume of volatile cc/g-mol

$M2$	the molar volume of polymer, cc/g-mol
P	the partial pressure of the volatile in the vapor phase
Q	the rate of heat transfer
R	the rate of change with time of the variable under study. It could be heat transfer, reaction rate, or volatile stripping
S	the monomer remaining in the polymer at 100 °F, lb
$S1$	the volatile solubility parameter $(cal/cm^3)^{0.5}$
$S2$	the polymer solubility parameter $(cal/cm^3)^{0.5}$
SD	the skeletal density of the polymer, lb/ft^3
T	the total volume of the hopper car, ft^3
t	the amount of time in the dryer, minutes
T_P	the absolute temperature of the process fluid in the tubes
T_g	the absolute temperature of the gas
T_m	the absolute temperature of the tube metal
U	the interaction parameter between the volatile and the polymer
$U*A$	the lumped parameter (C) or for heat transfer the heat transfer coefficient multiplied by the area
V	the monomer in vapor space of the hopper car, lb
$V1$	the volume fraction of the volatile in the polymer
$V2$	the volume fraction of the polymer
VP	the vapor pressure of the volatile at the dryer temperature
VT	the vapor volume of the hopper car, ft^3
$X1$	the concentration of the volatile in the polymer, ppm
X	the actual concentration of the volatile in a polymer
X_e	the equilibrium concentration of the volatile in the polymer
X	the concentration of component X
Y	the concentration of component Y
X_f	thc conccntration of thc volatile in the outlet polymer
X_o	the concentration of the volatile in the inlet polymer
Z	the lattice constant. In theory = 0.35 ± 0.1
Π	the total absolute pressure, psia

10

Application to Unsteady State

10.1 Introduction

While most process engineering courses deal with design and operations at steady-state conditions, industrial processes operate at constantly changing conditions. Frequently, the differences between steady and dynamic conditions can be ignored. However, an approach to considering unsteady operations will be of value to every process engineer. There are multiple instances of unsteady-state operations that a process engineer must often consider. Some of them are as follows:

- Start-ups are never steady state, and this is probably the most challenging phase of problem solving. This is often an area where the intuition of an experienced operator or engineer is more valuable than a detailed problem-solving activity. However, there are examples where a frequently encountered start-up activity that involves a problem (reactor start-up, for example) can be approached using some of the principles outlined in this book.
- Batch operations are always dynamic. If problems are uncovered during a batch operation, typical process engineering calculations based on steady-state operations will rarely be of any value. Thus, the industrial problem solver needs to have tools to allow him to approach these dynamic operations. Example Problem 10.1 is a typical batch operation real-life example.
- Unsteady-state upsets can often provide valuable information. For example, short-term upsets caused by feed impurities can be extrapolated to steady-state values using techniques that relate the time of the upset to the actual reactor residence time. This will allow the problem solver to determine the steady-state impact of these impurities, that is, the impact on the process that would occur if the short-term impurity concentration in the feed was at the upset level consistently. This information can be used to determine the specification of the impurity in the feed.

Problem Solving Approaches for Maintaining Operational Excellence in Process Plants, Second Edition.
Joseph M. Bonem, Nattapong Pongboot, and Wiroon Tanthapanichakoon.
© 2025 the American Institute of Chemical Engineers, Inc. Published 2025 by John Wiley & Sons, Inc.

10.2 Approach to Unsteady State Problem Solving

The most difficult aspect of solving unsteady-state problems in an industrial environment is the balance between formulating technically correct hypotheses and the need for expediency. This was referred to in Chapter 4 as *optimum technical depth*. Because of the difficulty of determining this balance, one of two extremes is often present. In one of these extremes, the problem solver begins developing a highly sophisticated dynamic model that requires a powerful computer to solve the differential equations. As the work slowly progresses, the management becomes impatient and cancels the project. At the other extreme, the problem solver often "gives up" and uses intuitive "gut feel" rather than developing a simplified dynamic model. *Optimum technical depth* is the approach used to achieve a balance between these two extremes. This balance will be a strong function of the environment surrounding the problem. For example, during a start-up, a balance slanted toward intuition is likely the best approach. On the other hand, a chronic problem which does not have any apparent answer will require a more sophisticated approach. The following paragraphs present guidelines for approaching unsteady-state problems.

If the problem requires an *immediate* recommendation such as during a start-up, the best approach will be to rely on the instinct of an operator or experienced start-up engineer. Medical research has shown that our minds work by both instinct and logical reasoning. Obviously, instinct is faster than development of a detailed logical path. Minds conditioned by years of experience with unsteady-state operations such as a start-up will react instinctively to provide an immediate response, as opposed to having to think about the situation.

The basic concept for considering unsteady-state operations is equation (10.1) shown below:

$$\mathrm{AD} = I - O - \mathrm{RD} \tag{10.1}$$

where

AD = rate of accumulation. This could be accumulation of anything – level, heat, or reactant

I = inflow of material or heat

O = outflow of material or heat

RD = removal/addition of heat or formation/destruction of the material by reaction

This relationship can be used as a building block for all unsteady-state considerations from something as simple as the change of level in an accumulator to the development of complicated dynamic models. The thrust of this chapter is the development of simple, but accurate dynamic models.

If it is considered desirable to build a dynamic model of the unsteady-state process, all physical components of the system must be considered. For example, with steady state, the heat capacity of the process vessel is correctly never considered. However, for unsteady-state heat balances, the heat capacity of the vessel (walls and contents) must be considered and will often be a moderating influence. This moderating influence will cause the rate of temperature change to be less than calculated if the heat capacity of the vessel is ignored. The dynamic model should be kept simple by use of the perfectly mixed vessel or plug flow assumptions where appropriate, use of lumped parameter constants (such as overall heat transfer coefficients), and an assumption of uniform metal temperature.

If the reaction is involved, it should be recognized that there are two models that can be used to simulate an unsteady-state reaction. Traditional process engineering indicates that the reaction rate is proportional to the concentration of the reactant. That is, the rate of reaction will be increasing as the concentration of the reactant in the vessel increases. This has been referred to as the "slow response model." For some reactants such as impurities in a polymerization reactor, the impurity reacts as soon as it enters the reactor. This "fast response model" is easier to develop because it allows assuming that the concentration of the impurity in the reactor approximates 0. In order to determine whether the traditional model or the "fast response model" should be used requires that the problem solver understand the technology. As a general rule, the "fast response model" will apply only to reaction impurities in a polymerization reaction. Most other reaction modeling will be more accurate if the traditional process engineering approach is utilized. Regardless of which model is used, short-term upsets and fundamental models of stirred tank or plug flow reactors can be used to estimate the full impact of an impurity.

10.3 Example Problems

The problem-solving techniques associated with unsteady-state operations are illustrated by the actual case histories described in the following paragraphs. Problem 10.1 illustrates the use of these techniques for solving a problem associated with a batch reactor that experienced a temperature "runaway" (such a rapid temperature increase that the emergency devices were activated). Problem 10.2 illustrates the use of these techniques to extrapolate short-term reaction upsets caused by an impurity to steady-state conditions.

10.4 Problem 10.1

A batch reaction was carried out in a reaction vessel with a cooling water jacket. The reaction was initiated when the primary reactant was added to the reaction vessel. The primary reactant was dissolved in hot hexane in a small vessel and then pressured into the reaction vessel with nitrogen as fast as possible. While the technology had been developed in the research facility of a major petrochemical company, the actual manufacturing of the material was being done by a contract manufacturer. His operation was designed and operated on an exceptionally low budget. There were no flow instruments to measure and control the rate of addition of the reactant. The average rate of reactant addition was obtained knowing the amount of material in the small vessel and the time that it took to add it to the reaction vessel.

The exothermic reaction had been conducted successfully many times. However, during a recent batch, a rapid increase in temperature was experienced as the primary reactant was added to the reactor. The rapid temperature increase was so fast and of such a magnitude that the safety release system on the reactor was activated. The actual manufacturing location was located several hundred miles away from the technology center. Operations personnel decided to solve the temperature runaway problem without asking for help from the technology center. In order to eliminate the rapid increase in temperature, the operations personnel decided to add the reactant at a very slow rate. The next batch was produced without a temperature runaway. However, the product produced in this fashion did not meet the product morphology specifications. After this batch failed to meet the morphology specifications, the operations personnel requested help from the technology center. The problem solver was charged with the responsibility of determining how to conduct this reaction with an 80–90% probability that a second temperature runaway would not occur and that the product morphology would be satisfactory.

The problem solver began accumulating technology information. It was known that the primary reactant reacted instantly when it entered the reactor (fast response model). It was desirable to add the reactant as fast as possible for two reasons. As indicated earlier, the morphology of the product was adversely impacted by a slow addition rate of the reactant. In addition, the heat of reaction provided heat input to increase the reactor temperature rapidly from 149 to 185 °F.

The heat of reaction was unknown. It was not determined in the laboratory during the development part of the project. Actually, the laboratory chemists believed that there was no heat of reaction. The small-scale reactor had an inherently high area-to-volume ratio. In addition, this laboratory-scale reactor was surrounded by a constant temperature sand bath. This combination essentially eliminated the chance to observe any heat of reaction in the laboratory. The nature of the reaction was such that the heat of reaction could not be determined by classical literature approaches. In addition, time and financial constraints did not allow determining this variable from laboratory data.

The problem solver began using the five-step problem-solving approach described earlier, as is shown below.

Step 1: Verify that the Problem Actually Occurred

Verification of this problem was easy since the hexane vapor from the safety release system on the reactor (a rupture disk) condensed in the air and "rained" on a local veteran's day parade. There were no injuries, but several uniforms and band instruments required cleaning and reconditioning. In addition, a review of the reactor temperature and pressure data indicated that the rupture disk did not release prematurely. That is, the pressure and temperature had really gotten out of control.

Step 2: *Write* Out an *Accurate* Statement of What Problem You Are Trying to *Solve*

The problem statement developed by the problem solver was as follows:

> "An exothermic reaction conducted in a batch reactor had a temperature runaway even though the same reaction had been conducted many times previously with no temperature runaway. In a subsequent batch it was demonstrated that the temperature could be controlled by using a slow rate of addition of the primary reactant. However, it is necessary to add the primary reactant at a rapid rate in order to obtain the desired product morphology. Determine what caused the batch reactor to have a temperature runaway In addition, develop procedures that insure that there is at least an 80 to 90 percent confidence level that this temperature runaway will not reoccur when producing a product that meets the morphology specifications."

Step 3: *Develop* a Theoretically Sound Working Hypothesis that Explains as Many Specifications of the Problem as Possible

Table 10.1 shows the questions from Chapter 6 along with appropriate comments.

A review of the comments associated with these questions and a review of all previous runs indicated several points of interest that would require additional considerations. These were as follows:

- While the batch size was the same, the rate of reactant addition varied greatly from run to run. As indicated earlier, this rate of addition was only determined after completion of the addition since there were no flowmeters on the transfer line. There were no procedures given for the rate of reactant addition. Some operations personnel pressured the drum used to mix the reactant and hexane with nitrogen prior to opening the transfer valve. Others simply started the nitrogen flow to pressurize the drum and then immediately opened the transfer valve. There were times when the mixing drum was hotter than in other times, which impacted the solution viscosity, which would affect the rate of reactant transfer.

Table 10.1 Questions/comments for Problem 10.1.

Question	Comment
Are all operating directives and procedures being followed?	For the batch that had a temperature runaway, all appeared to have been followed. However, there were no guidelines given for how fast to add the primary reactant
Are all instruments correct?	The instruments had allegedly been calibrated
Are laboratory results correct?	The product morphology on the second batch was confirmed by two independent techniques
Were there any errors made in the original design?	Yes. The laboratory chemists indicated that there was essentially no heat of reaction based on their studies. In hindsight, this was obviously in error
Were there changes in operating conditions?	Yes. Rate of addition of the reactant was highly variable even though batch sizes were the same
Is fluid leakage occurring?	Not applicable
Has there been mechanical wear that would explain the problem?	This is possible, but it would cause the heat removal capability to decrease. This was not occurring based on measured coolant flows and temperatures
Is the reaction rate as anticipated?	The reaction rate was highly dependent on the reactant rate of addition
Are there adverse reactions occurring?	Adverse reactions were likely the cause of poor morphology. However, this is exceptionally complicated chemistry. Thus, solving the morphology problem while maintaining slow addition of reagents is unlikely
Were there errors made in the construction?	Not applicable. Some batches worked fine

- While the laboratory chemists reported that the heat of reaction was insignificant, there had to be a significant heat of reaction to cause the vessel contents to increase from 149 to 185 °F in a short period of time. If the heat of reaction was known or could be determined, it would be possible to estimate the rate of temperature increase for any batch size and/or rate of reactant flow.
- In addition to the problem batch, there were several other runs where the temperature increased *almost* as rapidly; however, the reactor temperature did not become unstable. The only difference in these runs was the rate of reactant addition.

While it could be theorized at this point that the temperature runaway was due to the rate of reactant addition, this theory would only allow one to specify that an addition rate equivalent to that of the first failed batch would cause a temperature runaway. It would not allow determination of the maximum rate of reactant addition. A more fundamental approach was required to allow specifying the target reactant addition rate.

The problem solver decided to approach the problem by using a fundamental approach. He felt that if he could express the factors involved in the temperature runaway mathematically, he could understand better how to prevent it. A temperature runaway occurs when the rate of heat generated is greater than the maximum rate that heat can be removed. This can be expressed mathematically as follows in equations (10.2) through (10.4):

$$Q_g > (Q_r)_{max} \tag{10.2}$$

where

Q_g = the rate of heat generation, BTU/hr
$(Q_r)_{max}$ = the maximum rate of heat removal, BTU/hr

$$Q_g = \Delta H_R{}^{*}R \tag{10.3}$$

$$(Q_r)_{max} = U^{*}A^{*}(T_R - T_C) \tag{10.4}$$

where

ΔH_R = the heat of reaction, BTU/lb
R = rate of reaction. For the case of an instantaneous reaction, it is the rate of reactant addition, lb/hr
U = the heat transfer coefficient, BTU/hr-F-ft^2
A = the heat transfer area, ft^2
T_R = the temperature of the reactor, °F
T_C = the *minimum* temperature of the coolant, °F

Note that equation (10.4) has been simplified from the traditional heat transfer equation that involves use of a logarithmic relationship. This is valid for this particular case since the coolant is flowing through the reactor with a minimal increase in temperature. The temperature to be used is the minimum coolant temperature that can be obtained. It is assumed that the control system will react to provide the minimum coolant temperature possible. In addition, because the reactor vessel is well mixed, the reactor temperature throughout the vessel is constant at any point in time.

For a batch reactor, equations (10.3) and (10.4) can be combined, and a heat accumulation term can be added to take into account the unsteady-state nature. In addition, any heat accumulation will cause an increase in the temperature of the reactor. The equations below represent this situation.

$$\text{AC} = \Delta H_R{}^*R - U^*A^*(T_R - T_C) \tag{10.5}$$

$$\text{AC} = W^*C_P{}^*dT/dt \tag{10.6}$$

where

AC = the heat accumulation, BTU/hr
W = the weight of material (metal and reactants), lb
C_P = the average specific heat of material, BTU/lb-°F
dT/dt = the rate of temperature rise, °F/hr.

Since the reaction of the reactant being pressured into the reactor is instantaneous, the heat generated (ΔH_R*R) depends only on the rate of reactant addition. Thus, equations (10.5) and (10.6) could be combined and modified to allow solving for the heat of reaction.

$$\Delta H_R = \left(W^*C_P{}^*dT/dt + U^*A^*(T_R - T_C)\right)/R \tag{10.7}$$

In the above equation, "U" and "A" can be estimated from physical dimensions and typical vessel heat transfer coefficients. Since all previous batches and future plans called for using the same vessel, obtaining exact values of "U" and "A" was not important. The product of U*A was considered a "lumped parameter constant."

Equation (10.7) was used to determine the heat of reaction for all previous runs. Since a reasonably constant heat of reaction was calculated over a wide range of reactant addition rates (R), this approach seemed valid. After determining the heat of reaction using data from the batch runs, calculations were done to determine the maximum rate of reactant addition. Referring back to equation (10.2), a typical approach is to provide a 10% safety factor so that the maximum rate of heat removal is 10% greater than the maximum rate of heat generated. Thus, equation (10.2) can be modified as follows:

$$1.1^*Q_g < Q_{r\ max} \tag{10.8}$$

In order to estimate the maximum rate of reactant addition, equation (10.8) along with equation (10.9) shown below, which is a rearrangement of equation (10.7), was used.

$$dT/dt = (R^*\Delta H_R - U^*A^*(T_R - T_C))/(W^*C_P) \tag{10.9}$$

Equation (10.9) was used to estimate the maximum rate of reactant addition (R) that would allow the temperature to rise from 149 to 185 °F without causing a temperature runaway. This maximum rate of addition was determined using an iterative procedure along with a spreadsheet. This iterative procedure consisted of running multiple cases at different addition rates. Each case generated a curve similar to Figure 10.1. From these cases, the maximum addition rate that would increase the temperature to 185 °F was determined. After the maximum rate was determined, it was reduced by 10% to be consistent with the requirements shown in equation (10.8).

Using this approach, it was estimated that the maximum rate of reactant addition was 25% less than utilized in the batch on which a temperature runaway occurred. The problem solver developed the following working hypothesis:

> "It is theorized that the temperature runaway was caused by rapid addition of the primary reactant. This addition rate was so high that the temperature could not be controlled even when the coolant was at the maximum rate. Calculations indicate that if the primary reactant rate addition was reduced by 25 percent the reaction could be controlled."

Step 4: Provide a Mechanism to Test the Hypothesis

Developing a mechanism to test the hypothesis was easy from an engineering standpoint. As described above, the maximum rate of reactant addition to achieve a peak temperature of 185 °F was easily estimated. However, convincing the operations personnel to add the reactant at this rate was difficult since they recently had good success controlling the temperature while adding the reactant at a much lower rate and allowing the reactor to slowly heat up. As indicated earlier, this resulted in a product with poor morphology. In order to convince them to test the hypothesis, it was necessary to give them a complete explanation for the importance of the morphology of the product and to stand by the reactor when the reactant was added at a rapid rate. The problem solver had estimated the nitrogen pressure that would be required to transfer the reactant to the drum at the desired rate. He also had calculated the rate of level change in the reactant storage drum that would result if the reactant was added at the correct rate. The test was successful in producing a product with good morphology, and the temperature increased as predicted without a temperature runaway.

Step 5: *Recommend* Remedial Action to Eliminate the Problem without Creating Another Problem

The hypothesis test (Step 4) proved that the reactant could be added at a high rate. In order to prevent another temperature runaway, additional controls were considered to assure that the target reactant addition was maintained. Since as indicated earlier, the only means used by the operations personnel for knowing the addition rate was the amount of time that it took to add the reactant, a more positive technique was mandatory. The project did not have funds for the installation of a flowmeter nor was the required shutdown time immediately

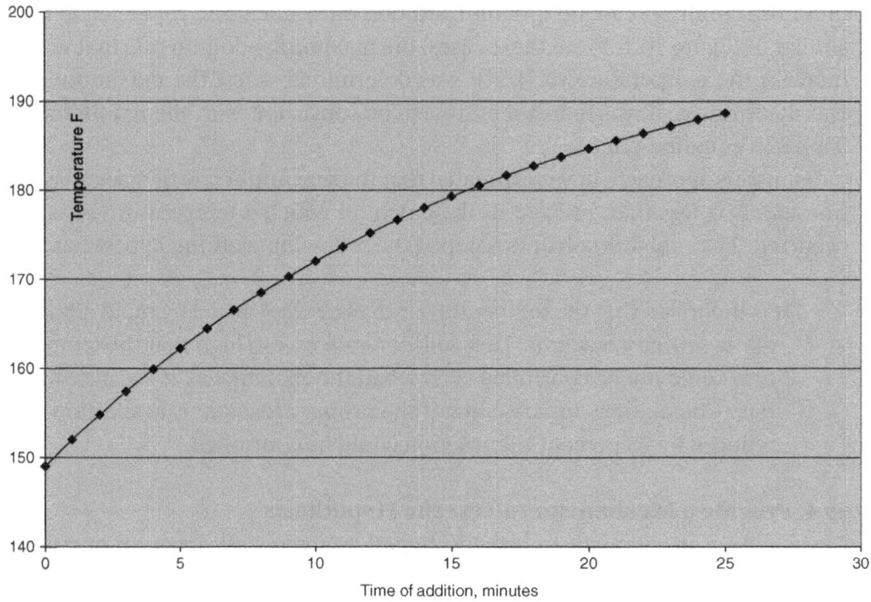

Figure 10.1 Reactor temperature versus time.

available. Three approaches were used to ensure that the reactant was being added at the correct rate. The pressure to be reached and maintained on the reactant storage drum was specified in the operating procedures. In addition, the rate of level change that would be anticipated was specified. As a final check, the curve shown in Figure 10.1 was provided and was to be used to monitor the rate of temperature increase in the first few minutes. For example, if the initial temperature increase was too fast, it would indicate that in spite of the other techniques, the rate of addition was too fast. The rate of addition would then be reduced.

10.5 Lessons Learned

Although it has nothing to do with direct problem solving, one of the lessons learned was that it is very difficult to determine the heat of reaction from laboratory experiments that are not specifically designed to obtain this variable. A typical laboratory experiment will be designed to obtain reaction rate variables or product quality attributes. The best way to do this will be at a constant temperature. This is usually done by utilizing small equipment that has a high area-to-volume ratio and is surrounded by a constant temperature bath. With this facility, it will be impossible to determine the heat of reaction or even to notice whether there is a heat of reaction or not.

The utilization of the unsteady-state relationship, equation (10.1), for heat accumulation proved to be a reliable tool for determining the heat of reaction. As indicated, the key to being able to successfully use this equation is the full inclusion of all of the heat content (vessel and reactants) of the process under consideration. The qualification that time and money were not available to determine the heat of reaction in the laboratory is often the case in industrial problem solving. In order to obtain the heat of reaction in the laboratory, it would have taken time and funds to build a laboratory reactor specifically designed to obtain heat of reaction data.

In problem solving, the engineering calculations are often the easiest part of the job. Convincing either the operations management or the hourly workforce of operators and mechanics that the recommend solution is correct is often more difficult. The process of convincing the operations organization to adopt the answer will be easier if they are convinced that the problem solver is there with them to assist them if something was wrong with the calculations. Thus, in this example problem, the problem solver was present when the operator added the reactant at the calculated rate. While things went as planned, the operator felt much more comfortable in setting the pressure to give the high rate of flow since the problem solver was present with him.

10.6 Example Problem 10.2

A CSTR (continuous stirred tank reactor) in a polymerization plant was plagued by short-term "loss of reaction" events. While these events were small in magnitude, they created product quality upsets, as well as a loss of catalyst efficiency. These upsets appeared to be associated with spikes of carbon dioxide (CO_2) in the monomer feed. The reactor residence time was approximately 3 hours. The short-term spikes of CO_2 lasted for only 30–60 minutes. Some batch data were available from the laboratory indicating that the CO_2 specification of the monomer feed currently at 10 ppm needed to be reduced. However, the management would like to be convinced that the laboratory data were consistent with the plant experience. The management also believed that there was probably another impurity that was not being analyzed that was causing the problem. This belief was based on previous experience in another plant that utilized a similar, but not identical catalyst system. The management was also concerned that a project to reduce the CO_2 specification and to eliminate spikes would take several months to implement, as well as requiring significant investment. Thus, they did not want to proceed with the project unless there was a high probability of success.

The problem solver was asked to determine if these upsets were really caused by feed impurities, what the likely feed impurity was, and what the impurity's real impact on the process was. It was believed that this approach would lead to determining and setting realistic impurity specifications on the feed monomer.

The problem solver approached the problem using the five-step problem-solving technique as shown in the following paragraphs.

Step 1: Verify that the Problem Actually Occurred

Verification of this problem was done by two different techniques. The polymer was analyzed to determine the concentration of the catalyst in the polymer leaving the reactor. From this result, the catalyst efficiency was calculated, and it was clear that during CO_2 spikes, the catalyst efficiency did decrease. However, this approach only gave an average catalyst efficiency at any point in time. In addition to this analytical approach, the calculated production rate (as determined by the process control computer using a heat balance) was used to determine the instantaneous production rate.

Step 2: *Write* Out an *Accurate* Statement of What Problem You Are Trying to *Solve*

The statement written by the problem solver was as follows:

> "Frequent losses of reaction episodes are occurring in a polymer plant. These episodes cause a decrease in average catalyst efficiency and inconsistent product quality. These episodes appear to occur at the same time as short-term spikes of CO_2 concentration in the monomer feed. Determine the cause of the frequent 'loss of reaction' episodes. Any conclusions based on plant data must be consistent with available laboratory data on the impact of CO_2 on the polymerization reaction. This evaluation should also indicate whether or not there is any indication that another impurity besides CO_2 is present. Based on these data, the monomer feed specification should be reviewed."

Step 3: *Develop* a Theoretically Sound Working Hypothesis that Explains as Many Specifications of the Problem as Possible

In order to fully assess the possible causes of the upsets, the questions shown in Chapter 6 were utilized and Table 10.2 was developed.

The comments in Table 10.2 tended to confirm the suspicion that the upsets were caused by the presence of CO_2. However, there was still a residual doubt that possibly there was another impurity that was causing the problem. The problem solver determined that if the batch laboratory data could be shown to be consistent with plant data, this doubt would be removed. In order to develop this comparison, a technique was required to allow extrapolating short-term plant upset data to steady-state operations. The steady-state operation values would

Table 10.2 Questions/comments for example Problem 10-2.

Question	Comment
Are all operating directives and procedures being followed?	All appeared to be correct and being followed
Are all instruments correct?	The instruments had allegedly been calibrated. In addition, the heat balance was checked by manual calculations
Are laboratory results correct?	Yes. The X-ray machine used to determine the catalyst concentration was cross-checked with another machine. No other impurities besides CO_2 were observed on continuous analyzers during upsets. In addition, samples of the feed taken during upsets and analyzed in the laboratory confirmed the presence of CO_2
Were there any errors made in the original design?	No
Were there changes in operating conditions?	No
Is fluid leakage occurring?	The potential for reaction quench fluids leaking into the reactor was checked, and it was concluded that they were not leaking into the reactor
Has there been mechanical wear that would explain the problem?	No
Is the reaction rate as anticipated?	Yes, except during upsets
Are there adverse reactions occurring?	Yes. During upsets, the product quality attributes decreased
Were there errors made in the construction?	Not applicable

allow comparison to the batch laboratory data. After some literature research, the problem solver developed the four important concepts described in the next few paragraphs.

In most polymerization reactions, highly reactive impurities such as CO_2 react immediately upon entering the reactor. That is, the rate of catalyst deactivation is not dependent on the concentration of the impurity in the *reactor*, but depends on the concentration of the impurity in the *feed*. As discussed earlier, this is known as the "fast response model." This is the first important concept.

The second concept is that for industrial problem solving, the polymerization reaction can be represented by simple first-order kinetics with respect to the catalyst and monomer concentration. The equation shown below was utilized to approximate the polymerization kinetics in a reactor with a fixed volume.

$$R = k^* C^* M \tag{10.10}$$

where

R = the polymerization rate or monomer consumption rate, lb/hr
k = the kinetic rate constant including reactor volume, ft^6/hr-lb
C = the catalyst concentration in the reactor, lb/ft^3
M = the monomer concentration in the reactor, lb/ft^3

Note that the units of the kinetic constant include a fixed reactor **volume**. The absolute value of the kinetic constant is not important since, as will be shown later, all that is considered is the relative change in this constant.

The third concept involves the modification of equation (10.1) to include a catalyst deactivation term and development of an expression for catalyst concentration in the reactor. The rate of accumulation of the catalyst in the reactor can be expressed as shown in equation (10.11):

$$A_C = I_C - O_C - D_C \tag{10.11}$$

where

A_C = the rate of accumulation of the catalyst in the reactor, lb/hr
I_C = the inflow of the catalyst, lb/hr
O_C = the outflow of the catalyst, lb/hr
D_C = the deactivation of the catalyst, lb/hr

Based on the "fast response model," any CO_2 entering the reactor will immediately deactivate some of the catalyst. This relationship can be expressed as follows:

$$D_C = Z^* B \tag{10.12}$$

where

Z = the ratio of the catalyst deactivated to CO_2, lb/lb
B = the rate of CO_2 entering the reactor, lb/hr

Equation (10.11) then becomes

$$A_C = I_C - O_C - Z^* B \tag{10.13}$$

If the flow of CO_2 continues for an extended period, the steady state will be reached. At the steady state, $A_C = 0$ and $O_C = I_C - Z * B$. This steady-state value will be the full impact of the particular rate of CO_2 entering the reactor. While the deactivation ratio (Z) is of theoretical interest, it can be eliminated from pragmatic considerations as described later.

As indicated in equation (10.10), the important variable is the catalyst concentration in the reactor. Equation (10.13) can be transformed into equation (10.14) that uses the catalyst concentration by the appropriate substitutions.

$$dC/dt = (I_C - O_C - Z^*B)/V \tag{10.14}$$

where

dC/dt = the rate of catalyst concentration change in lb/ft^3-hr
V = the volume of the reactor, ft^3

Equation (10.15) can be used along with equation (10.14) and the initial catalyst concentration in the reactor to determine the concentration at any point in time.

$$C_i = C_0 - (dC/dt)^*t \tag{10.15}$$

where

C_i = the catalyst concentration in the reactor at any point in time, lb/ft^3
C_0 = the original catalyst concentration in the reactor, lb/ft^3
t = the elapsed time from the start of the upset to C_i

The fourth concept involves developing a relationship between the concentration of the active catalyst in the reactor at any point in time, the initial concentration, and the final concentration. This relationship assumes that a step change is made to an equilibrium condition and that the step change is continued until the steady state is reached.

This relationship will allow translating upset data into steady state for comparison to the batch laboratory data. For a perfectly mixed CSTR, the concentration of a component at any point in time after a step change is made can be expressed as shown in equation (10.16). In the specific case under consideration, the component is the catalyst and the step change is an increase in the CO_2 concentration in the feed.

$$(C_i - C_o)/(C_f - C_o) = 1 - e^{-T} \tag{10.16}$$

where

C_i = the concentration of the catalyst in the reactor at any point in time
C_o = the concentration of the catalyst in the reactor at time = 0
C_f = the concentration of the catalyst in the reactor after an infinite amount of time
T = the number of reactor displacements. This is simply the time after a step change divided by the average residence time

As indicated in equation (10.12), the amount of the catalyst deactivated by each pound of CO_2 is not known and was set equal to "Z." However, the instantaneous reaction rate is directionally proportional to the catalyst concentration in the reactor (assuming a constant monomer concentration) regardless of the value of "Z." Thus, the reaction rate can be substituted for the "C" values in equation (10.16). Equation (10.16) then becomes

$$(R_i - R_o)/(R_f - R_o) = 1 - e^{-T} \tag{10.17}$$

where

R_i = the reactor production at any point in time
R_o = the reactor production at time = 0
R_f = the reactor production after an infinite amount of time
T = the number of reactor displacements at any point in time

Thus, if reactor production at time = 0 (R_o), the reactor production at any point in time (R_i), and the number of reactor displacements at any point in time (T) are known, the steady-state impact (R_f) of a step change in impurity level can be estimated. The number of reactor displacements is the time since the step change in impurity level occurred divided by the reactor residence time. The use of equation (10.17) also eliminates some of the concerns associated with the theoretically imperfect derivation of eq. (10.14) and equation (10.15).

Unfortunately, the impurity upsets observed by the problem solver did not occur as step changes. He simulated the approximate sinusoidal curve of concentration versus time that often occurred during upsets as a step change. A typical impurity curve along with the technique used to simulate a step change is shown in Figure 10.2.

The problem solver recognized that selecting the magnitude of the upset level to use was somewhat subjective. He tried to select it so that the areas below and above the increasing part of the actual curve were equal. He also used the peak in the concentration versus time curve as the termination point of the upset (R_i or C_i).

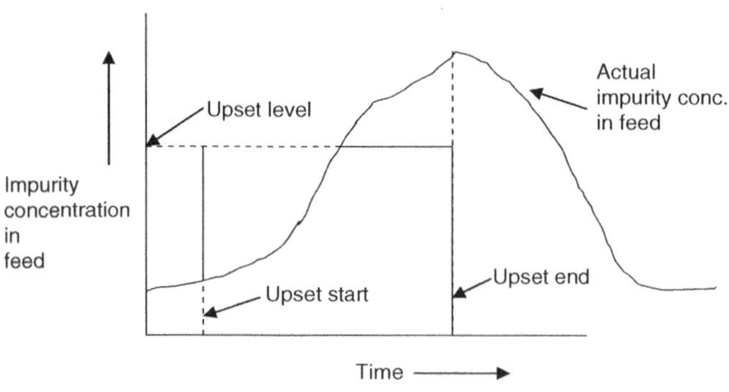

Figure 10.2 Simulation of actual change.

The problem solver used this four concept technique because it provided a consistent method for evaluating upsets of varying magnitudes and durations. For example, an upset where the CO_2 concentration increased from 0 to 25 ppm and lasted for 15 minutes might have a less actual impact than an upset where the CO_2 concentration increased from 0 to 5 ppm and lasted for 45 minutes. However, the projected steady-state impact of the first upset was greater. It was obvious that a steady-state concentration of 25 ppm of CO_2 in the monomer feed would have a much larger impact than 5 ppm However, the exact magnitude of this difference could not be known without using the techniques described earlier.

In order to completely formulate the working hypothesis, several reaction upsets apparently caused by CO_2 of varying severity and duration were followed, and the techniques discussed above were used to extrapolate to the steady-state effect. These data along with the batch laboratory data are shown in Figure 10.3.

The problem solver then developed the following working hypothesis based on the data shown above:

> "It is theorized that the short-term reaction upsets are due to the presence of CO_2 in the reactor feed. The plant data are consistent with laboratory data that indicate that CO_2 is a potent reactor impurity. The reaction upsets are likely to cause product quality upsets due to the unsteady nature of the reactor during these transients. In addition, the presence of a potent reactor impurity is anticipated to affect polymer quality attributes. There were no indications of any other impurities in any of the upsets that were followed."

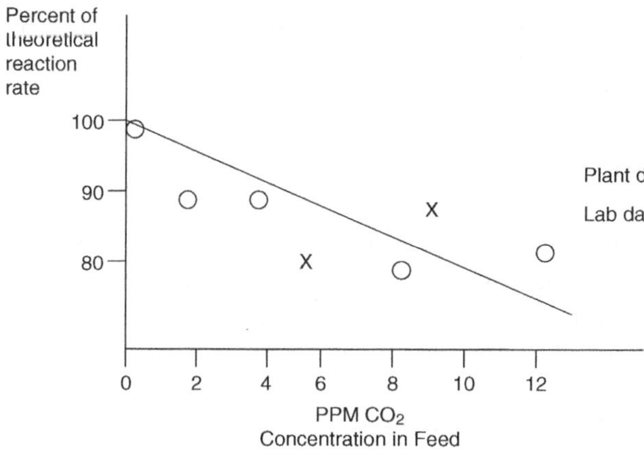

Figure 10.3 Projected steady-state reaction rate versus CO_2 concentration.

Step 4: Provide a Mechanism to Test the Hypothesis

In order to test this hypothesis, a technique had to be developed to ensure that the reactor feed was free of CO_2 for an extended period of time. If no reaction upsets occurred during this time period, agreement was reached that this would be considered a successful test and proof of the hypothesis. While it would involve additional operating costs, it was agreed to operate the monomer production facilities in a fashion that there would be minimal probability of CO_2 upsets. While this additional operating cost was not acceptable in the long range, it was acceptable for a 1-month trial. During this 1-month period, there was no measurable CO_2 present in the monomer feed nor were there any reaction upsets. Thus, this appeared to be a successful test, and the working hypothesis was proven.

Step 5: *Recommend* Remedial Action to Eliminate the Problem without Creating Another Problem

Based on the successful test that proved the hypothesis, the following remedial actions were taken:

- An evaluation was made to determine if the specification should be reduced from the existing 10 ppm to a lower level. Because of the high cost of the catalyst and the fact that operating at 10 ppm would cause the reaction rate to decrease from 100% of theoretical to about 80% of theoretical (see Figure 10.3), it was concluded that the specification should be reduced to 1 ppm
- Additional facilities were installed to allow reducing the specification from 10 to 1 ppm and to eliminate the occurrences of impurity upsets.
- A new highly accurate continuous analyzer to monitor the reactor feed for CO_2 was installed.
- Contingency plans were developed for the possibility of a CO_2 upset. These plans included shutting down the reactor if the concentration reached such a high level that serious product quality problems would be encountered.

10.7 Lessons Learned

While this problem might be considered an isolated example that has minimal application to the industrial world, it should be recognized that upsets do occur on a frequent basis. These upsets are often ignored because they are only short term. When they are considered, it is often with a minimal amount of data analysis, and wrong conclusions are often reached.

In the real-life example problem, pursuing the presence of another impurity besides CO_2 would have been an easy trap to fall into because of the presence of a previous experience and the persuasiveness of the proponent of the idea.

However, it would have been the wrong route to pursue. It was only when data were developed and analyzed that it became obvious that the impurity upsets were associated with CO_2. The problem also illustrates the value of quantitative data. The relationship between the concentration and reaction kinetics (Figure 10.3) allowed determination of the optimum monomer impurity specification.

The problem illustrates the type of shortcuts that can be taken in an industrial environment. While theoretically, polymerization reaction kinetics involve several steps such as catalyst activation, polymerization initiation, polymerization propagation, and polymerization termination, there is no need to include all of these reactions. Essentially, all reactions can be modeled by simple first-order kinetics with respect to the catalyst and monomer concentration. In addition, since in this specific problem the reactor volume is fixed, there is minimal need to include it in the kinetic relationship. What is important is that the correct model be used for the reaction of the impurity (fast response or slow response) and the reactor fluid flow (CSTR or plug flow). In this case, the problem description includes the idea that the impurity CO_2 reacts immediately as it enters the reactor and thus can be simulated with a fast response model. The problem description also indicated that the reactor was a CSTR.

The fact that the impurity concentration versus time relationship was simulated as a step change was a decision made by the problem solver. It would have been possible to simulate this as a sinusoidal relationship. However, it is doubtful that this would have improved the accuracy of the solution and it would have increased the complexity of the approach significantly.

10.8 Final Words

The primary purpose of this chapter is to illustrate both the need to have and apply techniques for the unsteady state. The engineering concept given in equation (10.1) is simple.

$$\mathrm{AD} = I - O - \mathrm{RD} \tag{10.18}$$

The techniques required to apply the equation are much more complex. However, the problem solver must not be overcome by the complexity. Like all problems and/or projects, the complexity must be approached one step at a time. For example, what is the input, what is the output, and how can the reaction or addition or removal of heat be estimated. The benefit of doing the work required to utilize the techniques described is that quantitative problem solutions are possible. For example, in the first example (Problem 10.1), it was possible to specify the rate of reactant addition based on the data analysis.

The two example problems are actual problems. The time pressures to develop solutions rapidly were present. But there was also a need to work the problem right the first time as opposed to multiple "trial and error" attempts often characterized by the phrase "We have got to try something."

Nomenclature

A	the heat transfer area, ft^2
A_C	the rate of accumulation of the catalyst in the reactor, lb/hr
AC	the heat accumulation, BTU/hr
AD	rate of accumulation. This could be accumulation of anything—level, heat, or reactant
B	the rate of CO_2 entering the reactor, lb/hr
C	the catalyst concentration in the reactor, lb/ft^3
C_f	the concentration of the catalyst in the reactor after an infinite amount of time, lb/ft^3
C_i	the catalyst concentration in the reactor at any point in time, lb/ft^3
C_0	the original catalyst concentration in the reactor, lb/ft^3
C_p	the average specific heat of material, BTU/lb-°F
D_C	the deactivation of the catalyst, lb/hr
dC/dt	the rate of catalyst concentration change in lb/ft^3-hr
dT/dt	the rate of temperature rise, °F/hr
I	inflow of material or heat
I_C	the inflow of the catalyst, lb/hr
k	the polymerization kinetic rate constant including reactor volume, ft^6/hr-lbs
M	the monomer concentration in the reactor, lb/ft^3
O	outflow of material or heat
O_C	the outflow of the catalyst, lb/hr
R	the rate of reaction. For the case of an instantaneous reaction, it is the rate of reactant addition, lb/hr. For a polymerization reactor, it is the polymerization rate or monomer consumption rate, lb/hr
R_f	the reactor production after an infinite amount of time
R_i	the reactor production at any point in time
R_o	the reactor production at time = 0
RD	removal/addition of heat or formation/destruction of the material by reaction
T	the number of reactor displacements. This is simply the time after a step change divided by the average residence time
T_R	the temperature of the reactor, °F

T_C	the *minimum* temperature of the coolant, °F
t	the elapsed time from the start of the upset to C_i
U	the heat transfer coefficient, BTU/hr-°F-ft^2
V	the volume of the reactor, ft^3
W	the weight of the material (metal, water, and reactants), lb
Z	the ratio of the catalyst deactivated to CO_2, lb/lb
ΔH_R	the heat of reaction, BTU/lb

11

Application to Other Plant Improvements

11.1 Introduction

The emphasis of this book is on detecting, specifying, developing theoretically correct hypotheses, and providing problem solutions. However, there are instances where the chemical engineering need is well known, and what is required is a study to understand how to provide a solution for the need. This chapter is devoted to these needs.

The format for this chapter is somewhat different than that of other chapters. The format for problem solving has been replaced by a step-by-step format describing what was done to eliminate the problem. As such, several case studies of various types are presented. Regardless, of the change in format, the reader will observe the careful application of chemical engineering fundamentals is provided. Thus, one of the key emphases of the book has been maintained. These case studies also indicate the range of details required. Some cases are only for a conceptual or scoping study. Others are more detailed and typical of studies that need to be done prior to final implementation in operating plants.

The type of needs that this chapter addresses are as follows:

- A scoping study of debottlenecking known reactor limitations
- A detailed debottlenecking study of a known hydraulic limitation problem
- Two detailed studies of improved operating procedures

Debottlenecking is defined as increasing capacity by eliminating limitations in the process with a minimal amount of investment. It has sometimes been referred to as "brownfield" investment as opposed to "greenfield" investment. Without a doubt, debottlenecking where possible is less expensive and hence more economical than building a grass roots greenfield plant However, the engineering and construction phases are generally more complicated and require good knowledge

Problem Solving Approaches for Maintaining Operational Excellence in Process Plants, Second Edition.
Joseph M. Bonem, Nattapong Pongboot, and Wiroon Tanthapanichakoon.
© 2025 the American Institute of Chemical Engineers, Inc. Published 2025 by John Wiley & Sons, Inc.

of the existing plant and engineering fundamentals. Brownfield projects and the associated test runs are discussed in more detail throughout this chapter and later on in Chapters 14 and 15.

11.2 Debottlenecking Reactors

As a general rule, unit operations such as fractionation or heat transfer are very well understood. Thus, debottlenecking these operations is relatively easy compared to the complexity of reactor debottlenecking. Reactor debottlenecking involves the following:

- *Reaction kinetics*: Many reactor debottlenecking projects involve reduced reactor residence times. Thus, reaction kinetics must be considered. The compensation for reduced residence time might be a new catalyst with increased activity, increased reactor temperature, or reduced once through conversion to name a few techniques.
- *Heat transfer*: Essentially, all reactions have a heat of reaction and 85–90% are exothermic. The discussion in this section assumes that the reaction is exothermic. Heat removal must be considered on a macro basis and on a micro basis. The term macro basis refers to the overall heat that must be removed, as determined by the heat balance. On the other hand, the micro basis considers a localized temperature. This might occur as hot spots in a liquid-phase reactor, in the catalyst particle of a fixed-bed reactor, or in individual particles in a polymerization reactor. A micro heat balance is required to determine if this is occurring.

The need to increase reactor capacity occurs for both continuous and batch reactors. For the purpose of this section, only continuous reactors are considered.

There are multiple types of reactors such as the following:

- Gas-phase reactors
- Liquid-phase reactors—these could be either diluent or nondiluent reactors.
- Two-phase reactors—these could be gas–liquid, gas–solid, or liquid–solid reactors
- Fixed-catalyst bed reactors—in the reactors described above, the catalyst is injected into the reactor. Any of these described could be conducted over a fixed bed except those that contain or produce solids.

11.2.1 Kinetics

As a general rule, kinetics are the first consideration for debottlenecking reactors. There are multiple means to overcome this limitation. The most common are increasing catalyst activity by either a more active catalyst, increasing reactor temperature, or increasing reactant concentration. In the special case of a polymerization reactor, residence time can be increased by increased polymer concentration in the reactor. There are potential heat removal problems associated with all of these approaches.

11.2.2 Heat Removal

Macro heat removal occurs by indirect heat transfer and/or sensible heating of feed streams. The most common is the combination of sensible heat and indirect heat transfer. The relative amount of indirect heat removal to sensible heat removal depends on the temperature of both the feed streams and the reaction. There are multiple ways to accomplish this indirect heat transfer. The equipment can be vessel jacket, pump around shell and tube exchangers, internal heat exchange tubes, or vaporization/condensation of a reactant or diluent.

In addition, micro heat removal must be considered. Micro heat removal involves the heat removal from individual particles or stagnant areas of the reactor. Inadequate heat removal from reactants, products, or catalyst can cause excessively high temperatures at spots in the reactor or in individual solid particles. This can occur in at least two different modes. In one mode, the distribution of the catalyst/reactants or heat transfer area is such that a "hot spot" develops. One example is a fixed-bed reactor where reactants are poorly distributed or where the heat transfer surfaces are not correctly placed. A second example is the injection flow rate of a highly active catalyst. In this example, the active catalyst causes a high rate of reaction at the catalyst inlet due to poor catalyst distribution.

A second mode occurs in a reaction that produces solids (polymerization reactor). An assessment of heat transfer that occurs from these individual particles to the bulk liquid or gas phase is required to determine if the particle temperature will exceed some critical level. Above this critical temperature, the individual particles may agglomerate and/or foul the reactor. Many catalysts have high initial reactivity so that this is often a problem with reactor startup.

11.2.3 Stepwise Procedure

Regardless of the type of reactor, there are five steps to the procedure to determine if the reactor capacity can be increased. The steps are as follows:

- Step 1: Determine the target debottlenecked rate. A reasonable estimate might be an increase of 10–20%. A level of 50% increase is likely the maximum.
- Step 2: Determine how the kinetic limit will be eliminated.
- Step 3: Determine the macro heat removal requirements at the new production rate. Determine if the existing equipment is adequate for the increased rate.

In these calculations, allowance should be made for temporary upsets that might increase the heat load by 10–15%.

- Step 4: Consider the micro heat removal requirements. This will likely be a good place to utilize CFD (calculational fluid dynamics). However, since this is an expensive step that likely requires the use of a specialized consultant, a preliminary analysis might be of value.
- Step 5: Consider what other reactor limits might be involved. In a polyethylene gas-phase reactor, the macro heat removal might be increased by using spare capacity in the recycle gas compressor. However, this might increase the probability of excessive entrainment of solids. In the highly competitive field of polyolefins, the possible existence of intellectual properties must also be considered.

11.2.4 Real Life Example

An example of this debottlenecking approach is the scoping study of the expansion of a polypropylene plant at a minimal cost. The example of a polymer plant is used because it represents one of the more complicated types of debottlenecking. In this particular case, it was determined that the equipment downstream of the reactor could be modified at a minimum cost and that a new catalyst with 30% more activity was available. Thus, the primary question was reactor heat removal capacity. The reactor was a bulk (no diluent) boiling reactor. Heat was removed by the sensible heat of the feed, the water-cooled reactor jacket, and the condensation of the boiling propylene in an elevated condenser. The propylene condensate was drained back to the reactor from the exchanger.

11.2.5 Macro Heat Removal

The focus of this example begins with Step 3 since Steps 1 and 2 were essentially given. Step 3 involves considering the macro heat removal. The heat removal by sensible heat or by cooling water in the jacket of the reactor are fixed. The additional heat load associated with debottlenecking must be removed in the elevated condenser. The calculation steps to confirm that the elevated condenser has sufficient capacity are as follows:

1) Estimate the heat load on the elevated condenser by a heat balance around the reactor at the expanded rate. This should consider the lowest reactor temperature, the highest feed temperature, and the highest coolant temperature that are reasonable to consider. Add 10–15% to this value to allow for slight control excursions. This is necessary to prevent an uncontrollable reaction temperature from occurring.
2) Determine by a plant test the heat transfer coefficient of the elevated condenser.

3) Determine if the elevated condenser has sufficient capacity to handle the increased heat load.
4) Check the pressure balance in the vapor and liquid between the reactor and condenser. The velocity in the liquid drain line from the condenser to the reactor should be low enough to allow vapor to disengage rather than being carried down to the reactor. The pressure drop in the vapor line from the reactor to the elevated condenser should be low enough so that vapor does not flow from the reactor to the condenser through the liquid drain line.

Assuming that these calculations indicate that there is sufficient capacity to remove the macro heat of polymerization, it is necessary to consider the removal of heat from the individual particles (micro heat removal). If the heat created by the reaction cannot be removed from the individual particles, they will likely overheat, forming agglomerates and/or fouling the heat transfer surfaces.

11.2.6 Micro Heat Removal

This overheating can occur by two separate mechanisms, both of which are complicated and require a fundamental analysis. One of these mechanisms is associated with mixing of the catalyst being injected into the reactor. Inadequate dispersion of the catalyst will cause a high catalyst concentration in the proximity of the injection nozzle. This high rate of reaction may cause the individual particles to overheat. Typically, the catalyst is dispersed in propylene and injected into the reactor at a high velocity so that it was reasonable to assume that the catalyst was well dispersed.

The second mechanism is associated with the very high activity of the catalyst as it is initially injected into the reactor. When considering this very active catalyst, the reaction rate might be high enough so that the heat transfer capability from the small particle is inadequate to remove the heat of reaction. If this occurs, the particle temperature would increase to provide the necessary temperature driving force to allow the heat removed to equal the heat generated. This new temperature might be above the softening point of the polymer being formed, leading to agglomeration or fouling of the heat transfer surfaces.

Considering the second mechanism, it is known that the catalyst used in the polymerization of propylene is a heterogeneous catalyst. As insoluble catalyst is injected into the reactor, the high-porosity spherical catalyst serves as a template for the growing polymer particle, which polymerizes in the pores as well as on the surface of the catalyst. Over time, the particle grows, and the porosity decreases. The result of this growth and decrease in porosity is both a reduction of catalyst activity and particle heat transfer area-to-volume ratio.

The prediction of the temperature of the single particle as a function of time then requires simulation with a dynamic model that takes into account both the constants and the changing variables of catalyst activity and particle heat transfer area. There are at least two approaches for developing this model. The simplest one makes use of literature predictions of the particle heat transfer coefficient and assumes a perfectly mixed vessel. With this approach, the temperature of the boiling liquid is assumed to be the same throughout the reactor. In addition, each particle being considered is assumed to be at a uniform temperature. This approach provides no projections of entrainment by the vaporizing liquid.

A more sophisticated approach utilizes computational fluid dynamics (CFD), but requires significantly more time and as a result is more costly. This approach requires specialized software and almost always specialized expertise. However, it is likely more accurate and provides an approach to estimating entrainment.

Both approaches utilize a dynamic model for predicting the temperature of the individual particles. The model is a basic heat accumulation model around a single particle. To provide some understanding of the dynamics involved, the simplified approach is shown below:

$$\text{Heat Accumulation} = \text{Heat Generated} - \text{Heat Removed} \tag{11.1}$$

$$(dT/d\theta)^{*}W^{*}\text{Cp} = R^{*}\text{Hr} - U^{*}A^{*}(\text{Tp} - \text{Tr}) \tag{11.2}$$

This can be transformed to the following:

$$dT/d\theta = (R^{*}\text{Hr} - U^{*}A^{*}(\text{Tp} - \text{Tr}))/W^{*}\text{Cp} \tag{11.3}$$

where

$dT/d\theta$ = rate of temperature change
W = weight of particle (catalyst and polymer)
Cp = average specific heat of particle
R = rate of polymerization
Hr = heat of polymerization
U = heat transfer coefficient
A = total surface area of particle including pore area
Tp = temperature of particle
Tr = temperature of reactor liquid

The maximum particle temperature is determined by solving equation (11.3) in an iterative fashion using very small increments of time ($d\theta$). Initially, the contents of the reactor will be at 158 °F. From this starting point, the rate of temperature change can be calculated, and the new temperature can be developed. This procedure is repeated until enough time has elapsed so that the maximum temperature has been achieved.

An examination of equation (11.3) indicates that there are both variables that are known or readily available as well as critical variables that must be estimated.

These critical variables and how they are estimated are described as follows:

- *Polymerization rate*: The rate of reaction was determined from simplified kinetic models developed from both plant and laboratory data. The simplified model takes the following form:

$$\mathrm{CE} = A^{*}e^{-(B/RT)*}\ M \tag{11.4}$$

- *Heat transfer coefficient*: While there have been several approaches suggested for evaluating the overall heat transfer coefficient from a particle to the surrounding fluid, the one that seemed to give the most reasonable results is

$$U = 36.2^{*}d^{\wedge}0.68^{*}\mu^{\wedge} - 0.79^{(1)} \tag{11.5}$$

1) *Chemical Engineering Kinetics* by J. Smith (page 413 Third Edition). McGraw-Hill This empirical relationship gives heat transfer coefficients between 50 and 100 BTU/ft^2-hr-F

- *Polymer particle surface area*: Because of the large amount of pore volume in the typical polypropylene catalyst and the fact that the catalyst serves as a template for the polymer, the initial polymer has a large amount of pore volume. This means that the surface area for heat transfer is much higher than the calculated surface area based on the particle diameter. As the polymerization continues, this surface area is reduced as the pore area is coated with the polymer. The model includes this loss of heat transfer area.

The model treats the particle as a constant temperature particle with no temperature differential from the surface to the center. While this is not exactly correct, it is believed to be an adequate representation because of the large amount of porosity of the particle during the early phase of polymerization. The large amount of porosity greatly reduces the thermal resistance since the average thickness for heat transfer is much less than the radius of the particle.

In the above:

CE = catalyst efficiency lb polymer/lb. of catalyst
A and B = Typical Arrhenius constants determined from laboratory, plant, or literature data
M = monomer concentration, lb/ft^3
R = gas constant, 1.987
T = absolute temperature, °Rankine
θ = time of reaction, hours
U = heat transfer coefficient BTU/hr°F-ft^2
d = particle diameter, microns
μ = slurry viscosity, centipoise

Since this relatively simplified model calculates the critical variables as a function of time, the temperature profiles of the two catalyst systems can be compared. This comparison is shown in Figure 11.1. As shown in this figure, the new

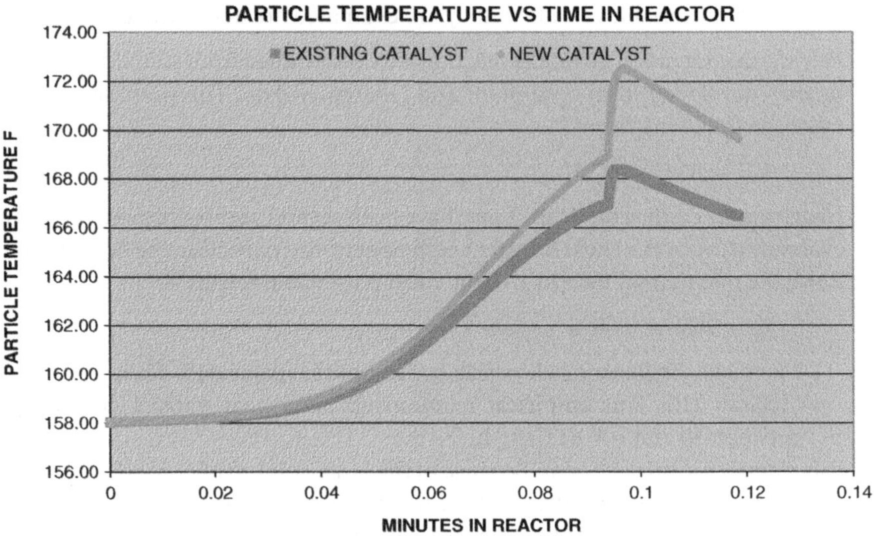

Figure 11.1 Temperature profile with existing and new catalyst.

catalyst is anticipated to have a peak temperature 4 °F higher than the existing catalyst. While neither the difference in peak temperatures or the absolute peak temperatures themselves should be considered highly accurate, they do indicate that additional work is required. This additional work might consist of the use of a more sophisticated CFD model and/or an evaluation of polymer morphology that might occur with higher peak temperatures. Consideration should also be given to all grades of polypropylene being produced to ensure that different operating conditions or different morphologies do not impact any conclusions.

This simplified approach can be used on other reactor styles keeping in mind the words of one famous scientist "All models are wrong. But some are useful." Simplified models based on fundamental concepts are of value to point out the direction of future work.

11.3 Real-World Hydraulic Debottleneck

11.3.1 Introduction

A comprehensive hydraulic review is a critical part of a successful revamp of any plants in chemical process industry (CPI)—to ensure the process fluid can smoothly move from one place to another. Depending on the remaining hydraulic margin, a certain degree of modifications may be required to accommodate a new unit throughput/configuration.

In revamps, a design engineer should exploit every bit of design margin to avoid costly equipment modifications. While hydraulic design guidelines (i.e., fluid velocity limit) are intended for greenfield problems (also see Chapter 15 for problem definition) to ensure an economical design, it does not mean the designer cannot conflict with them, as long as integrity/reliability criteria (e.g., erosion limit) are not violated.

It must be noted that the economic justification is different here, as the plant owner already invested in the existing equipment. Utilizing the existing hardware is always a better choice from financial/practical perspectives, thus requiring a careful hydraulic assessment.

Another essential part of a successful revamp is a good test run. There are countless disastrous revamp projects out there that failed to meet the revamp objectives just because the designer did not incorporate actual plant data and equipment performances during the design phase.

Moreover, people tend to overly focus on rotating equipment. This notion is not always the case and could lead to a costly solution, e.g., replacing a large compressor. Ask yourself if you would still replace a pump or compressor if you found that the inlet block valve was partially/fully closed. In many cases, pump/compressor issues are not from the equipment itself.

This guideline explains helpful tips to significantly reduce the man-hours required for hydraulic review, based on real-world chemical engineering experiences. Furthermore, unconventional yet practical ways of equipment sizing will also be demonstrated here as fast-track solutions to unlock hydraulic constraints.

11.3.2 Always Account for Non-idealities

While theoretical calculations are good starting points, one must remember that engineering assumptions or even the hydraulic model itself (e.g., Darcy–Weisbach or Hazen–Williams) might not perfectly represent the real world. For example, the hydraulic design assumptions listed in Table 11.1 might be reasonable for a new design. However, it does not mean that the hydraulic model developed using the same assumptions will 100% reflect what happens in the existing hydraulic system.

This is the primary reason why a design margin is always there when designing a new piece of equipment/device to account for any nonidealities and provide additional operational flexibility. Almost all process piping, pumps, and compressors are designed based on rated flow rates, typically 110–120% of normal. Therefore, 10–20% spare capacity may be available for the revamp operation.

Some examples of nonidealities in hydraulic systems include fouling, corrosion, and erosion. A piping system or equipment might be fouled (or corroded), thus

Table 11.1 Examples of equivalent lengths for piping components based on major design/operating companies.

Piping Component	Detail	Equivalent Length
Butterfly valve	8″ and larger	20–40 D
Elbow	90°, $R = 1.5\ D$	13–20 D
Tee	Flow through branch	60–72 D
Gate valve	Full bore	8–13 D
Check valve	Swing type	92–135 D

(a) (b)

Figure 11.2 (a) A tube bundle heavily fouled by asphalt and coke. (b) The inner pipe wall of the severely fouled fuel oil line.

affecting both actual pipe roughness and flow area to a certain degree, as illustrated in Figure 11.2. These nonidealities are reasons why ideal hydraulic models deviate from reality.

In many cases, some revamp projects are just a remedy of a failed grassroots design. This further emphasizes the need to augment theoretical calculations with real-world data when one handles a revamp work.

11.3.3 A Quick Guide for a Successful Hydraulic Test Run

The main objective of a unit test run is to find actual performances and constraints of the existing system. Process calculations and solution findings based on only theoretical equipment performance may be misleading. A test run is a perfect opportunity to reveal all hidden design margins, which might not be explicitly shown in the equipment datasheet. Moreover, it is also highly beneficial to use the test run data to establish a base case for your revamp and improve the calculation accuracy.

A good example is a typical pump sizing workflow. Several parties often add design factors (usually arbitrary and based on institutional knowledge) to each design stage:

(1) Basic Design
(2) Front-End Engineering Design (FEED)
(3) Detailed Engineering Design

By assuming a typical design margin of 10%, the compounded design margin could be as large as 33% (e.g., 1.10^3) when the pump is installed. It is not uncommon to see a pumping system with a highly throttled flow control valve. In some cases, one might need to replace or trim the impeller to improve pumping operation and energy efficiency.

Neglecting the actual equipment and system performances can lead to an undesired situation, where the maximum capacity of the existing equipment is substantially underestimated or overestimated. Test-run data and observations should be reviewed and discussed with the involving parties to avoid recommending a modification that conflicts with the operating data. For example, the control valve was too small considering the past actual data, but it is not replaced during the revamp. In the worst-case scenario, the revamp could fail, as the designer may not address the real issues.

Comprehensively conduct test runs at high unit throughputs to ensure that the pressure drop between the measurement locations is more significant than the error of the pressure instruments. As a best practice, preliminarily check remaining equipment margins before the actual test run. Simulating expected process conditions as a test-run guideline is always helpful.

There is no rule preventing the revamp designer from exceeding the design unit capacity. The real constraints are what is being observed in the field, e.g., valve travel or motor load, rather than design values based on the paper. As mentioned earlier, the goal is to utilize all hidden margins and define a new baseline here. If the existing hardware limits the maximum throughput, some actions can be accomplished during the test run, e.g., open a bypass valve or start a spare pump. Altering process conditions, e.g., raise pressure or level of the suction vessel to increase pump discharge pressure and available net positive suction head ($NPSH_a$), can also help. It is a good practice to incorporate past equipment inspection and reliability records into the analysis.

The new performance baseline can be developed by measuring flows, temperatures, and pressures with well-calibrated instrumentation and analyzing process streams. When deemed applicable, field measurement/check, i.e., a single-gage pressure survey, is highly recommended.

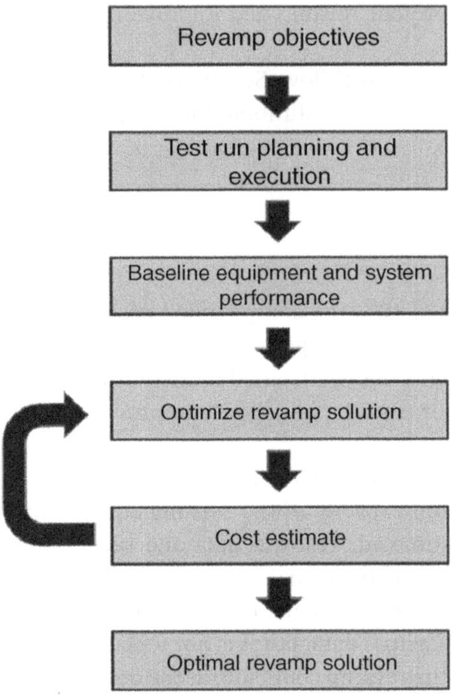

Figure 11.3 A pathway to maximizing success rate and minimizing CAPEX.

Always verify, correct, and reconcile the test run data to improve data quality before further use. For more details on conducting plant test run and data reconciliation, please see Chapter 15.

11.3.4 A Recommended Workflow for an Effective Hydraulic Revamp

As earlier emphasized, relying on theoretical calculations can lead to a costly revamp or even a total failure. The best practice illustrated in Figure 11.3 is suggested for all revamps. This figure shows a pathway to maximizing success rate and minimizing CAPEX.

Mostly, the revamp solution of hydraulic debottlenecking projects can be developed by the following approaches:

11.3.4.1 Utilize Equipment/Device's Characteristics

As a first step, try this simple yet effective technique before setting up the full-hydraulic model. Normally, the performance characteristics of equipment/device are usually well-defined if not subjected to severe fouling or damages and practically independent of the hydraulic system (i.e., valve characteristic

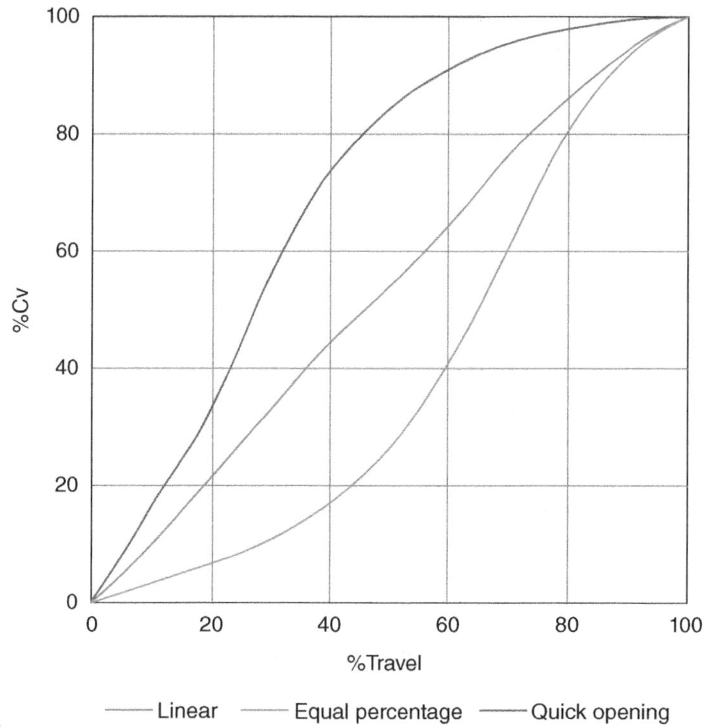

Figure 11.4 Typical inherent valve characteristics.

curves in Figure 11.4). Together with good baseline data, these characteristics are generally adequate to check the hydraulic adequacy of the existing system or even design a new control valve or pump. This technique employs the simplest mathematical equations, i.e., flow coefficient (C_v) equation, and few pieces of process data, thus requiring minimal effort. Figure 11.4 shows typical inherent valve characteristics—the relationship between C_v and valve travel (*%Travel*). This relationship remains virtually the same for any hydraulic system and thus is a good measuring stick for hydraulic availability.

11.3.4.2 Use a Calibrated Hydraulic Model

If the first approach does not work, it usually means it requires something more complicated to achieve the goal—a full hydraulic model.

First, a design engineer must select a suitable hydraulic model. For example, the Hazen–Williams equation is usually good enough to handle a water distribution system with a relatively smooth pipe surface. However, this equation is not valid for other fluids, therefore requiring the Darcy–Weisbach equation.

Second, a designer should systematically calibrate the hydraulic model. Typically, the modeler has to adjust multiple tuning parameters to match the actual data, such as

- *Resistance coefficients/equivalent lengths for piping components*: Different design guidelines suggest different values (as highlighted in Table 11.1). There is no harm in tuning these parameters.
- *Pipe roughness*: The values from design guidelines rarely match the reality due to fouling and corrosion. This parameter is particularly crucial for cast iron or aged piping systems.
- *Effective pipe diameter*: If the hydraulic system is subject to fouling, the modeler can also adjust the effective pipe diameter (also see Figure 11.2b for example).

A well-calibrated hydraulic model should be able to fit two sets of data. If it fails to do so, it usually means the tuning parameters are unfit, therefore requiring recalibration. One must note that a model calibration does not have to be perfect but should be at least good enough for decision-making.

To better appreciate how both approaches work, the following real-world case studies will be described here.

11.3.5 Maximize Diesel Production by Revamping the Draw-off System

The main objective here is maximizing the diesel production from the hydrocracker fractionator. In this case, the product cut point between kerosene and diesel was already minimized, but limited by the existing diesel pump capacity at 199.4 m^3/hr from a high motor load. It must be noted that a separate hydraulic review had already been conducted to confirm that the draw-off valve and piping were not the hydraulic constraints here. According to the economic study, the diesel production capacity should be increased to 226.0 m^3/hr (at the pumping temperature). Some potential revamp solutions include the following:

1) Replace both existing diesel pumps (main and spare) with larger ones
2) Add the 3rd larger pump

Intuitively, installing one pump is normally a cheaper solution than replacing two pumps. Eventually, the refinery decided to install the 3rd larger pump, as illustrated in Figure 11.5. This new pump will be the main workhorse, with the existing pumps as spares, and will only be offline when diesel demand is low.

As earlier described, diesel is desirable in this refinery. Therefore, the refinery often ran both existing pumps in parallel to maximize its production. From a reliability perspective, this operating scenario should be avoided as the existing pumps cannot be offline for preventive maintenance. Additionally, the liquid level inside

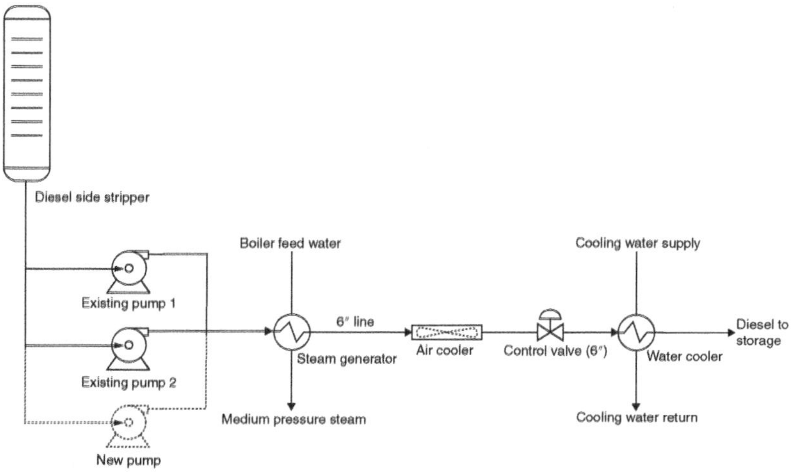

Figure 11.5 The scheme of the existing diesel draw-off system with the 3rd pump highlighted by dotted symbols.

the diesel side stripper was sometimes unstable when the diesel hydrotreater was run at high throughput (two diesel streams from both units join together, causing higher back pressure), forcing the refinery to reduce the diesel production from the hydrocracker fractionator—as the level control valve becomes unstable.

As the worst-case scenario, the design engineer conducted the test run at the design revamp flow rate (Q_{revamp}, 226.0 m^3/hr) with the diesel hydrotreater at its maximum throughput. *Both diesel pumps were in parallel operation during the test run* with %Travel of the level control valve at 95 (preferably less than 80). Based on Q_{revamp}, the baseline pump's head (ΔH_{base}) is 159.7 m, as graphically demonstrated in Figure 11.6. The composite (both pumps in operation) pump curve (dotted) is constructed based on a single pump curve as a representation of parallel operation.

This diesel draw-off system consists of long and complex pipelines plus multiple heat exchangers, thus requiring challenging and expensive demolition works during the turnaround if modified. Hence, the new diesel pump will be sized, so %Travel of the level control valve is reduced to 60 by including a design margin (ΔH_{margin}) as per equation (11.6).

$$\Delta H_{revamp} = \Delta H_{base} + \Delta H_{margin} \tag{11.6}$$

According to the hydraulic study, the revamp pump's head (ΔH_{revamp}) must be 175.7 m to ensure good operability. By increasing the pump discharge pressure, the downstream control valve needs to reduce its opening to maintain the same flow rate. In this case, a new relief valve is unavoidable due to higher shut-off pressure, but the additional cost is easily outweighed by the benefits from more diesel sales.

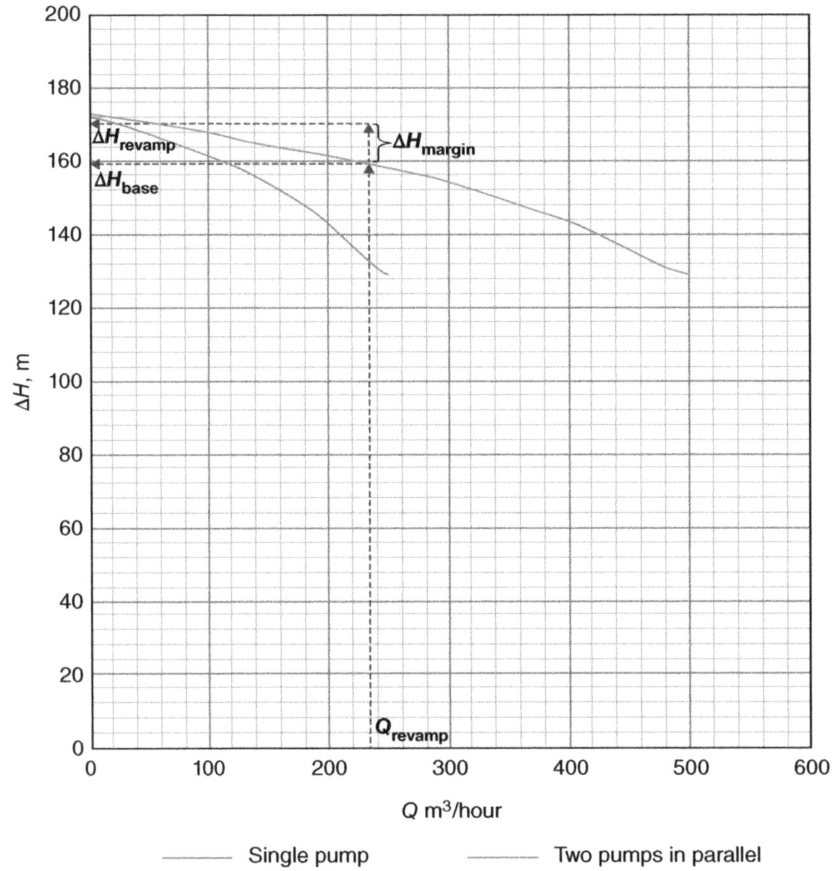

Figure 11.6 Graphical illustrations of ΔH_{base} and ΔH_{revamp} estimation.

11.3.6 A New Feed Preheater for Energy Saving

To improve the energy efficiency, a new feed preheater was designed and installed during the refinery turnaround to reduce the energy consumption of a heavy oil hydrocracker. As shown in Figure 11.7, a mix of two liquid feed streams is heated up by a hot diesel product from a diesel hydrotreater. Subsequently, this mixed liquid feed stream must go through two stages of filtration to remove solid particulates, which might end up plugging the very first beds of the reactor.

This new feed preheater must fit into the existing system without causing flow control issues. As such, one should ensure that Valve A and Valve B in Figure 11.7 have %Travel less than 80 (a typical maximum %Travel for reasonable flow control).

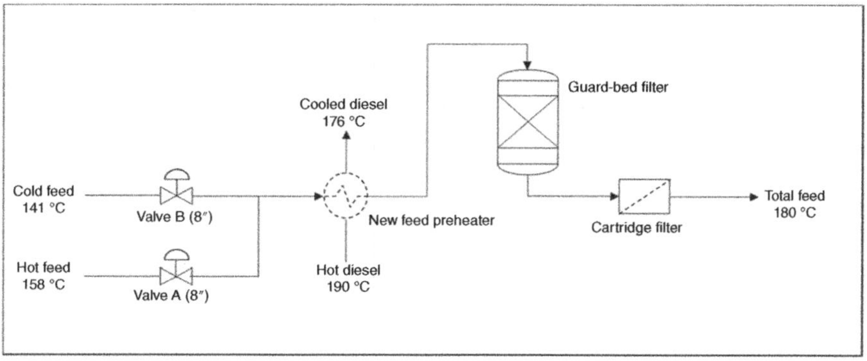

Figure 11.7 Feed system with its new feed preheater.

Table 11.2 Test run data for a new preheater design.

Stream	Q_{revamp} (m^3/hr)	%$Travel_{base}$	Temperature (°C)	S.G.
Cold feed	78.2	10.0	141	0.820
Hot feed	237.4	26.4	158	0.818

As a worst-case scenario, the refinery conducted a test run at desired revamp flow rates (Q_{revamp}) and maximum filter's pressure drops (see Table 11.2). It must be noted that Q_{revamp} for each stream is based on the specific gravity (S.G.) at the corresponding operating temperature.

For the hydraulic review, one can utilize characteristics of control valves described as follows:

1) Estimate the baseline flow coefficient ($C_{v,base}$) from the baseline valve travel (%$Travel_{base}$), as graphically illustrated in Figure 11.8 Mostly, these C_v data can be obtained from vendor's literature, e.g., Fisher's Catalog 12.
2) Find a baseline pressure drop (ΔP_{base}) using equation (11.7).

$$\Delta P_{base} = \frac{Q^2_{revamp}\text{S.G.}}{C^2_{v,base}} \tag{11.7}$$

3) After obtaining ΔP_{base} from Step 2, calculate a revamp pressure drop (ΔP_{revamp}) from equation (11.8) based on an expected pressure drop of the new heat exchanger and associated piping ($\Delta P_{new\ equipment}$).

$$\Delta P_{revamp} = \Delta P_{base} - \Delta P_{new\ equipment} \tag{11.8}$$

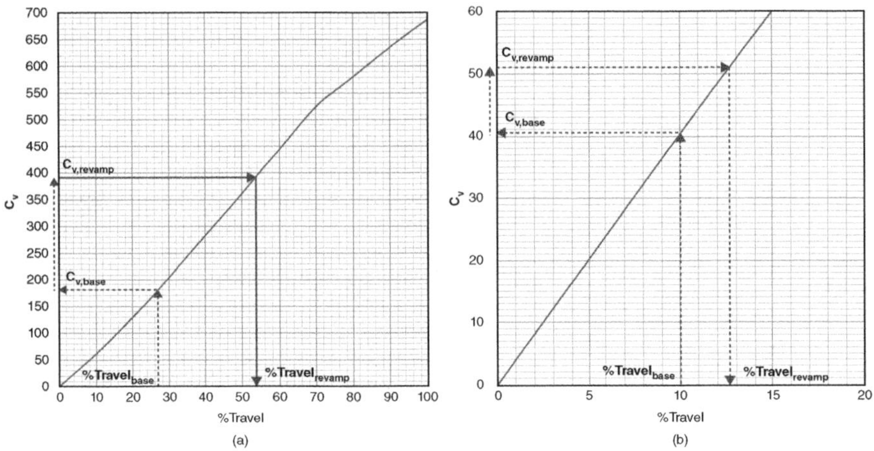

Figure 11.8 Graphical determinations of $C_{v,base}$ and %Travel$_{revamp}$: (a) Valve A (linear – full scale), (b) Valve B (equal percentage – zoomed).

4) Calculate a revamp flow coefficient ($C_{v,revamp}$) from equation (11.9) based on ΔP_{revamp}

$$C_{v,revamp} = Q_{revamp}\sqrt{\frac{S.G.}{\Delta P_{revamp}}} \tag{11.9}$$

5) Finally, estimate the revamp valve travel (%Travel$_{revamp}$), as demonstrated in Figure 11.8

With a typical $\Delta P_{new\ equipment}$ of 1.5 bar, the final results can be summarized as per Table 11.3. The existing feed system has adequate hydraulic availability to accommodate the new feed preheater with %Travel$_{revamp}$ less than 80 for both Valves A and B.

This technique can also be applied to similar cases, e.g., adding a new adsorber/filter into the existing system.

Table 11.3 Hydraulic calculation results based on the valve performance characteristics.

Step	1	2	3	4	5
Component	$C_{v,\ base}$	ΔP_{base}	ΔP_{revamp}	$C_{v,\ revamp}$	%Travel$_{revamp}$
Valve A	180.0	1.90 bar	0.40 bar	391.8	54.0
Valve B	40.6	4.06 bar	2.56 bar	51.1	12.6

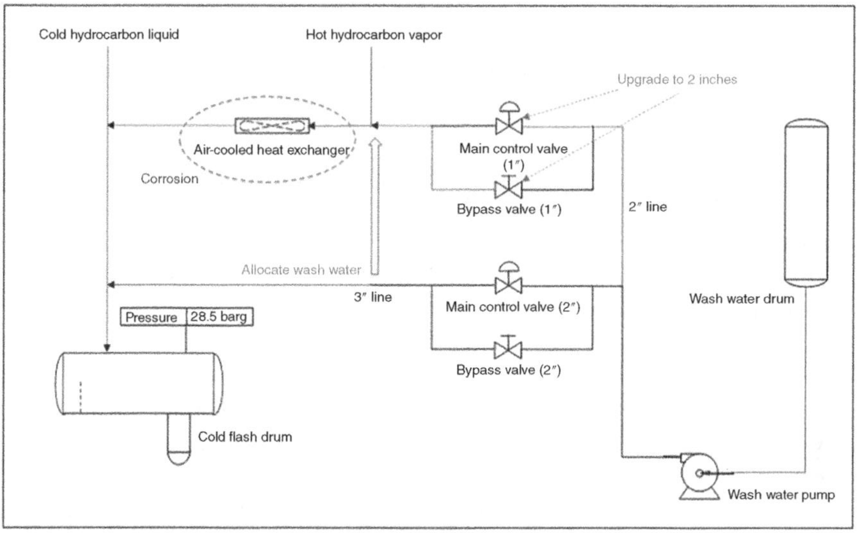

Figure 11.9 Illustration of a wash water injection system of a heavy oil hydrocracker.

11.3.7 Increase Wash Water Flow to Reduce the Corrosion Rate

An air-cooled heat exchanger of a heavy oil hydrocracker unit suffers from ammonium salt corrosion. Based on a corrosion study, the wash water flow rate should be increased from 150 to 200 m^3/day to ensure at least 25% of the injected wash water remains in the liquid phase (only liquid can dissolve salt) by allocating wash water from another parallel injection point (no corrosion issue)—as illustrated in Figure 11.9. As a precaution, the wash water flow rate must be below 215 m^3/day to prevent excessive erosion. This injection point has a relatively constant operating pressure of 28.5 barg. The piping system is made of a sour-service steel grade (expensive plus a long-lead time) with a complicated piping route, making piping replacement unattractive.

The control and bypass valve are of 1 in. (½ inch trim), with the line size of 2 in. for the concerned piping section. This suggests that one can still upgrade the control valve to increase the wash water flow. It is not uncommon to see a piping system with a smaller control valve (and even smaller trim size).

To maximize the success rate and simplify the work process, the test run data and the mathematical relationship proved below were used to size the new control valve. As previously suggested, one can always consider opening a bypass valve to overcome hydraulic limitations during the test run.

Principally, the total flow to the injection point (Q_{total}) is a summation between the flow across the main control valve (Q_{main}) and the bypass valve (Q_{bypass}) as per equation 11.10.

$$Q_{total} = Q_{main} + Q_{bypass} \tag{11.10}$$

By employing a liquid-phase flow coefficient equation, one can rewrite equation (11.10) as follows:

$$C_{v,total}\sqrt{\frac{\Delta P_{total}}{S.G.}} = C_{v,main}\sqrt{\frac{\Delta P_{main}}{S.G.}} + C_{v,bypass}\sqrt{\frac{\Delta P_{bypass}}{S.G.}} \tag{11.11}$$

In practice, all pressure-drop terms (ΔP_{main}, ΔP_{bypass}, and ΔP_{total}) are identical. Accordingly, one can estimate the total flow coefficient for each scenario ($C_{v,total}$) from equation (11.12).

$$C_{v,total} = C_{v,main} + C_{v,bypass} \tag{11.12}$$

As demonstrated in the previous example, the flow coefficient for the main control valve ($C_{v,main}$) and bypass valve ($C_{v,bypass}$) can be estimated based on the valve performance characteristic.

In the test run, the bypass valve was opened to determine the maximum wash water rate of this injection point. Table 11.4 summarizes the test-run data and calculation results.

Opening a bypass valve can lead to undesired situations, i.e., vessel overflow. As such, the existing 1-in. control valve was upgraded. In this case, a full-bore 2 in. valve was selected. From Table 11.4, the total flow coefficient of the revamp case ($C_{v,total,revamp}$) is 9.14. Consequently, the travel of the new control valve ($\%Travel_{revamp}$) would be 47, as demonstrated in Figure 11.10.

11.3.8 Improve Diesel Recovery from a Crude Distillation Column

A crude distillation unit experienced high column pressure, resulting in lower diesel recovery. It must be noted that the crude heater outlet temperature was

Table 11.4 Test-run data and calculation results for the wash water system revamp.

	Base	Revamp	Maximum
Q, m^3/hr	150.2	200.6	210
$\%Travel_{main}$	48	91	100
$\%Travel_{bypass}$	100	100	100
$C_{v,main}$	1.00	4.23	4.91
$C_{v,bypass}$	4.91	4.91	4.91
$C_{v,total}$	5.91	9.14	9.82

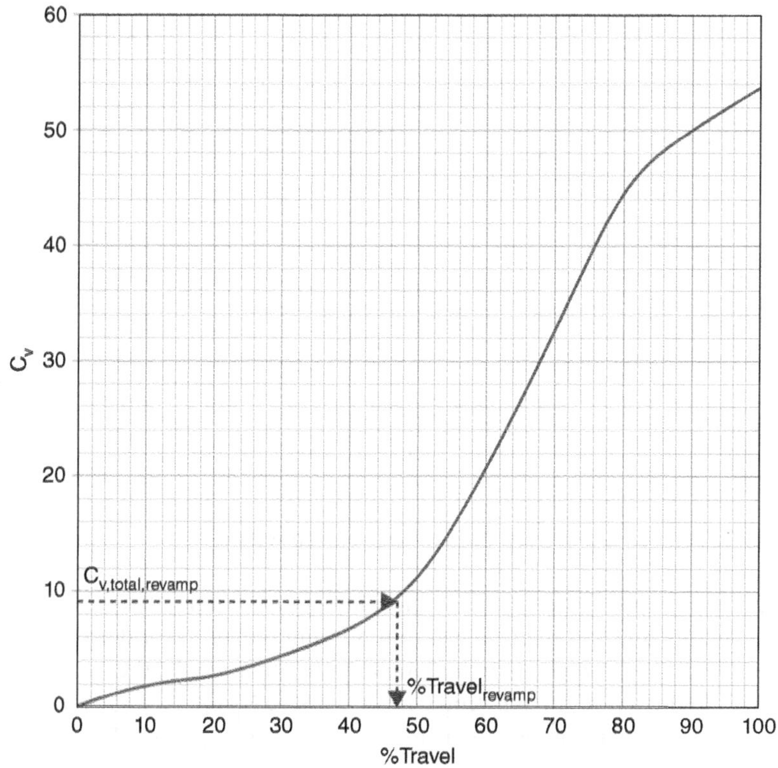

Figure 11.10 Graphical determinations of $\%Travel_{revamp}$ for a full-bore 2-in. control valves.

already at the maximum. As per the design, the suction pressure of the overhead compressor directly sets the column pressure (Figure 11.11). The pressure control system varies %Travel of the discharge control valve to maintain a constant accumulator pressure. When the %Travel reaches 100%, e.g., at a high vapor flow rate, the accumulator and column pressure become uncontrolled and increase with even higher vapor flow. In this case, two possible options to improve diesel recovery are as follows:

1) Replace the overhead compressor with a larger one
2) Reduce suction piping pressure drop

A brute force solution could be replacing the overhead compressor, but it would cost multi-million dollars. To avoid such a costly revamp, one may consider reducing suction piping pressure drop to free up compressor capacity. Increasing suction pressure is the most effective way to unload compressor capacity, compared with discharge pressure and suction temperature reduction, especially

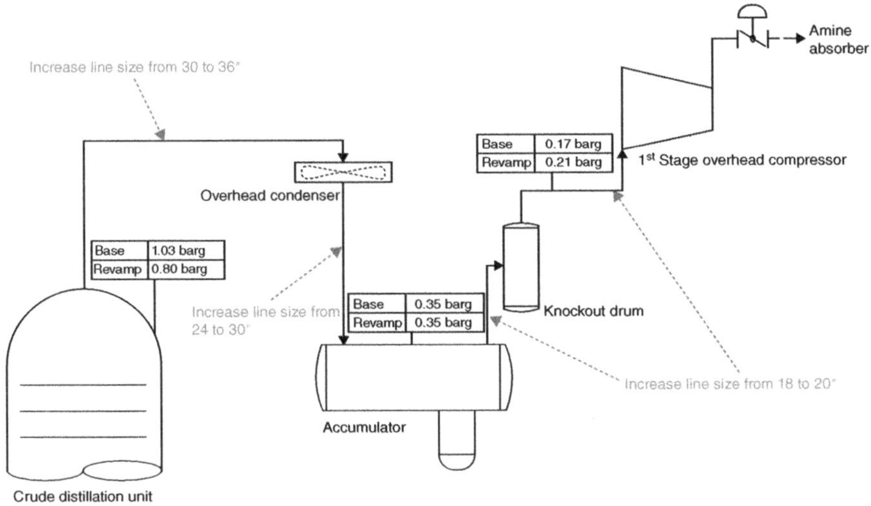

Figure 11.11 Illustration of the crude distillation overhead system.

in lower pressure applications, e.g., atmospheric distillation. This capacity unloading is mainly due to a reduced gas volume from higher vapor density and increased condensation at higher suction pressure. In other words, the operating point is shifted to the left of the polytropic head curve ($\Delta H_{\text{polytropic}}$, Figure 11.12) so that the discharge control valve can retake control, e.g., with %Travel less than 80.

As recommended earlier, the design engineer conducted multiple test runs at high throughputs. Everything from flow rate, temperature, pressure, and stream composition were recorded and verified. In this case, a full-hydraulic model is unavoidable. Before optimizing a revamp solution, the baseline hydraulic model was calibrated and tested against multiple test run data to ensure its validity.

In this case, it is possible to reduce the column's top pressure by 0.23 bar at the same accumulator pressure (with a total pressure drop reduction of 0.27 bar) by upgrading piping sections on the suction side, as illustrated in Figure 11.12. It must be noted that the piping replacement costs six times less than an overhead compressor revamp.

11.3.9 Other Considerations

- Where applicable, the design engineer should reuse equipment/device that is no longer fit for its original service in a different service to minimize CAPEX.
- When one exploits design margins, the equipment/device will be closer to its limits, thus having a narrower operating range. This risk may be managed by

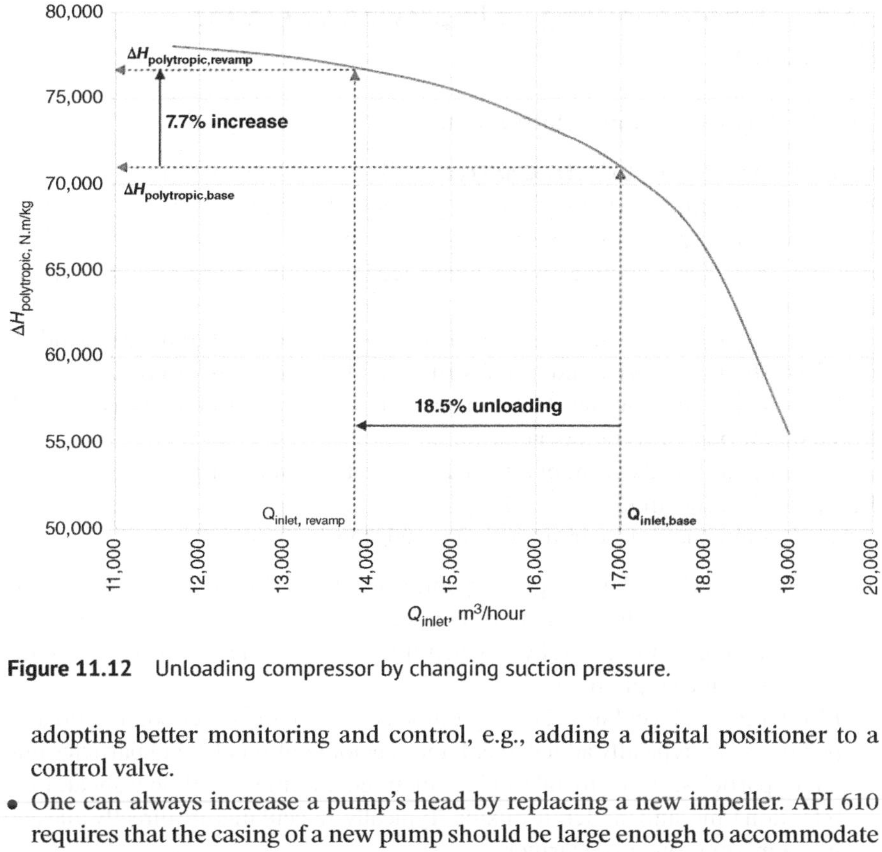

Figure 11.12 Unloading compressor by changing suction pressure.

adopting better monitoring and control, e.g., adding a digital positioner to a control valve.

- One can always increase a pump's head by replacing a new impeller. API 610 requires that the casing of a new pump should be large enough to accommodate a larger impeller, so a 5% head increase is possible.
- In contrast with a general belief, the head rise to shutoff (from the rated point) could be as low as 5% for a single-pump operation (and could be even lower as per Chaware and Joshi 2014). According to API 610, a minimum of 10% head rise is for pumps in parallel. Nonetheless, many EPC contractors adopt 10% as a design criterion for all scenarios. This overconservative mindset could result in unnecessary modifications. For example, a design engineer might need to add a relief valve into the existing system as the new shutoff pressure exceeds the piping design pressure from a suboptimal head rise to shutoff. For revamp scenarios, the operating point tends to be on the far right of the head curve, e.g., near the maximum throughput. In other words, the operating point will not change that often as the plant owner always tries to push the production unit to its limit. Therefore, a bit flat head curve should be acceptable.
- Always conduct a pulsation analysis (per API 674 and 618) before adding new parallel reciprocating equipment to avoid possible piping and structural failures.

11.4 Debottlenecking By Improving Operating Procedures

11.4.1 Improving Water Clarification

11.4.1.1 Introduction to Water Clarification

Water is an essential part of virtually all plants in the chemical process industries (CPI) and power generation business. Some examples of its applications include makeup water for a cooling tower, boiler feedwater for steam production, and firewater.

As a first step, raw water must be treated or clarified to remove suspended/dissolved solids, bacteria, and other forms of impurities to help prevent system scale, corrosion, and fouling. This can be achieved by clarification processes, followed by various forms of media filtration.

A simplified flow diagram of a typical raw water treatment plant is illustrated here as an example in Figure 11.13.

Clarification takes place in three sequential steps as follows:

1) *Coagulation*: Coagulants, e.g., inorganic salts of aluminum or iron, are usually added to accomplish coagulation. This step prefers high shear and rapid mixing. Coagulants are generally injected into the feed line upstream of a static mixer to induce rapid mixing.
2) *Flocculation*: Flocculants, i.e., high-molecular-weight, water-soluble organic polymers, are typically added to enhance agglomeration of small particles into larger particles. It ideally takes place in an environment with low shear, high retention time, and moderate mixing. Typically, flocculants are directly injected into a mixing zone of a clarifier.
3) *Sedimentation*: The physical removal of agglomerated particles from suspension or settling, usually under low water velocity in a clarifier's settling zone.

In addition to coagulants and flocculants, disinfectants (such as sodium hypochlorite) are also applied to suppress microbial growth. Sometimes, pH adjusters, i.e., sodium hydroxide, can also be used for neutralization. In most raw

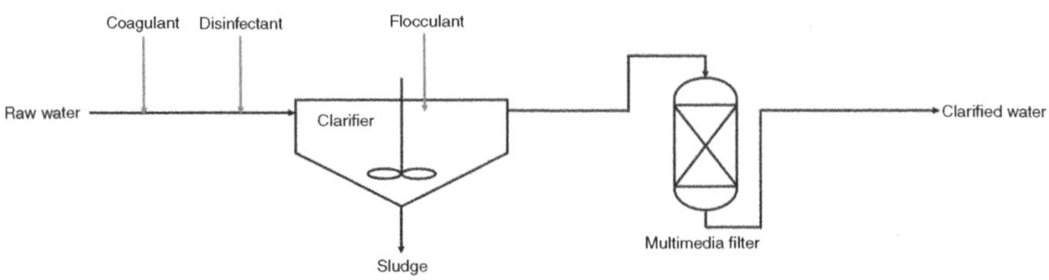

Figure 11.13 A simplified raw water treatment process.

water clarification, at least a portion of the clarified water is filtered by the aid of a sludge blanket in the settling zone. Clarifier effluents of 2–10 NTU may be improved to 0.1–1.0 NTU by conventional sand/media filtration.

In general, raw water clarification is a highly manual process and often requires intensive human monitoring, so it is prone to process upsets. For example, a sludge blanket might disappear from excessive sludge draining during the nightshift where visibility is an issue, causing high turbidity in effluent water. When this happens, a downstream sand filtration system is the last resort to ensure acceptable suspended solid concentrations in the finished water.

This troubleshooting guideline will provide readers with proven field tuning and troubleshooting techniques to ensure reliability and robustness of the sand/media filtration system. While pressurized sand filtration is used as an example here, one can also apply the main ideas to other types of sand filters or similar equipment.

11.4.1.2 Sand Filtration Basics

Pressurized sand filters are typically filled with multiple filter media, e.g., sand or anthracite, to enable the deep-bed filtration mode. The use of different filter media is the main reason why they are also sometimes called multimedia filters. As presented in Figure 11.14, anthracite with a larger diameter is placed on top of smaller diameter sand to trap larger-sized flocs/particulates, followed by gravel as a support. For efficient operation, flow distributors are equipped to aid flow distribution in all operating steps.

Over the operating cycle, the pressure drop across these filter vessels will gradually increase. Once the pressure drop reaches the maximum limit (typically 1.0–1.2 bar for pressurized filters), these filter vessels need to be cleaned (backwashed) to remove trapped particulates so that they can again return into service. Some operators use a turbidity breakthrough as the main criterion, while others just set a fixed-timer to trigger backwash operation, e.g., every 24–36 hours.

Equipment designs are relatively simple, as illustrated in Figure 11.15. They usually employ a set of automatic on/off valves to enable each operating step given as an example in Table 11.5, along with other

Figure 11.14
Mixed-media filter and simplified loading diagram.

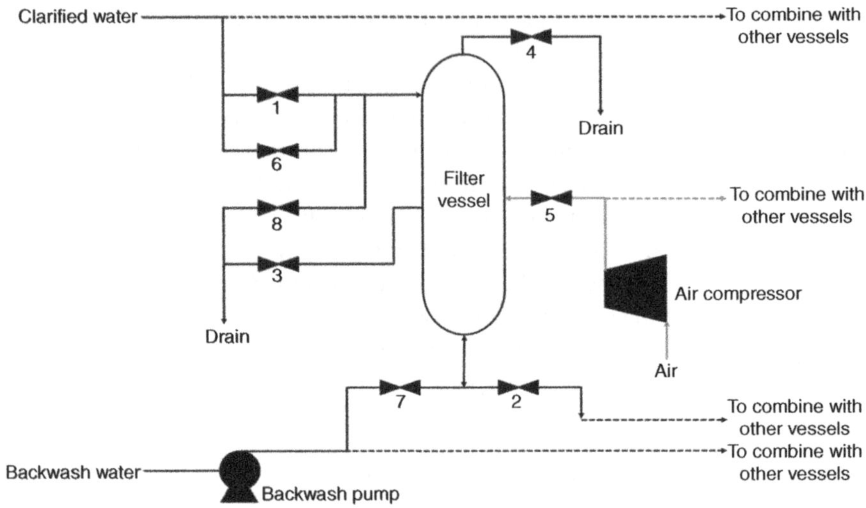

Figure 11.15 A typical sand/media filter setup.

Table 11.5 Typical operating steps for automatic sand/media filters.

Step	Duration	Open	Close	Backwash Pump	Air Injection Compressor
Service	24 hr	1,2	3,4,5,6,7,8	Off	Off
Drain down	20 min	3,4	1,2,5,6,7,8	Off	Off
Air injection	15 min	4,5	1,2,3,6,7,8	Off	On
Refill	15 min	4,6	1,2,3,5,7,8	Off	Off
Backwash	10 min	7,8	1,2,3,4,5,6	On	Off

peripheral equipment, e.g., backwash pump, air injection compressor. Generally, a programmable logic controller or PLC controls these operating sequences, e.g., which valves to be open/close or when the backwash pump shall be started/stopped.

Periodically, these filter media need to be fluidized and expanded to a certain extent by a counter-current backwash flow for cleaning. In general, the degree of bed expansion during any backwash operation should be greater than 10% for efficient cleaning but less than 20% to prevent filter media loss. According to industrial surveys, poor backwashing is the most frequent cause of filter failures.

In addition to the degree of bed expansion, the design backwash rate also depends on the size and density of filter media as well as backwashing techniques but generally falls in the range of 37–61 m/hr (superficial velocity). Various air injection strategies can be applied to aid backwashing, e.g., air first/water second or simultaneous air and water. It is utmost important to use filter media with consistent size and density to ensure the validity of the design backwash rate.

11.4.1.3 Evaluate Backwash Performance

Generally, if backwash water is relatively clean at the start with insignificant turbidity rise throughout the backwash operation, it often means the backwash flow rate is inadequate to clean filter media. Conversely, if there is a significant amount of anthracite (or sand) in backwash water, the backwash flow rate is too high.

Figure 11.16 shows real backwash profiles of two different backwash operations with 1- minute sampling frequency. The trend with a high turbidity peak implies that the backwash operation was highly efficient and completed after 10 minutes.

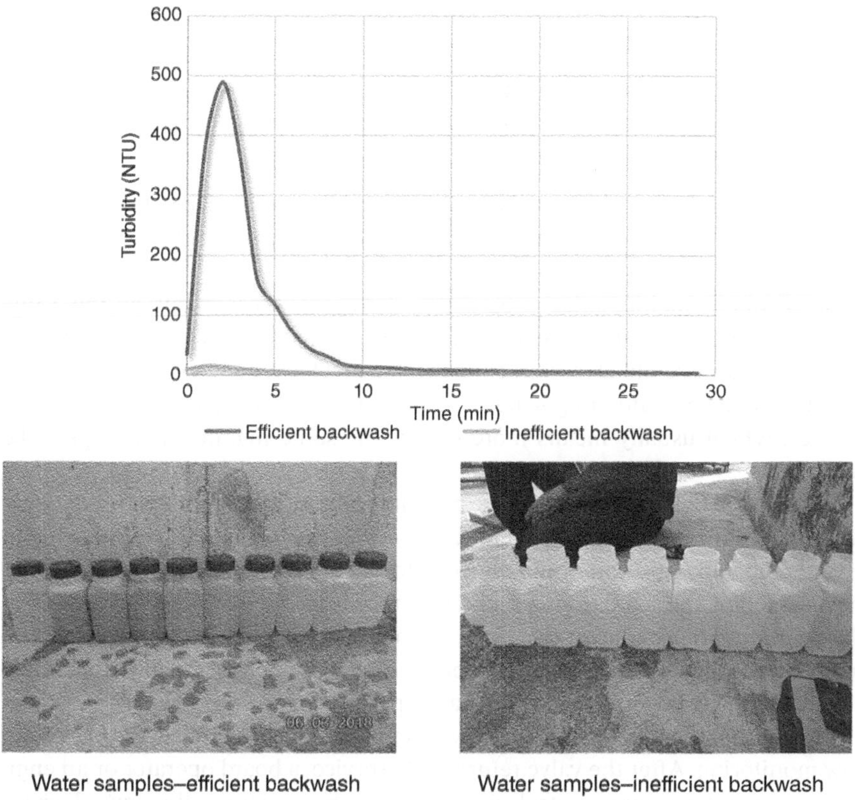

Figure 11.16 Water turbidity profiles during backwash operation.

Additionally, the turbidity changes in a backwash cycle happened abruptly, with a significant peak in the second minute.

On the other hand, the flat turbidity trend implies an inefficient backwash operation as the turbidity profile was relatively flat. Regardless of the backwash duration, this filter vessel would never be thoroughly cleaned.

As a general recommendation, American Water Works Association (AWWA) has recommended terminating backwash when the turbidity is in the range of 10–15 NTU (AWWA Research Foundation n.d.). In addition to water loss, excessively long backwashing is also detrimental to the filtration process, contributing to post-backwash turbidity breakthroughs and requiring an extended ripening period.

11.4.1.4 Operational Issues Related to Inefficient Backwash

Many of these sand/media filtration systems do not have a dedicated flowmeter for backwash flow monitoring, especially old or budgetary units. Most of them rely on the valve opening of these on/off valves. These on/off valves serve two purposes in this type of application:

1) As an on/off valve for sequential control
2) As a fixed-opening control valve for flow regulation by optimizing the opening position—fully or partially open.

The functionality of these on/off valves is crucial to a successful operation. Their rotational movement should be smooth, while the seat tightness should be acceptable.

An adequate maintenance program is a key to success but at the same time could also be a burden. As part of preventive maintenance, these valves are periodically disassembled, repaired, and tested at a local maintenance workshop. Sometimes, an operator might observe malfunctions and request valve maintenance, which usually means more or less the same maintenance steps to be performed.

Based on the field experience, these good intentions often turned into a disaster. Surprisingly, many instrument technicians do not realize how important it is to have the opening position optimally adjusted. More frequent than people think, they just follow the standard maintenance procedure and put these valves back in place without proper adjustment. Quite often, the engineer in charge does not even notice if the valve was removed for repair, as an area operator or shift supervisor is typically the one who gives maintenance permission.

For a well-equipped system, there will probably be a flowmeter for backwash flow monitoring. After the valve returns into service, a board operator or an engineer can monitor from a control room whether the backwash flow rate deviates

Figure 11.17 Plugged filter media.

from the original value. Depending on luck, the backwash flow rate might be the same as before, thus not needing further adjustment. If that is not the case, the field adjustment would be required if there are any significant deviations. More often than people think, there is no flowmeter available for backwash flow monitoring.

Failure to detect these anomalies can lead to performance issues of filtration systems. For example, an inadequate backwash flow results in premature plugging of filter media, and it could be as bad as shown in Figure 11.17. The pressure drop across filter vessels will sharply increase over a short time and result in an unacceptably short cycle length with possible premature turbidity breakthrough. Additionally, too frequent backwash operation resulting from premature plugging often causes water balance issues. When the backwash flow is really low, the operator/engineer might be able to tell by just visually observing on site.

In this figure, the top surface of anthracite was fully covered by carried-over sludge from an upstream clarifier. This phenomenon can also occur when the anthracite layer is inadequate to enable deep bed filtration, e.g., anthracite loss.

A good question here is how to estimate/measure the backwash flow rate, when most sand/media filters in the field are not equipped with a dedicated flowmeter for backwash operation.

One might think about using nonintrusive flow sensors, e.g., ultrasonic flowmeter. That is a good option, provided there is an adequate pipe straight-run.

Unfortunately, these filter vessels and backwash systems are usually located in a cramped space, close to each other. As such, the piping systems typically consist of many bends and relatively short pipe sections, rendering the ultrasonic flowmeter useless. Some piping materials, such as low-density plastics and rubber (or even thick paint), can attenuate signals or possibly create false echoes, causing inaccurate flow measurements.

11.4.1.5 Field Techniques for Backwash Tuning

If there is no flowmeter available on site, one may consider the following recommendations given below:

1) Set a flow rate target (Q), and then determine a corresponding differential head (ΔH) from the pump characteristic curve, as illustrated in Figure 11.18. A good starting point is the design backwash flow rate provided by the design company (Q_{design}).

   This corresponding ΔH can be converted into a differential pressure target across the main pump using the following formula.

$$P_{discharge} - P_{suction} = \rho_{water} g \Delta H \tag{11.13}$$

   For more understanding of how to use field data, consider the system depicted in Figure 11.19. The main pump is in service delivering water for a backwash

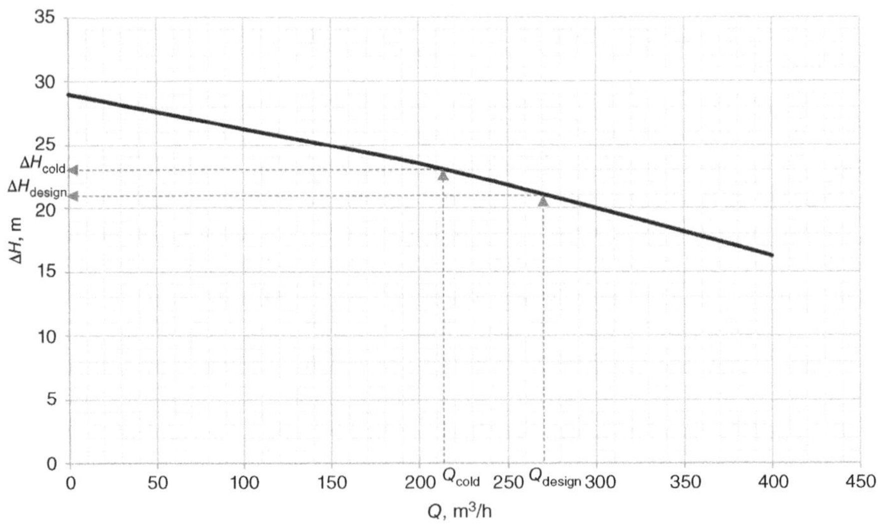

Figure 11.18 Determining ΔH_{design}.

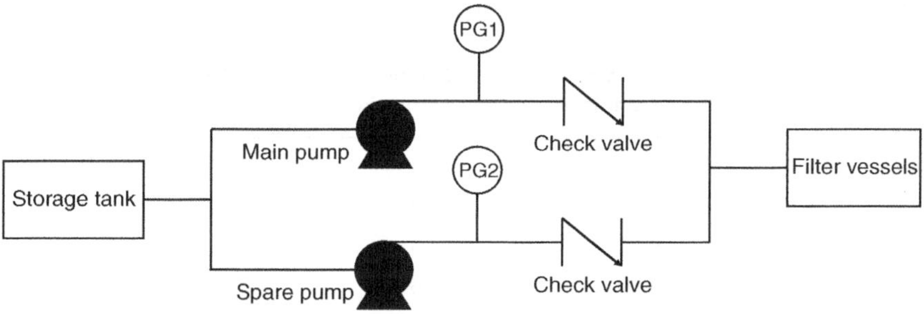

Figure 11.19 A typical backwash system.

operation. In this particular case, $P_{discharge}$ is a pressure reading of PG1, PG2 is an indirect indication of the suction pressure of the main pump.

If there is no spare pump available, just read a pressure gage located at the discharge line to estimate $P_{suction}$ *before commencing* the backwash operation. This is an acceptable assumption and usually good enough for practical purposes—as the pump suction pressure should be largely governed by the tank level and pressure.

2) Try to adjust the opening position of the valve controlling the backwash flow rate (valve no. 7 in Figure 11.15). A good strategy is to start at a low flow rate to prevent filter media loss, e.g., at 50% of the opening, and then gradually increase until $P_{discharge}$-$P_{suction}$ (or PG1-PG2) matches the expected value. In most cases, a design value is good enough to achieve efficient backwashing.
3) If the performance is still unsatisfactory, consider first increasing the backwash flow rate by 10%. In case the backwash performance is still unsatisfactory after increasing the backwash flow rate, the real root cause of inefficient backwash operation might be something else and should be investigated, e.g., an air injection step might be ineffective. Going beyond 10% is not recommended, as it might result in excessive loss of filter media.
4) Water temperature can also affect the degree of bed expansion. The colder the water is, the more filter media will expand at the same backwash flow rate due to higher drag forces, as a result of higher water viscosity. A best practice is using the design backwash flow rate during the warm season and periodically checking for filter media loss during the cold season, i.e., by observing backwash water. These seasonal variations are another reason why adjusting the backwash flow rate beyond 10% of the design value (a typical design factor) is not recommended. Table 11.6 provides correction factors for standard sand and anthracite at different temperatures as a starting point for tuning.

Table 11.6 Correction factors for typical filter media at various temperatures.

Water Temperature, °C	1.0 mm Anthracite	2.0 mm Anthracite	0.5 mm Sand	1.0 mm Sand
30	1.10	1.06	1.13	1.09
25	1.00	1.00	1.00	1.00
20	0.91	0.93	0.90	0.91
15	0.80	0.86	0.79	0.83
10	0.70	0.79	0.69	0.73
5	0.61	0.71	0.61	0.64

Source: Adapted from AWWA Research Foundation (n.d.).

When the water temperature is colder, one can employ equation (11.14) to estimate the required backwash flow rate. For better clarity, this will be demonstrated in the subsequent example.

$$Q_{\text{cold}} = C_{\text{d}} \times Q_{\text{hot}} \tag{11.14}$$

5) To ensure good backwashing over a comprehensive range of water levels in the storage tank, tune the backwash flow rate when the tank is at the minimum controlled level, and then observe filter media losses at the maximum controlled water level (graphically illustrated in Figure 11.20). Ideally, there should be minimal anthracite (or sand) loss over the entire controlled range. In the case of excessive media loss at the maximum water level, the minimum controlled level should be raised and then the backwash flow rate is readjusted, so it will not increase too much at the maximum controlled water level.
6) Failure to reach the design flow rate even at the full valve opening could be a sign of some other issues. Figure 11.21 shows an example where fine dirt/particles almost 100% blocked the bottom distributor.

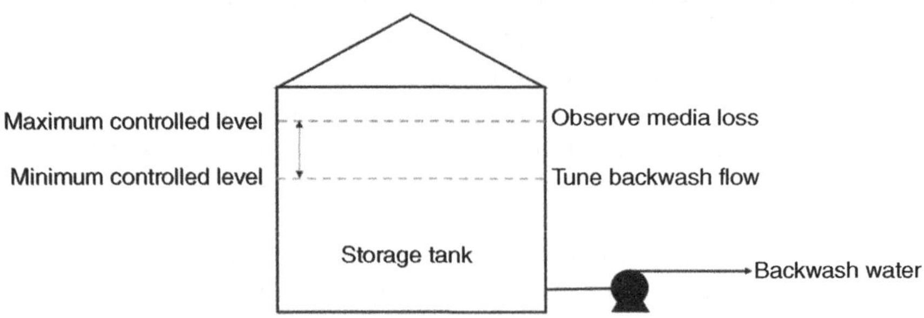

Figure 11.20 A recommended tuning strategy for varying water level.

Figure 11.21 Blocked distributor.

As shown in this figure, the bottom distributor has too narrow water passages, making it susceptible to plugging and extremely difficult to unblock.

11.4.1.6 A Real-World Example

An engineer tried to optimize the backwash flow rate for four identical dual-media filter vessels containing anthracite and sand. Based on a design reference book, the design backwash flow rate (Q_{design}) is 270 m^3/hr at 25 °C. This water treatment plant was suffering from high-pressure drop across multimedia filters. It must be noted that there is no flowmeter for backwash flow monitoring.

1) **Field backwash tuning** From Figure 11.18, the corresponding differential head (ΔH) for 270 m^3/hr of backwash flow is 21.0 m. As it was a warm day, the water temperature was approximately 25 °C with a ρ_{water} of 997 kg/m^3.
   By applying equation (11.14), the differential pressure target can be determined as follows:

$$P_{discharge} - P_{suction} = 997\frac{kg}{m^3} \times 9.81\frac{m}{s^2} \times 21.0\ m = 205{,}392\ Pa\ or\ 2.05\ bar$$

   This calculated differential pressure will be a target for the field tuning.
   Before commencing the backwash operation, the opening position of the backwash inlet valve (valve no. 7 in Figure 11.15) was pre-adjusted at approximately 50% as a preventive measure to prevent excessive filter media loss.
   Based on the initial field pressure data, $P_{discharge} - P_{suction}$ was 2.50 bar (approximately 150 m^3/hr of backwash flow), while the target is 2.05 bar. Furthermore, a visual check of backwash water also confirmed that the backwash flow rate was too low to expand the filter media as the water appearance was quite clear.
   The opening position of the backwash inlet valve was then adjusted in such a way that the differential pressure became approximately 2.05 bar. Once the differential pressure approached 2.05 bar, a significant amount of sludge and silt had been observed in the backwash water, with a negligible amount of

anthracite. A dirty appearance of backwash water implied an efficient backwash operation (also see Figure 11.16). It was also proof that the differential pressure target of 2.05 bar worked well, although a full backwash performance test would still be required.

2) **Effects of water level** On the day the backwash flow rate was adjusted, the water level in the storage tank was intentionally set at 85%, the minimum controlled level.

   To ensure the system's robustness, the backwash operation when the water level was at 95%, the maximum controlled level, was also monitored.

   As per the field monitoring, minimal anthracite in backwash water was observed with the estimated backwash flow rate of 280 m^3/hr (or 4% above design) using the same technique described earlier.

3) **Effects of water temperature** At 85% water level and 25 °C, the optimal backwash flow rate during the warm day (Q_{warm}) is 270 m^3/hr. This system utilizes 1.0 mm anthracite and 0.5 mm sand.

   Historically, the temperature could drop significantly to 15 °C (ρ_{water} is 999 kg/m^3), thus requiring a lower backwash flow rate (denoted by Q_{cold} in Figure 11.18). From Table 11.6, the derating factor (C_d) for 1.0 mm anthracite is 0.80/1.00 (a ratio between correction factors at 15 and 25 °C) or 0.80. In the same fashion, C_d for 0.5 mm sand is 0.79/1.00 or 0.79.

   As a conservative approach, use the lower C_d for correction to minimize the risks of filter media loss. From equation (11.14), Q_{cold} can then be calculated as follows:

$$Q_{cold} = 0.79 \times 270\ \text{m}^3/\text{hr} = 213.3\ \text{m}^3/\text{hr}$$

   From the pump characteristic curve in Figure 11.18, the corresponding ΔH for cold water (denoted by ΔH_{cold} in Figure 11.18) is 23.0 m water.

   Thus,

$$P_{discharge} - P_{suction} = 999\frac{\text{kg}}{\text{m}^3} \times 9.81\frac{\text{m}}{\text{s}^2} \times 23.0\ \text{m} = 225{,}404.4\ \text{Pa or } 2.25\ \text{bar}$$

   When the water temperature is 15 °C, one can use 2.25 bar pump differential pressure as a first guess; then observe the actual backwash performance and adjust as needed.

11.4.1.7 Other Tips

- Install a dedicated backwash flowmeter for the new unit or retrofitting the existing plant for early detection of any anomalies. Although it still requires a field check to ensure problem-free backwash operation, the proposed technique is nonetheless useful to prove the accuracy of the flowmeter.
- Create a transparent logging system to ensure everyone involved in process monitoring can track the status of valve maintenance, e.g., the unit engineer

can create a dedicated section in the shift report for any critical valves currently under maintenance. Also, communicate with all instrument technicians in charge to ensure they fully understand the effects of opening valve position on filter operation.

- When sludge carryover from the upstream clarifier is frequent, a preventive measure could be setting the backwash flow rate at 10% above design value to handle potential clarifier's upsets. One may consider installing a real-time sludge level detector for better sludge level monitoring.
- Seasonally or quarterly conduct backwash performance evaluation. The optimal backwash flow rate for each season should be appropriately documented for future references.
- Be careful when extending the time interval between each backwash, e.g., from 24 to 36 hours, as trapped flocs/particulates will become more difficult to remove. A backwash flow rate or duration should be accordingly adjusted to ensure efficient backwashing.
- Create a task for a field operator in charge to monitor the backwash operation, at least once a month for every filter.

11.5 Trust Creating a Disaster—Heat Exchanger Corrosion from Improper Cooling Water System Operation

11.5.1 Background

A major chemical company hired an Engineering Procurement and Construction (EPC) contractor on a lump sum basis to design and build a new olefin complex. The scope also included commissioning and start-up of the whole complex, including commissioning of the cooling water system.

It was observed during the pre-commissioning phase that the majority of water-cooled heat exchangers experienced corrosion issues. The first indication was the high iron content and turbidity in the cooling water during the pre-commissioning of the cooling water system, as illustrated in Figure 11.22.

After continuous observation of high turbidity and a high total iron content, the plant owner (a major chemical company) requested the EPC contractor to isolate and remove some heat exchangers for inspection, as they suspected corrosion. It is well known in the industry that high turbidity and total iron are signs of corrosion products.

Some of the inspection photos are illustrated in Figures 11.23 and 11.24. Clearly, all removed heat exchangers (six in total) experienced severe fouling and moderate under deposit corrosion regarding shown inspection results. As a

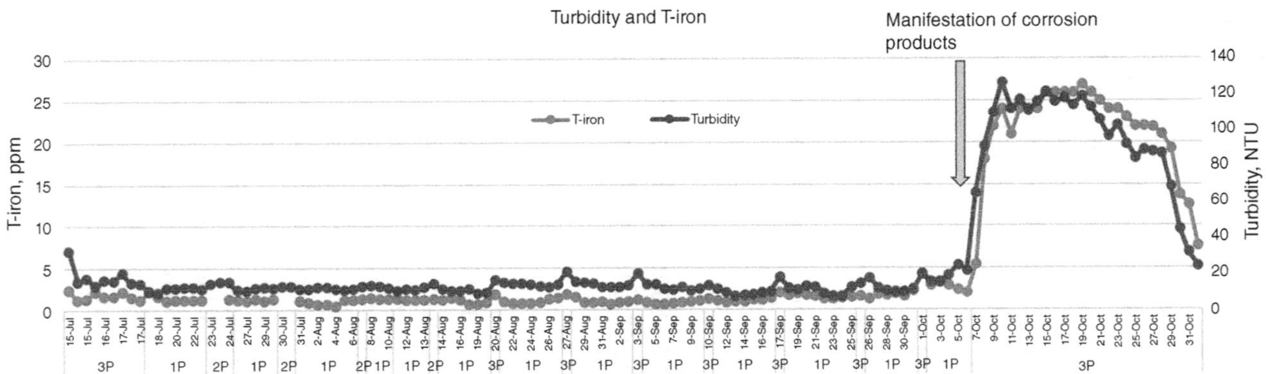

Figure 11.22 Turbidity and total iron (T-iron) during the pre-commissioning of the cooling water system.

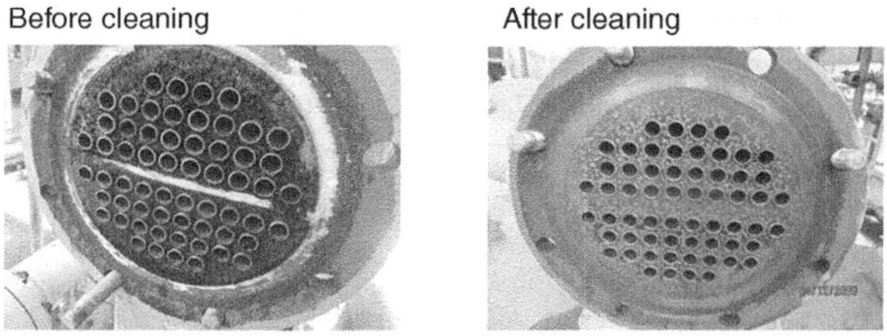

Figure 11.23 Fouling and under deposit corrosion of the heat exchanger's tube sheet.

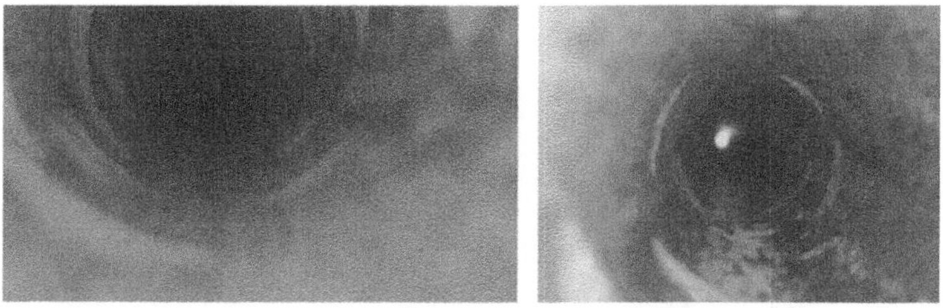

Figure 11.24 Under deposit corrosion in heat exchanger tubes.

further confirmation, scale samples were analyzed, revealing that the collected scale samples constitute of more than 55 wt% iron. This scale analysis along with pitting corrosions observed emphasized the occurrence of under deposit corrosion. A detailed mechanism of under deposit corrosion will not be elaborated here and can be found elsewhere (Flynn 1979; NALCO 2004, 2015).

Consequently, the plant owner decided to temporarily stop the pre-commissioning and clean all heat exchangers with carbon steel tubes. It must be noted that all mild steel corrosion coupons being removed just prior to the incident also experienced under deposit corrosion.

This incident resulted in a 2-month start-up delay and other costs associated with reduced heat exchanger tube thickness.

11.5.2 Penny Wise, Pound Foolish–Misconception of Cooling Water System Operation

By design philosophy, this cooling water system has four running pumps with one spare (two steam-driven, three motor-driven). Upon completion of the passivation program (a general practice to create the protective film on heat exchanger tubes by mean of chemical treatments), the EPC contractor, proposed to run only one out of four available cooling water pumps on a regular basis with normal cooling water

Table 11.7 A typical control parameter for the cooling water system.

Parameter	Unit	Target	KPI	Control	Monitor
pH		7.8–8.2		x	
Conductivity	uS/cm	<1500		x	
Free residual chlorine	ppm	0.3–0.5		x	
Turbidity	NTU	<20			x
Chloride	ppm	<400			x
Soluble iron	ppm				x
Total iron	ppm	<3		x	
Soluble phosphate	ppm	8–10		x	
Total phosphate	ppm				x
Delta PO_4	%	<30%	x		
Soluble zinc	ppm	>0.5		x	
Total zinc	ppm				x
Dispersant	ppm	20–25			x
Total hardness	ppm	<300			x
Ca hardness	ppm	<200			x
M alkalinity	ppm				x
Silica	ppm	<150			x
Suspended solid	ppm				x
Sulfate	ppm				x
Bacteria plate count	colonies/ml	<100,000	x		
Sulfate reduction bacteria	col/ml	0	x		
Cooling cycle		>4			x
Corrosion rate from coupon					x
Mild steel	mpy	<3			x
Copper	mpy	<0.3			x
Stainless steel	mpy	<0.5	x		

treatment program described in Table 11.7. On a weekly basis, they proposed to run two pumps during the daytime. The EPC contractor believed that any fouling which might have occurred during the periods with one pump operation would be periodically flushed away by intermittently running two pumps. It must be noted that there was no steam available during the cooling water system commissioning, so only three motor-driven pumps were available.

Table 11.8 Minimum tube velocities for cooling water used by different companies.

Description	European Oil and Gas Operator	North American Oil and Gas Operator 1	North American Oil and Gas Operator 2	Major Technology Licensing Company	Major EPC Company
Minimum tube velocity	1.0 m/s	1.1 m/s (3.5 ft/s)	0.9 m/s (3 ft/s)	0.9 m/s (3 ft/s)	0.9 m/s (3 ft/s)

Prior to the onset of the high iron incident, the cooling water system was subjected to the periods with only one pump in operation for more than 80% of the time prior to the observation of continuously high-iron content in the cooling water (also see Figure 11.22).

It is a generally accepted practice that tube velocities of cooling water shall be maintained above a certain minimum value, as too low velocities can result in accelerated fouling (which usually leads to under deposit corrosion), as a result of a reduction in turbulence. Minimum tube velocities of cooling water adopted by different organizations are reviewed and summarized here in Table 11.8.

From Table 11.8, every major organization agrees that the minimum tube velocity for cooling water should lie somewhere between 0.9 and 1.1 m/s. A mean value of 1.0 m/s is taken as a criterion to decide whether the actual tube velocity is acceptable.

As an explanation why it is essential to adhere with this best practice, the relationship between the corrosion rate and flow velocity is illustrated in Figure 11.25.

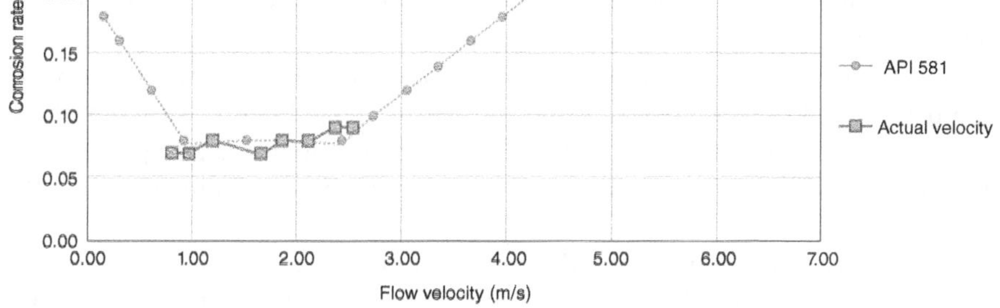

Figure 11.25 Corrosion rate versus flow velocity Mokhtar et al. (2018).

From Figure 11.25, the flat "bottom of the horseshoe" is the 1.0–2.0 m/s velocity range, which is ideal for minimal corrosion rates. Also note that as the flow increases above 2.0 m/s, corrosion increases due to erosion corrosion. More than that, NALCO (2004) also suggests that anything below 0.3 m/s is considered stagnant environment and should be avoided. Examples of negative effects that relate to less than 1.0 m/s tube velocity include the following

- Biocides, scale, and corrosion inhibitors will not be replenished as rapidly at the metal surface when low flow exists, e.g., thicker diffusion film. Control of microbial growth, scale dispersion, and corrosion protection are reduced under low flow conditions, and the corrosion and fouling rates become extremely high.
- Low flow or stagnant conditions allow the suspended material to settle onto the bottom of tubing and bottom surfaces of other components. This promotes under-deposit corrosion.
- At low flow, iron oxide deposits are generally loose and nonprotective.
- Tubercle formation on iron is often more severe in low flow situations. This is a form of under-deposit corrosion.

Despite conflicting with the best practice, the EPC contractor wanted to minimize the electricity consumption from running three pumps continuously and convinced the plant owner to follow this false recommendation to run only one pump for the majority of pre-commissioning. It must also be noted here that the plant owner initially objected to the operating strategy proposed by the EPC contractor but at the end allowed it to be commenced since any liabilities resulted from this decision will fall under the EPC contractor.

In this project, the design tube velocity ranges between 1.0 and 2.0 m/s with the assumption that all four cooling water pumps are in service.

With only one pump in operation, one can presume the tube velocity to be approximately 1/4 of the design value. Considering the fact that most heat exchangers were designed with 1.0–2.0 m/s tube velocity, the tube velocity of most heat exchangers was likely to be much less than a recommended value, 1.0 m/s, for the majority of the pre-commissioning period. This hypothesis can be confirmed by the actual tube velocity measured by the plant owner that the tube velocity was mostly below 0.3 m/s.

Even when two pumps were in service, the tube velocity would only be approximately 50% of the design value and hardly above 1.0 m/s. Although there is no actual velocity data being reported here, intermittent two pump operation was not likely to significantly alleviate fouling, as previously claimed by the EPC contractor.

Another issue accompanied with lower-than-design cooling water velocities is water quality control. The water quality control parameters in Table 11.7 are intended for normal operation where tube velocities are above the minimum recommended value, e.g., 1.0 m/s. When tube velocities are below the minimum safe value, it is very difficult to determine the optimal range of each quality control parameter.

Initially, only two pumps were online during the intermittent flushing (1 day per week, during the day time). This means the tube velocity of most heat exchangers was well below 1.0 m/s. Therefore, minimal amount of fouling was likely to be removed by the intermittent flushing, as also evidenced by the fact that the turbidity of cooling water was relatively constant over that particular period of time, e.g., no turbidity spike when the 2nd pump was in service, refer to Figure 11.22. Another explanation for insignificant turbidity and T-iron spikes during intermittent flushing periods might be attributed to the fact that the operation was still at its initial stage. As a result, less fouling and corrosion products were removed from heat exchanger surfaces during the intermittent flushing.

Later on, the EPC contractor decided to change their operating approach and used three pumps every time the intermittent flushing was performed. This change in operation was pushed by the plant owner concerned that their heat exchangers might have experienced fouling by inadequate cooling water circulation.

As expected, almost every time three pumps were put into service, significant turbidity and T-iron spikes were accordingly observed, as clearly highlighted by the arrow in Figure 11.26. Interestingly, the magnitude of these spikes increased over time. This implies that fouling and corrosion had become more severe as the pre-commissioning progressed, e.g., more fouling and corrosion products were removed during the intermittent flushing.

The level of turbidity and T-iron continuously increased to the point they brought serious concerns to the plant owner. At the end, the plant owner requested the EPC contractor to continuously operate all three available cooling water pumps with the hope to flush away any fouling or corrosion products and continue the pre-commissioning without interruptions.

It must also be noted that the EPC contractor also recommended to temporarily increase the level of active polymer (indication of fouling dispersant) from 6 to 10 ppm to above 10 ppm with the expectation to minimize re-deposition of suspended solid and T-iron during the continuous flushing period. Unfortunately, these deposits would not be 100% removed by just increasing the water circulation rate.

Once fouling initiates, it is nearly impossible to completely remove it by just increasing the tube velocity, e.g., by running multiple pumps. Only newly formed layers will be removed by this procedure, providing the tube velocity is actually above 1.0 m/s during the flushing period. The remaining deposits will stay on heat exchanger surfaces, grow thicker, and harden as time progresses.

The ability of high-water velocities to minimize fouling also depends on the nature of the foulant. Clay and silt deposits are more effectively removed by high water velocities than aluminum and iron deposits, which are stickier and form interlocking networks with other precipitates.

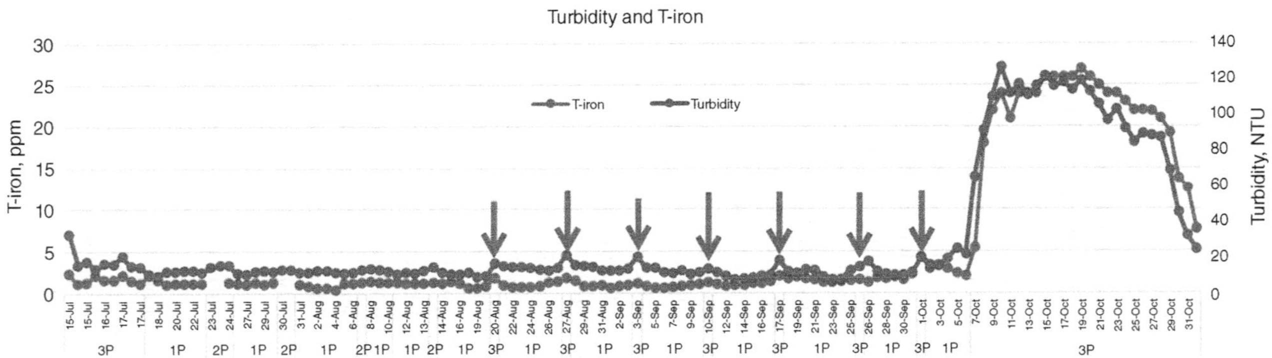

Figure 11.26 Spikes in turbidity and T-iron when 3 pumps were in service.

The phenomenon explained above is actually the main reason why mechanical cleanings are required during heat exchanger maintenances, as these hardened fouling layers could not simply be removed by shear stress induced by cooling water circulation.

The fact that some corrosion products had already been observed, as indicated by a very high level of T-iron and turbidity, implies that the majority of passivation films might have already been damaged. Simply re-passivating these heat exchangers would fail to protect heat exchangers from further corrosion, as these remaining deposits would block treating chemicals and other essential elements to access heat exchanger surfaces, to form a new passivation film.

Under deposit corrosion would still continue, with even faster rate if these heat exchangers are exposed to heat load without being cleaned and re-passivated.

11.5.3 Take Ownership

Deep down, the plant owner was aware that the EPC contractor conflicted with the industrial best practices. However, the contract seemed to prevent them from overriding the decision of the EPC contractor. In principle, the plant was still under the EPC contractor's account until the completion of commissioning. Contractually, the plant owner could only comment/review but had no legal right to intervene during the commissioning.

While this dilemma is understandable, the plant owner probably should have been more aggressive at an earlier point as abnormal operating techniques were observed. At the end, the EPC contractor agreed to pay penalties as a result of a 2-month delay. Nonetheless, the compensation is nothing compared to revenue losses, especially with this large operational scale.

This example emphasizes the importance of ownership when dealing with real-world chemical engineering problems. While good engineering fundamentals/judgments are essential to successful plant design and operation, there is always a risk of relying on contractors/vendors, which may not be always competent.

Nomenclature

%Travel	valve travel
$\%\text{Travel}_{\text{base}}$	valve travel for the base case
$\%\text{Travel}_{\text{revamp}}$	valve travel for the revamp case
A	total surface area of particle including pore area
A and B	Typical Arrhenius constants determined from laboratory, plant, or literature data

Cd	derating factor, a ratio between two corresponding correction factors of two different water temperatures for the same size and type of filter media
Cp	average specific heat of particle
C_v	flow coefficient
$C_{v,base}$	flow coefficient for the base case
$C_{v,bypass}$	flow coefficient for the bypass valve
$C_{v,main}$	flow coefficient for the main control valve
$C_{v,revamp}$	flow coefficient for the revamp case
$C_{v,total}$	total flow coefficient across the main control and bypass valve
$C_{v,total,revamp}$	total flow coefficient for the revamp case
CE	catalyst efficiency lb polymer/lb. of catalyst
d	particle diameter, microns
$dT/d\theta$	rate of temperature change
D	nominal pipe diameter
g	gravitational acceleration
Hr	heat of polymerization
M	monomer concentration, lb/ft^3
$NPSH_a$	available net positive suction head
$P_{discharge}$	a pressure reading of the pressure gage located at the discharge of the main pump
$P_{suction}$	a pressure reading of the pressure gage located at the discharge of the spare pump
Q	volumetric flow rate of backwash water
Q_{bypass}	actual volumetric flow rate across the bypass valve
Q_{cold}	an estimated backwash flow rate for colder water
Q_{design}	design volumetric flow rate of backwash water
Q_{hot}	a tuned backwash flow rate based on warmer water
Q_{inlet}	actual inlet volumetric flow rate
$Q_{inlet,base}$	actual inlet volumetric flow rate for the base case
$Q_{inlet,revamp}$	actual inlet volumetric flow rate for the revamp case
Q_{main}	actual volumetric flow rate across the main control valve
Q_{revamp}	actual volumetric flow rate for the revamp case
Q_{total}	total actual volumetric flow rate across the main control and bypass valve
R	rate of polymerization
R	gas constant, 1.987
S.G.	specific gravity at the corresponding operating temperature
T	absolute temperature, °Rankine
Tp	temperature of particle
Tr	temperature of reactor liquid

U	heat transfer coefficient
U	heat transfer coefficient BTU/hr°F-ft^2
W	weight of particle (catalyst and polymer)
ΔH	differential head of backwash pump
ΔH_{base}	pump's head for the base case
ΔH_{margin}	design margin for the pump's head
$\Delta H_{\text{polytropic}}$	compressor's polytropic head
$\Delta H_{\text{polytropic,base}}$	compressor's polytropic head for the base case
$\Delta H_{\text{polytropic,revamp}}$	compressor's polytropic head for the revamp case
ΔH_{revamp}	pump's head for the revamp case
$\Delta P/100\text{ m}$	pressure drop per 100 m
ΔP_{base}	pressure drop across the control valve for the base case
ΔP_{bypass}	pressure drop across the bypass valve
ΔP_{main}	pressure drop across the main control valve
$\Delta P_{\text{new equipment}}$	pressure drop across the new equipment
ΔP_{revamp}	pressure drop across the control valve for the revamp case
ΔP_{total}	total pressure drop across the main control and bypass valve
ρ_{water}	water density
θ	time of reaction, hours
μ	slurry viscosity, centipoise

12

Applications of Novel Process Engineering Fundamentals to Plant Problem Solving

12.1 Introduction

This chapter deals with applying novel approaches to plant problem-solving. To solve plant problems effectively, we need to be flexible and creative to apply different approaches to the same problem. There were so many times where the author could solve difficult chronic problems of a new never-experienced process after only a few weeks or a few months of learning about the problem, while experts or plant people who had been directly involved in that plant operations for so many years and were supposedly very familiar with the plant could not solve that chronic problem for years.

What are the keys to the author's successful plant problem-solving? The answer is the author's flexibility and creativity to apply additional principles and theories outside his own familiar field or outside the typical range of college chemical and mechanical engineering coverage. Those principles and theories may often come from high school science subjects such as physics, chemistry, and biology. This chapter will totally change your perspectives on plant problem-solving.

12.2 Novel Approaches to Plant Problems

There are times when plant problems may involve not only operation but also maintenance and other miscellaneous plant issues. The problem-solver needs to find another solution to the same problem that is both technically and economically superior.

These novel approaches to plant problems can be identified in two main categories:

1) Application of additional principles and theories outside your own familiar field

Problem Solving Approaches for Maintaining Operational Excellence in Process Plants, Second Edition.
Joseph M. Bonem, Nattapong Pongboot, and Wiroon Tanthapanichakoon.
© 2025 the American Institute of Chemical Engineers, Inc. Published 2025 by John Wiley & Sons, Inc.

2) Holistic view to not only consider the problematic equipment or process but also surrounding components making up that equipment or process such as equipment accessories, piping, orientation, elevation, environment, and heat loss/gain.

Even though we studied much useful engineering knowledge in college, we must not forget the fact that many principles and theories we studied in schools can also be applied to plant problem-solving. Common science subjects in high school like physics, chemistry, biology, and simple mathematics can be applied to real-life plant problem-solving. This chapter will show you real-life examples where fundamental knowledge that we have learned from our high school days can be used to successfully solve several problems in the plant even without deep college chemical or mechanical engineering knowledge.

Another important tip to double or triple the success rate of problem-solving is to not jump to the conclusion that the equipment itself is always the root cause of the problem. For example, a shell-and-tube heat exchanger using cooling water to condense a process vapor may be malfunctioning or under-performing just because the cooling water inlet piping nozzle is wrongly placed on the upper side instead of correctly on the lower side of the channel cover on the tube side. Placing the cooling water inlet nozzle on the lower side will allow the water to be completely and uniformly filled inside the heat exchanger from bottom to top before leaving the exchanger from the upper side of the channel cover.

Another good example is when a piece of equipment needing sufficient heat removal to avoid heat accumulation can have a longer trouble-free life by placing inside a building with good ventilation (e.g., high air change rate) and good layout design to prevent putting equipment in dead zones where air is stagnant and flowing too poorly to take away heat from the equipment sufficiently.

12.3 Mostly used Engineering Fundamentals to Solve Plant Problems

1) Fluid flow and hydraulics
2) Mass and energy balances
3) Physics and forces
4) Chemistry and chemical reactions
5) Biology and biological conversions
6) Other engineering fundamentals
7) Ventilation design fundamentals

12.4 Application of New Engineering Fundamentals to Plant Maintenance Problems (Example Problem 12.1)

There are times when plant problems may involve not only operation but also maintenance issues. The problem-solver needs to find another solution to the same problem that is both technically and economically superior.

Maintenance problems may not be so obvious when there is already a potential solution on the table. However, the question that the problem-solver needs to ask is not "Can we solve the problem by using this method?", but should rather ask "Is there no better solution to solve the same problem?" The problem is whether we should tackle the same problem from another perspective or should we think more out-of-the-box or not.

This is based on a true story without disclosing a specific name and location. A crude oil refinery in Asia has several storage tanks in the same tank dike. However, due to changes in demand for each product, the designed volume of each tank is no longer suitable for the demand, causing frequent operating problems to pump in and pump out the products for truck or marine loading.

There are four tanks (A, B, C, and D) in the same dike, and they want to switch the location of tank A with tank C and tank B with tank D.

Let us assume that an engineer wants to relocate a storage tank with a diameter of 36.6 m by floating the tank in water. The tank total weight is 460 tons. How can we tackle this problem and find an easier and cheaper solution than using lifting cranes? The tank farm consists of a dike with bund walls on all four sides of the dike (Figure 12.1).

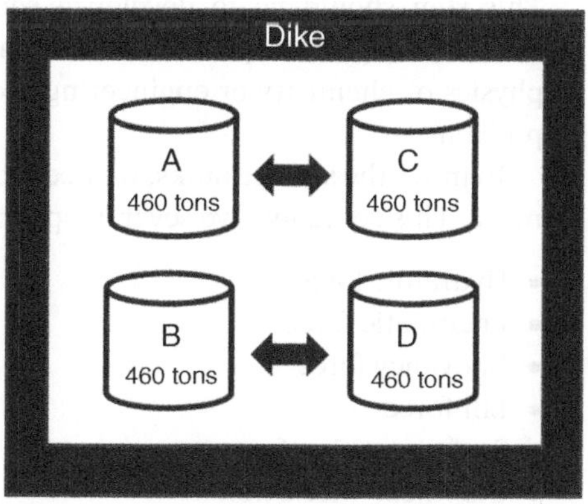

Figure 12.1 Storage tanks.

To relocate tanks in the same dike, it may be logical to think that using multiple lifting cranes should solve the problem already, and that problem-solving is not needed. However, we should always enter step 1 to find if there is a better solution or not.

Step 3 will not be to find the root cause of the problem, but rather the root cause of the desired solution, which is to cause the tanks to move in the easiest and cheapest way. Step 4 will involve calculations to prove the validity of the proposed solution hypothesis. Step 5 will involve recommending remedial actions and further steps to ensure that the same type of problem shall be handled properly in the future.

12.5 Application of the Disciplined Problem-Solving Approach

Step 1: *Verify* that the Problem Actually Occurred

In this case, if the problem-solver believes that using lifting cranes are already the best solution, then the problem does not exist. However, the problem-solver thinks that using lifting cranes is difficult and expensive.

Step 2: *Write* Out an *Accurate* Statement of What Problem You Are Trying to *Solve*

In this example, the problem should be stated as follows.

"Tanks in the same dike need to be relocated between them. Determine the easier and cheaper way than using lifting cranes to solve this problem."

Step 3: Develop a *Theoretically Sound* Working Hypothesis that Explains the Problem

This step should be to develop a sound working hypothesis to move the tanks. The problem-solver should have broad engineering knowledge on physics or chemistry or engineering theories that may be applied to the same problem.

To move the storage tanks, we need to use a kind of force to cause the tank to move. This could involve several types of forces like

- Hydraulic force
- Pneumatic force
- Buoyancy force
- Lift force
- Surface tension force
- Chemical interaction
- Reactive force, etc.

Table 12.1 Questions/comments for Problem 12.1.

Potential Types of Forces	Validation of the Proposed Force to Use
Hydraulic force	No idea was found to move the tank with hydraulic force directly, except using indirectly through equipment like lifting crane
Pneumatic force	No idea was found to move the tank with pneumatic force directly, except using indirectly through equipment like lifting crane
Buoyancy force	Possible if a certain level of water within the height limit of the dike can create sufficient buoyancy force
Lift force	Cannot create sufficient lift force by moving air at high velocity above the tank
Surface tension force	No applicable idea was found. Too small force available
Chemical interaction	No applicable idea was found
Reactive force	No applicable idea was found

Since the storage tanks are quite large and heavy, the force to use should be large enough to move a gigantic piece of equipment while not causing another new problem. Therefore, we narrow down our options to hydraulic force or pneumatic force or buoyancy force (Table 12.1).

From the table above, buoyancy force is a potential candidate because an emptied storage tank can act like a large steel ship, where fluid can create the buoyancy force due to its head and density. The force is equal to the total displaced volume.

Storage tanks are normally installed in tank farms with bund wall to positively contain the liquid in case of spill and tank leakage. The key question is "How high is the water level needed to cause the storage tank to start floating?"

Step 4: Provide a Mechanism to Test the Hypothesis

While testing a hypothesis often involves experimental work, using theoretically correct engineering calculations can also test the hypotheses. An engineer wants to relocate a storage tank with a diameter of 36.6 m by floating the tank in water. Tank total weight is 460 tons, and calculate the minimum water level to sufficiently float the tank so that it can be moved to a new location.

$$\text{Buoyancy force} = \text{weight of fluid displaced by the solid} \quad (12.1)$$

$$\text{Displaced volume} = (\pi D^2 h/4) = 460\ \text{m}^3 \quad (12.2)$$

For water 1 ton = 1 m^3

$$h = (4 \times 460)/\pi D^2 = (4 \times 460)/\pi (36.6)^2 = 0.437\ \text{m}$$

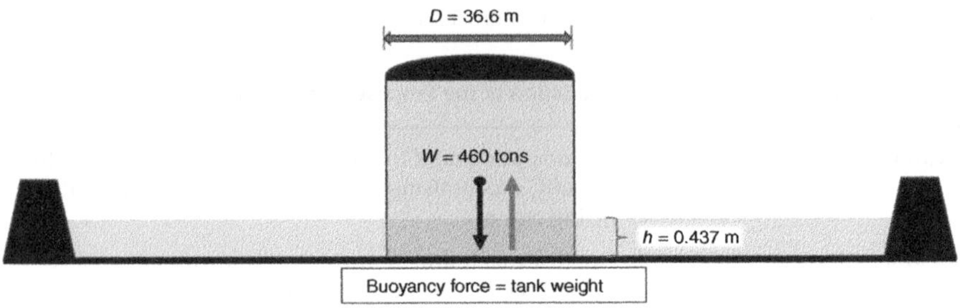

Figure 12.2 Tank relocation.

This height is within a typical bund wall height of 1–2 m, which was determined by the concept of holding the stored liquid volume of the single largest storage tank in the dike. The dike normally does not need to cover all tanks leaking simultaneously (a too remote scenario), but only a single tank leak at a time (a reasonably practical scenario).

Step 5: Recommend Remedial Action to Eliminate the Problem without Creating Another Problem

The required remedial actions developed by the problem-solver consisted of removing the tank from its foundation and filling water into the dike according to what the calculation suggested. Then, use a boat to pull the tank to the new position (Figure 12.2).

12.6 Lessons Learned

Frequently in situations like this, the problem-solver may tend to jump to the most obvious solution of using lifting cranes. The question of what should be done to improve the solution and further reduce costs might not have been considered. In this particular case, as the problem-solver investigated the problem in detail, he uncovered a new technique to solve this maintenance issue.

12.7 Tank Roof Raising for Maintenance Example

This is based on another true event. A process plant in the Middle East has several floating roof storage tanks in its facility. The maintenance crew had some activities to do inside the floating roof tank, but the roof needed to be raised to the top first.

12.8 Application of the Disciplined Problem-Solving Approach

Step 1: *Verify* That the Problem Actually Occurred

In this case, if the problem-solver believes that using lifting cranes is already the best solution, then the problem does not exist. However, the problem-solver thinks that using lifting cranes is difficult and expensive to raise the roof to the top.

Step 2: *Write* Out an *Accurate* Statement of What Problem You Are Trying to *Solve*

In this example, the problem statement should be as follows.

"A floating roof tank needs raising of its roof to the top. Determine an easier and cheaper way than using lifting cranes to solve this problem."

Step 3: Develop a *Theoretically Sound* Working Hypothesis That Explains the Problem

This step should be to develop a sound working hypothesis to raise the tank roof. The problem-solver should have broad engineering knowledge on physics or chemistry or engineering theories that may be applied to the same problem.

To raise the tank roof, we need to use a kind of force to achieve that. This could involve several types of forces like

- Hydraulic force
- Pneumatic force
- Buoyancy force
- Lift force
- Surface tension force
- Chemical interaction
- Reactive force, etc.

Since the storage tank roofs are quite large and heavy, the force to use should be large enough to move a gigantic piece of equipment while not causing another new problem. Therefore, we narrow down our options to hydraulic force or pneumatic force or buoyancy force (Table 12.2).

Possible solutions are narrowed down to only (1) buoyancy force; (2) lift force. However, due to the inconvenience and constraints of filling water into the tank to use the buoyancy force, the problem-solver opted to use the lift force as the first candidate.

Step 4: Provide a Mechanism to Test the Hypothesis

While testing a hypothesis often involves experimental work, using theoretically correct engineering calculations can also test the hypotheses.

Table 12.2 Questions/comments for Problem 12.2.

Potential Types of Forces	Validation of the Proposed Force to Use
Hydraulic force	No idea was found to move the tank with hydraulic force directly except using indirectly through equipment like lifting crane
Pneumatic force	No idea was found to move the tank with pneumatic force directly except using indirectly through equipment like a lifting crane
Buoyancy force	Possible to fill water to cause the tank roof to rise, but the problem is the difficulty of draining water and drying the tank internals after raising the roof
Lift force	Since the tank roof has a very large area, a small magnitude of pressure inside the tank can create a large lifting force on the roof by (Force) = (Pressure) (Area)
Surface tension force	No applicable idea was found. Too small force available
Chemical interaction	No applicable idea was found
Reactive force	No applicable idea was found

It is easier to consider this type of calculation by using the pressure unit in force per area such as kg/cm^2. Assume a tank roof diameter of 26 m having an area of 531 m^2 (5,310,000 cm^2) weighing 60 tons (60,000 kg). What should be the theoretically lowest minimum air pressure to raise the tank roof?

$$\begin{aligned}\text{Pressure}\,(\text{kg/cm}^2) &= (\text{roof weight in kg})/(\text{area in cm}^2)\\ &= (60,000)/(5,310,000)\\ &= 0.0113\ \text{kg/cm}^2\text{g}\\ &= 113\ \text{mmH}_2\text{Og}\end{aligned} \tag{12.3}$$

This air pressure is obviously a blower discharge pressure range. Thus, we should not use a compressor but a blower to raise this tank roof. This is the clear effect of large tank roof area decreasing the air pressure requirement to achieve the same magnitude of force. The large surface area effect on force is also a reason why many dam walls fail when the water level is high.

Step 5: Recommend Remedial Action to Eliminate the Problem without Creating Another Problem

The required remedial actions developed by the problem-solver consisted of closing the tank top manhole, using a blower to fill air inside the tank up to the calculated air pressure and monitor movement of the tank roof.

Another aspect of any potential problem analysis is the consideration of the potential safety problems associated with this operation. There are at least three potential safety problems to be considered in this proposed problem solution. They are as follows:

1) Over-pressuring the tank
2) Once the roof is raised, it should be maintained in the raised position by mechanical restraints. If this is not done, the differential pressure between the space above the floating roof and that below should be maintained at the design level. This should be done with highly accurate instruments and likely automated.
3) While it is obvious that the tank should be emptied of the liquid, it should be cleared of any vapor to prevent the formation of an explosive mixture when air is added. Fresh air masks may still be required depending on the specific material that was in the tank.

12.9 Lessons Learned

This is a great example of thinking out-of-the-box to solve a problem that had been solved by another more expensive method like lifting cranes. In this particular case, the problem solver uncovered a new technique to solve this maintenance issue by using high school physics (Figure 12.3)

$$(\text{Force}) = (\text{Pressure})(\text{Area})$$

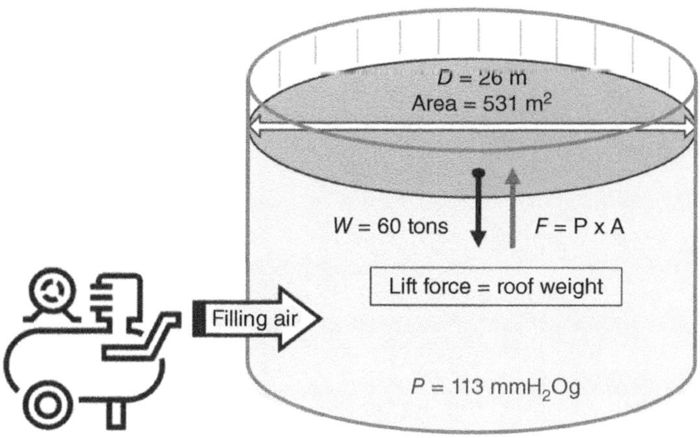

Figure 12.3 Schematic of physics involved.

Nomenclature

A	surface area
D	diameter
F	lift force
h	height. In this chapter, it refers to height of water level in the tank dike.
P	pressure
W	weight of the object

13

Verification of Process Instrumentation Data

13.1 Introduction

Data verification is a necessity of living for a successful problem solver. This is true whether the problem solver is working to discover a cure for cancer or working to get a process plant problem solved. In all cases, the verification involves use of both *human resources* and *technical resources*. While it may be possible to install multiple backup instrumentation, with frequent instrumentation calibrations and installation of elaborate communication devices to avoid the concept of data verification, this level of sophistication is rarely justified in a process plant.

Data verification can take many different approaches. This chapter aims to elucidate some of the techniques the authors have developed and/or used that are beyond the conventional "check out the instrument" approach. The approaches discussed in this chapter are certainly not all-inclusive. They are given for two purposes:

1) To illustrate some of the techniques that are available to check out suspect data. As technology growth occurs, newer techniques will cause this list to require continual updating.
2) To serve as an encouragement to the problem solver to take full responsibility for the area of data verification. Often, the problem solver is tempted to quit and say "I can't do any more until someone fixes the instrumentation."

13.2 Data Verification Via Technical Resources

The first step in verifying any instrument is a careful review of the instrument specification sheet and a comparison of the specification sheet to the actual field installation. The instrumentation specification sheet will have information such

Problem Solving Approaches for Maintaining Operational Excellence in Process Plants, Second Edition.
Joseph M. Bonem, Nattapong Pongboot, and Wiroon Tanthapanichakoon.
© 2025 the American Institute of Chemical Engineers, Inc. Published 2025 by John Wiley & Sons, Inc.

as instrument range, process fluid density, pressure, and temperature. If these values are not correct or if the range or zero point is set wrong in the field, the data cannot be expected to be correct.

A second area to consider when reviewing the instrument specification sheet is actual operating conditions compared to design conditions. That is, how close is the variable being monitored to the design level of flow, pressure, level, etc.? Many instruments provide a measurement that is based on differential pressure. If the actual measured value is significantly different from the design level, there may be inaccuracy introduced by this deviation. This inaccuracy is often overlooked in the era of digital data acquisition where the measured variable may be shown to four or five "significant" figures regardless of what percent of the range this is. An analog picture that may help in understanding this involves measuring the pressure of an automobile tire. The absolute accuracy of a pressure gage that is calibrated from 0 to 200 psig is much less than one that is calibrated from 0 to 40 psig. An error of 5% of maximum amounts to either 10 or 2 psig. This difference in the accuracy is quite significant when measuring tire pressures of 30 psig.

This "percent of range" problem is particularly significant for an orifice flow instrument. With an orifice flowmeter, the differential pressure across the orifice is proportional to the rate squared. Thus, a flow rate of 50% of maximum design flow rate only provides an output of 25% of the full meter range. The meter accuracy at 25% of the range will be much less than that at the full meter range.

The concept of reviewing the instrument data sheet first follows the idea of doing the easiest thing first. If errors are found or if the operating conditions are found to be only at a low percent of design conditions, these can almost always be rectified by simple adjustments of the instrument.

If the review of the instrument specification sheet fails to uncover any explanation for the suspected instrument being wrong, additional steps are required. In a continuous process plant, verification of instrumentation data is complicated by the fact that a shutdown of a process to verify an instrument is prohibitively expensive. Thus, data verification must be done by techniques that do not require a shutdown.

Essentially, all instruments consist of both a primary element and a display element. The primary element senses a process variable such as pressure or differential pressure. The display element takes that reading and converts it to the variable being measured such as flow rate, pressure, or level. The primary element is often not accessible without a partial or total plant shutdown. The display element can almost always be tested and replaced or adjusted while the process is in operation. For example, a flow instrument such as an orifice meter both creates and measures pressure drop. Since the orifice plate which creates the pressure drop is installed in the flowing fluid, it cannot be removed from service without at least a partial shutdown. However, the display element can be checked and adjusted or

Table 13.1 Typical sources of primary element errors.

Primary Measurement	Plugging in Element	Plugging in Tubing	Corrosion of Element	Condensation in Tubing
Flow	X	X	X	X
Level		X		X
Pressure	X	X		X
Differential pressure	X	X		X

X indicates that the primary element could likely be impacted by the condition shown at the top of the table.

replaced without a shutdown. The emphasis in this chapter is on data verification when the primary element is in question and cannot be removed from service.

Table 13.1 summarizes the typical sources of instrumentation errors of the primary elements. As an example of Table 13.1, the reader is referred to Figure 13.1, which shows a typical flow element and potential sources of errors.

As shown in Figure 13.1 and Table 13.1, the flowmeter can give erroneous results if one of the following occurs:

- The orifice could be partially plugged. This would create a greater-than-expected pressure drop for the given flow rate. This would result in the measured flow rate being higher than the actual flow rate.

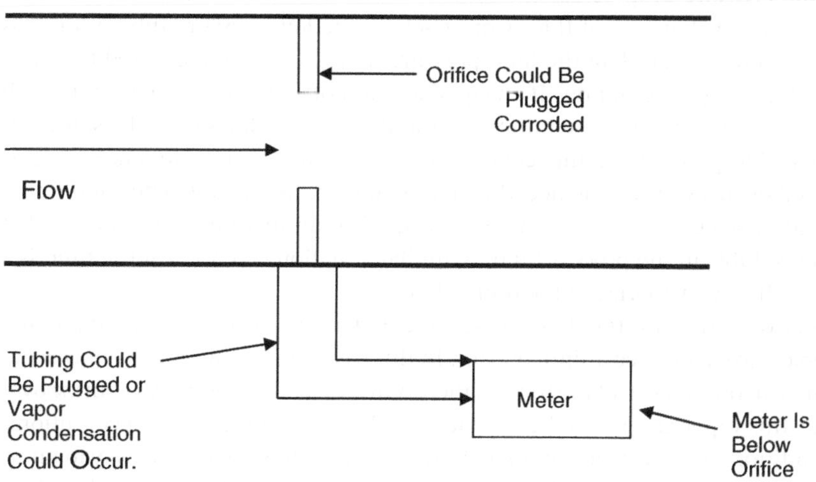

Figure 13.1 Potential flowmeter errors.

- The orifice could be corroded. This would create a lower-than-expected pressure drop for the given flow rate since the orifice hole would be bigger than the design. This would result in the measured flow being less than the actual flow. If corrosion of an orifice plate does occur, it might result in an orifice hole that was not uniform or was very rough, which might well create greater-than-expected pressure drop and a measured flow being greater than the actual flow.
- One of the tubing lines leading to the display element could be partially plugged. This would give a pressure reading at the display element that was different from the actual reading. The pressure sensors for the meter measure the pressure differential across the orifice. If the high-pressure side tubing is plugged, the measured pressure differential will be lower than the actual differential. This will cause the measured flow rate to be less than the actual flow rate. Conversely, if the low-pressure side of the tubing is plugged, the differential pressure will be higher than actual and the measured flow rate will be higher than actual flow. Possible sources of plugging are solids in the process fluid or freezing of the fluid in the tubing lines.
- Condensation in the tubing lines can also create a false differential pressure. The comments here assume that the meter is located below the orifice so that condensation will accumulate in the tubing lines. If the meter is above the orifice, any condensation will normally drain back into the process flow by gravity. This condensation can occur when the orifice is in hot vapor service or the ambient temperature is colder than the temperature of the flowing vapor. If this condensation occurs in the high-pressure tubing line and not in the low-pressure tubing line, the measured differential pressure will be higher than that created by the pressure drop across the orifice. This will result in the measured flow being higher than actual flow. Conversely, if the low-pressure tubing line has more condensation than the high-pressure tubing line, the measured flow will be less than the actual flow. If an equal amount of condensation occurs in both the high- and low-pressure tubing lines and the tubing lines have the same elevation change, then the impact on the measured variable is minimal. Since this is unlikely to occur in practice, the tubing lines are usually insulated and steam traced or sealed with either the process fluid or an instrumentation fluid that is not soluble in the process fluid. A similar situation can occur if the fluid is a liquid that can vaporize in the tubing lines.
- It is also possible for the flow measurement of a vapor stream entraining liquid droplets to experience a similar error. In this case, the liquid droplets may accumulate in one or both of the tubing lines. While these droplets may vaporize over time, this vaporization will be at a rate that depends on the ambient conditions. In addition, the rate of entrainment may also vary. Thus, there is no way to know that the tubing lines are full of liquid or vapor. The preventative steps described above are usually provided by the instrument designer in this case also.

Table 13.2 Data verification via technical resources.

Type of Instrumentation	Repair Mode[a)]	Data Verification Techniques
Flow measurement		
Primary element	Off-line	Insertable flowmeters, ultrasonic flowmeters, or process analysis
Display element	On-line	
Thermocouples with external wells		
Primary element	On-line	Thermometers, pyrometers
Display element	On-line	
Thermocouples without external wells		
Primary element	Off-line	Infrared pyrometers
Display element	On-line	
Pressure measurement		
Primary element with block valve	On-line	Pressure gauge
Display element	On-line	
Level measurement		
Primary element	Both off-line and on-line	Gauge glass, Geiger counter, or X-ray
Display element	On-line	

a) Off-line indicates that the process must be shut down to check or repair. On-line indicates that verification can be made while the process is in operation.

Table 13.2 shows possible techniques for validating instrumentation data. The emphasis in this table is verification without requiring a plant downtime. Thus, the most obvious approach of replacing an "in-line" instrument is not covered. The following paragraphs discuss verification techniques for the most common instruments.

13.3 Flow Measurement

If a primary flow measurement device (an orifice plate or venturi meter) is suspect and cannot be removed from service, it must be verified by either or both a noninvasive external flow measurement device or process analysis.

Table 13.3 External flow measuring devices.

Type of Device	Comments
Noninvasive ultrasonic instruments	These can be strapped onto a pipe and used to measure flow
Transit time	Clean homogenous fluids. Accuracy has been reported from 11 to 70%
Doppler shift	Used for suspended solids with concentrations from 200 ppm to the percent level
Invasive	These require a bleeder and a packing gland arrangement to insert them into the flow
Pitot tube	Can be highly accurate if sufficient readings are taken across the flow path and integrated to obtain the average flow. Generally for clean fluids only

The external flow measurement devices are generally not as accurate as the flow measurement devices installed in the process. There are some claims for high accuracy. However, this is usually based on devices that have been calibrated against a known flow rate of the suspect process fluid. If this field calibration against a reliable flowmeter is not available, then the claims for high accuracy are questionable. In addition, some of the external measurement devices (pitot tube) require insertion through a bleeder valve with a packing gland arrangement. This packing gland arrangement creates some safety risk especially in high-pressure service. Table 13.3 lists a few of the external devices that are available. This is not meant to be an exhaustive list. In addition, the changing technology world may well allow the development of different and/or more accurate devices.

Verification of a flow measurement by process analysis is a relatively simple, but often overlooked technique. It involves the use of heat and material balances to confirm the validity of a suspect flowmeter. It is probably best to consider it a "one-sided" test. For example, if the heat and/or material balances appear to check, the flowmeter is probably correct. If the heat and/or material balances do not check, the suspect flowmeter *or another* flowmeter is likely wrong. While these concepts are fundamental to all process engineering, some general guidelines are presented below to aid in the application of these principles:

- Accumulation may cause a system to appear to be out of material balance. Thus, it may be necessary to expand the traditional concept to utilize the principle of accumulation discussed previously in Chapter 10. That is,

$$\mathrm{AD} = I - O - \mathrm{RD} \tag{13.1}$$

where

AD = rate of accumulation. This could be accumulation of anything – heat or material
I = inflow of the material or heat
O = outflow of the material or heat
RD = removal of heat or material by reaction

- Beware of determining the flow rate as the difference between two large metered flows. Using the concept of concentration factor provides more reliable answers. Figure 13.2 illustrates this point. As shown in this figure, if the overhead distillate rate is calculated as the difference between the feed rate and the bottoms rate, a value of 500 lb/hr is obtained compared to a metered value of 1000 lb/hr. However, if concentrations are used to estimate the overhead rate, a flow rate of 909 lb/hr is calculated.
- Heat balances can be used to estimate the process or utility flow to an operation. For example, the steam rate to a reboiler for a simple fractionation column can be estimated knowing that the heat added by the reboiler and feed must be equal to the heat removed by the process streams and cooling water. Other examples are the cooling water flow to a tower condenser can be determined knowing the

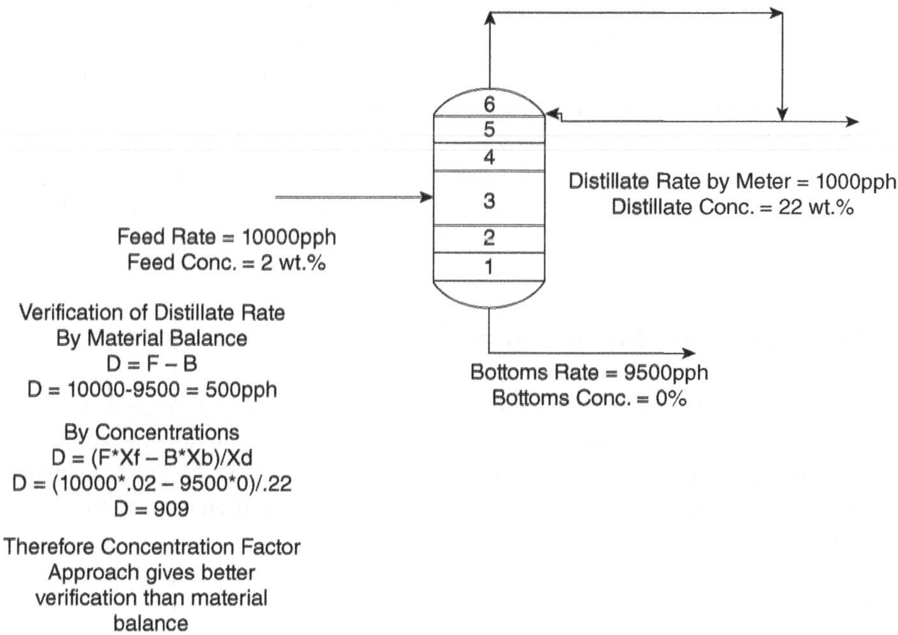

Figure 13.2 Estimating flow by concentrations.

heat removed in the condenser and the inlet and outlet water temperatures; and the "boil up" rate in an evaporator can be determined by the steam rate and the heat of vaporization of the process fluid.

13.4 Temperature Measurement

Often, a suspect primary temperature measuring element can be checked by removing it from a thermowell which is installed in the vessel or pipe. However, in some instances, the primary device is not removable. An example of this is the tube metal thermocouples in a furnace. These devices are used to measure the temperature of the furnace tube metal. In order to obtain this measurement, the thermocouples are attached to the tube itself and shielded from the flame as opposed to being installed in a thermowell. This shielding ensures that the thermocouple does not receive any radiant heat from the furnace flames. A high tube metal reading could be real or associated with a failure of the device or shielding equipment.

In the case of furnace tube thermocouples that are suspect, infrared temperature measurement can be utilized to measure tube metal temperatures through a furnace porthole if it can be accessed safely. Infrared temperature measurement is a noninvasive technique to determine the temperature of a small area. It can be used to measure reasonably and accurately the temperature of a particular spot on a furnace tube. It can also be used to measure the temperature of a flowing solid or to measure the temperature of a rotating part of the machinery.

Infrared temperature measurement requires a great deal of expertise, particularly the measurement of furnace tube temperatures. For example, measurement of furnace tube temperatures may require that the infrared detector be cooled by liquid nitrogen.

13.5 Pressure Measurement

A suspect pressure instrument can many times be removed and a gage installed to verify the pressure. However, if there is no way to isolate the pressure instrument or if the pressure tap is plugged, alternative methods of verifying the pressure instrument will be required. Fundamental process analysis can often be used to verify a pressure reading.

The classical chemistry phase rule can often be applied in this situation. This rule is as follows:

$$F = C - P + 2 \tag{13.2}$$

where:

F = the degrees of freedom (temperature, pressure, and composition)
C = the number of components in the mixture
P = the number of phases present

Thus, for a single-component system with two phases present, the degrees of freedom equal 1. This means if the temperature is known, the pressure is fixed. In a two-phase binary (two components) system, the degrees of freedom will be 2. If the binary system has a *fixed* composition, fixing the temperature also fixes the pressure since the *fixed* composition eliminates one degree of freedom. While this may seem very basic chemical/engineering, it is amazing how often this is overlooked in the data verification process. Example Problem 13.1 illustrates how this fundamental approach was used to solve a process problem rather than inventing more complicated theories.

13.6 Level Measurement

The verification of a level instrument can often be done simply by utilization of a gage glass if one is available. Other devices such as X-ray and Geiger counters can be used to determine the absolute level based on the fact that liquid is denser than vapor. Both of these techniques work in a similar fashion. A source of the radiation signal is located on one side of the vessel, and a detector is located on the other side. The amount of radiation that reaches the detector is inversely proportional to the length of the flow path and the density of the material in the flow path. The material in the flow path includes both the vessel wall and the process fluids. Thus, if the radiation absorption of the metal wall can be eliminated by a calibration technique, this technique can detect the interface between liquid and vapor. Key factors in the use of either of these techniques are as follows:

- Difference in densities between the liquid and vapor. As the density difference decreases, it is more difficult to measure level with this technique.
- Thickness of the vessel wall. The thicker the vessel wall, the more difficult it is for a signal to penetrate the wall of the vessel.
- Diameter of the vessel. At a fixed design pressure, larger-diameter vessels require thicker walls.
- Strength of the radiation signal. The stronger the source of the radiation signal, the easier it is to obtain a strong signal with the detector. However, stronger signals increase the radiation hazards and the design consideration difficulties.

Again, process analysis can be used to verify the range of the level instrument. In order to verify the absolute accuracy of the level instrument, the zero point and range must be known. The zero point is the level in the vessel that corresponds to a zero reading on the level instrument. This may or may not correspond to an empty vessel condition. The range is the difference (usually in inches) between the zero point and the 100% indication of the vessel level. For example, a 6-ft-high vertical drum with a zero-point set at the bottom tangent line of the vessel and a range of 72 in. would cover the entire height of the vessel. Thus, a reading of 50% on the level instrument would be expected to be 36 in. above the bottom tangent line.

If the vessel level instrument is in question, it may be due to an inaccurate zero point or an inaccurate range. The zero point and instrument range can often be checked without taking the vessel out of service. However, some types of level instruments cannot be checked without taking the vessel out of service. In addition, inaccuracies of level instruments can also be caused by vessel connection plugging.

If accurate flowmeters are available for the inlet and outlet flows, it may be possible to assess the accuracy of the range of the level instrument by accumulating in the drum or removing liquid from the drum and comparing the calculated inventory change from flowmeters to the measured inventory change. The approach to be used depends on the orientation of the drum. If the drum is a vertical vessel, the inlet and outlet flows should be set so that there is a significant difference between the two. The measured inventory change should then be compared to the calculated inventory change based on the flowmeters. If they compare well, this is a good indication that the range on the instrument is correct. However, the zero point could still be wrong. The zero point could be checked by reducing the inventory in the vessel until there was clear evidence that the drum was empty. The evidence could consist of "blowing through" as evidenced by the outlet flowmeter or by a change in temperature, which would indicate by the phase rule that there was only a single vapor phase present.

If the level instrument in question is on a horizontal vessel, the same technique can be utilized. However, determining the actual volumetric inventory change will be more difficult due to the curvature of the vessel walls. Tables and techniques are available to readily take this curvature into account. If an accumulation test is run with a horizontal vessel with a level zero point that is in error, it is possible to conclude that the level meter is correct due to the curvature of the drum. For example, a change in the level instrument from 65 to 70 (5%) might well be confirmed by the difference in flowmeters. However, if the actual level was 11% instead of 65%, a 5% level increase would seem in error when compared to the difference in flowmeters. To confirm the accuracy of the range of a level instrument on a horizontal vessel, it is best to run two tests at different vessel levels. If both of these tests indicate that based on differences in flowmeters the level changes appear correct, then it can be assumed that the range of the level instrument is correct. The techniques for determining the zero point discussed for a vertical vessel can also be utilized for a horizontal vessel.

13.7 Data Verification Via Human Resources

The role of the nonprofessional worker in problem specification was discussed earlier. It was implied that the operator or mechanic was cooperative and that all that was necessary was to solicit his data and/or the problem history. Obviously, an uncooperative worker can be a poor source of data by refusing to share his knowledge, presenting erroneous data, or simply acting as a "smoke screen." For example, an instrument technician that was known to have acrophobia probably did not check out an instrument as he claimed if it required him to be far above the ground.

While this book does not cover all aspects of interpersonal skills, the guidelines below will help the problem solver obtain the maximum assistance from the operating and mechanical workers.

- Respect the person that you are seeking information from as an equal whether he is college-educated or not.
- Learn people's names and do not forget them.
- Cultivate relationships by providing answers to questions, explaining your goals, reviewing the results of any tests, and listening to other's thoughts.
- Do not give the nonprofessional workers the impression that you are too busy to spend time with them.
- Be positive. Assume that people are always trying to do the right things. If a mistake has been made, assume that it was due to inadequate training or done accidentally. Do not take the position that someone did it on purpose. You will be wrong on occasions, but you will not destroy relationships.
- When it is occasionally necessary to look over someone's shoulder as they perform their job, try to convince them that they are educating you on their skills.
- Realize that everyone has bad days. If you get an undeserved "chewing out" by an operating or mechanical worker, take your licks and back away from the conversation without becoming defensive. Defending yourself to someone that is irrational is not possible and you will cut off a source of future data.

13.8 Example Problems

The example problems below illustrate some of the approaches described in this chapter. The first problem is a real-life problem that illustrates the value of investigating and verifying conflicting data, as opposed to simply assuming that one piece of the mysterious data is wrong. The second problem is a real-life problem that was solved by a careful review of the instrumentation specification sheet. The third problem is a fictitious problem that is included to illustrate how the technique of inventory change can be used to confirm a level instrument. The fourth problem is a fuel gas balancing problem.

13.9 Example Problem 13.1

A vessel in an ethylene refrigeration unit shown in Figure 13.3 served as a combination surge and knockout drum. Liquid ethylene flowed into the drum and flowed out to various heat exchangers. The mixture of vapor and liquid ethylene leaving the exchangers was returned to the drum where it was separated into vapor and liquid streams. The ethylene refrigeration system was used to cool liquid propane to a temperature of −145 °F. It was important to control the liquid level in the drum to ensure that minimal amounts of liquid were entrained with the vapor and also to ensure that liquid ethylene was always available for feeding the heat exchangers. The drum operated under a slight vacuum, and the normal temperature was −156 °F.

Current conditions were an anomaly. The pressure was at normal conditions. This slight vacuum was confirmed by a second independent pressure instrument. However, the temperature measured by two independent thermocouples was +11 °F. Operating personnel believed that the thermocouples were both wrong. The management had asked that technical resources determine what the real problem was.

The problem solver's approach to using the five-step problem-solving technique was as follows.

Step 1: Verify that the Problem Actually Occurred

Since there were two independent thermocouples that both indicated a much higher than expected temperature, it was obvious that something was wrong. This was a real problem.

Step 2: *Write* Out an *Accurate* Statement of What Problem You Are Trying to *Solve*

The problem statement that was developed was as follows:

> "The temperature on the ethylene knockout drum is reading much higher than normal. While the process appears to be operating normally, this higher temperature may be indicative of a potential problem that will occur in the future. Determine why the temperature on the ethylene knockout drum is reading much higher than the normal temperature of −156°F."

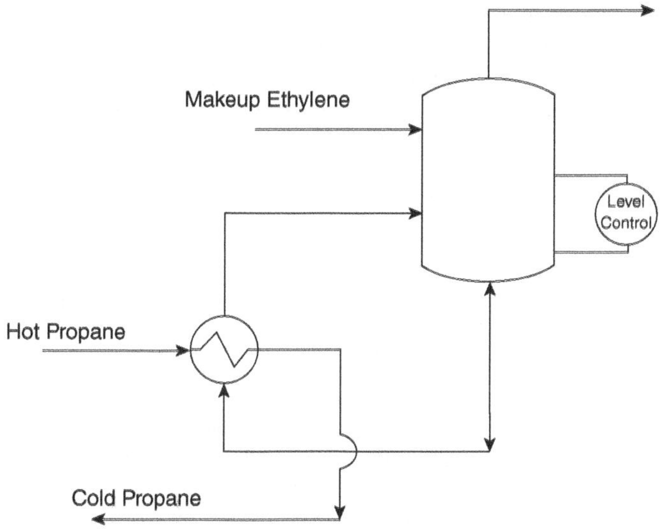

Figure 13.3 Ethylene surge/knockout drum.

Step 3: *Develop* a Theoretically Sound Working Hypothesis that Explains as Many Specifications of the Problem as Possible

Using the questions from Chapter 6, the problem solver developed Table 13.4. Based on the questions, four hypotheses were developed:

1) There is a leak in the exchanger, and the drum contains a mixture of propane and ethylene instead of pure ethylene.
2) The thermowell is so poorly insulated that the temperature instrument is really reading a mixture of the ethylene and the ambient temperature.
3) The new thermocouples were calibrated incorrectly.
4) The level instrument is wrong, and there is no liquid level in the drum.

Hypothesis 2 was ruled out by an inspection of the drum that showed the thermowells were well insulated. Hypothesis 3 was ruled out after discussions about the calibration techniques used by the instrument technicians.

The problem solver used the phase rule (equation (13.1)) shown below to conclude that either hypothesis 1 or 4 was a theoretically correct working hypothesis.

$$F = C - P + 2 \tag{13.3}$$

Based on the phase rule, he outlined the following three cases:

1) Under normal situations in the ethylene surge drum, there would be a single component ($C = 1$), two phases ($P = 2$), and the degrees of freedom (F) would be equal to 1. That is, the temperature in the surge drum would be fixed by the pressure and should be $-156\,°\text{F}$.

Table 13.4 Questions/comments for Problem 13.1.

Question	Comment
Are all operating directives and procedures being followed?	All appeared to be correct and being followed
Are all instruments correct?	The instruments had allegedly been calibrated
Are laboratory results correct?	Not applicable
Were there any errors made in the original design?	No. The process was an old one that was returning to service after repairs
Were there changes in operating conditions?	No
Is fluid leakage occurring?	The propane being cooled could leak into the ethylene refrigerant and concentrate in the knockout drum
Has there been mechanical wear that would explain the problem?	The unit had been down for repairs. The repairing of the insulation on the drum might not have been completed
Is the reaction rate as anticipated?	Not applicable
Are there adverse reactions occurring?	Not applicable
Were there errors made in the construction?	See comments above on mechanical wear. In addition, new thermocouples were installed

2) If there were two components ($C = 2$) in the drum, then the degrees of freedom would increase to 2 and the temperature and pressure in the drum would be independent.
3) If there was a single component in the drum ($C = 1$) but only a single phase ($P = 1$), then the degrees of freedom would also be equal to 2. Again, the temperature and pressure in the drum would be independent.

However, there are boundaries to this independence of temperature and pressure in case 2 shown above. For example, if there are two phases present and an exchanger leak had occurred (hypothesis 1), the temperature in the drum cannot be warmer than the boiling point of propane. If the liquid level instrument is correct (two phases are present) and as an extreme case all of the liquid in the drum is propane, the following analysis could be made:

$C = 1$ (nothing in the drum, but propane.)

$P = 2$ (the level instrument is correct, so there must be two phases.)

Therefore:

$$F = 1 - 2 + 2 = 1 \tag{13.4}$$

The boiling point of propane at atmospheric pressure is −44 °F. Thus, even if the liquid in the drum was 100% propane, the temperature would be slightly below −44 °F not +10 °F.

If there was no leak in the exchanger, but the level instrument was wrong and the drum was empty (hypothesis 4), the phase rule would indicate the following:
$C = 1$ (nothing in the drum but ethylene.)
$P = 1$ (nothing in the drum but vapor.)
Then:

$$F = 1 - 1 + 2 = 2 \tag{13.5}$$

In this case, the temperature of the ethylene vapor would be limited only by the amount of superheat added by the heat exchangers.

Therefore, the theoretically sound working hypothesis developed was as follows:

> "It is theorized that the mysterious temperatures in the ethylene knockout drum are due to the absence of a liquid level in the drum. Since there is no liquid level in the drum, the vapor can be heated to temperatures above the boiling point of ethylene at the pressure in the drum. In addition, the absence of this boiling liquid results in a very low heat transfer coefficient between the phase in the drum (vapor) and the thermowell. The normal situation in the drum is such that the thermowells are covered with boiling ethylene. The heat transfer coefficient in the normal case is likely as high as 200 BTU/hr-°F-Ft^2. The heat transfer coefficient from vapor to the thermowell is likely as low as 11 BTU/hr-°F-Ft^2. With this exceptionally low coefficient, the heat gained from the atmosphere flowing through the insulation and heating the thermocouple may create an additional crror sourcc."

Step 4: Provide a Mechanism to Test the Hypothesis

While the hypothesis could have been checked by simply raising the liquid level in the drum, this was deemed too risky. Since the drum served as a knockout drum, there was a risk of a high level causing liquid ethylene to be carried over into the compressors. The problem solver requested that the instrument technician (who had recently checked the level instrument and determined that the range and zero point were as they should be) accompany him to again check the instrument. He explained to the instrument technician why he thought that the level instrument was incorrect. When they jointly inspected the installation, they found that the steam tracing that was used to ensure that there was no liquid accumulation in the tubing lines was turned off. When this tracing was returned to service, the level instrument began to function correctly.

Step 5: *Recommend* Remedial Action to Eliminate the Problem without Creating Another Problem

In this case, the mechanism to test the hypothesis was the remedial action. In order to prevent future occurrences of this problem, a statement was added to the start-up procedure that called for the operators to check that the steam tracing had been put into service.

13.10 Lessons Learned

While it is true that an experienced engineer would not need to use the phase rule to analyze this problem, that is only because he knows by experience and intuition how to analyze similar situations. In applying his experience and intuition, he is using the phase rule whether he knows it or not. This problem might be considered too simple for an engineering analysis; however, it does illustrate two important points. In the first place, there is great value in doing a fundamental analysis instead of assuming that the first analysis (the thermocouples are both wrong) is correct. It would be unlikely that both thermocouples were incorrect, unless there was an external circumstance such as they were incorrectly calibrated or the heat transfer to the thermocouple was not what it should have been. As indicated, this could be due to excessive heat loss from the thermowell due to poor insulation or poor heat transfer (caused by the absence of liquid) from the vessel contents to the thermowell. The application of the phase rule was very helpful in indicating that the level instrument, which had been checked out, was truly in error.

The problem also indicates that it is often necessary to get the operator or mechanical personnel to recheck something that has already allegedly been checked. This requires tactfulness and a good working relationship. This good working relationship is developed by the principles indicated in Section 13.7. In this example problem, it was enhanced by the problem solver going with the instrument technician to look at the level instrument for a second time.

13.11 Example Problem 13.2

A new polymer was to be produced in a test run in a commercial polypropylene plant. This new polymer was to be a copolymer produced by using ethylene as a co-monomer in reactors normally polymerizing only propylene. Careful preparations were made for the test run, including reviews of batch and pilot plant operations producing the copolymer. These data indicated that the ethylene conversion should be about 85% during the planned operation. The test run plans called for careful monitoring of all variables, especially the ethylene conversion.

The ethylene conversion was to be monitored using the propylene and ethylene flowmeters, the ethylene content of the polymer, and the ethylene content of the unreacted gases. Because of the reliance on feed flowmeters, these were carefully checked out by the instrument technicians prior to beginning the operation.

After the test run was started and steady state was established, the calculated ethylene conversion based on flowmeters and the ethylene content of the polymer was essentially 110%. This was in contrast to the ethylene content of the unreacted gases, which indicated that the ethylene conversion was significantly less than 110%. While the unit ran well, the problem solver was greatly concerned about the deviation from the anticipated conversion based on calculations made using the feed meters. He convinced the management that even though this discrepancy between the anticipated conversion and the apparent actual conversion did not seem to be a problem on this run, it might be a problem on future runs. They agreed that he should spend some time investigating the problem.

The problem solver's approach to using the five-step problem-solving technique was as follows.

Step 1: Verify that the Problem Actually Occurred

Since the test run had been carefully planned and the instruments checked out ahead of time, there was no doubt that there was a problem. There was some doubt about whether the problem merited attention. The problem statement developed by the problem solver had to address the incentive for working the problem.

Step 2: *Write* Out an *Accurate* Statement of What Problem You Are Trying to *Solve*

The problem statement that was developed was as follows:

> "The calculated ethylene conversion based on the ethylene feed flowmeters and the ethylene content of the polymer is above 100 percent. This is in conflict with the pilot plant data and also in conflict with the analysis of the unreacted gases which shows the presence of ethylene. If the flowmeters are correct, then either the ethylene content of the polymer or the ethylene content of the unreacted gases is wrong. The process control strategy requires that all three variables (feed flowmeters, ethylene content of the polymer and ethylene content of the unreacted gases) be measured correctly. Determine why the ethylene conversion is significantly higher than anticipated based on feed flowmeters."

Step 3: *Develop* a Theoretically Sound Working Hypothesis that Explains as Many Specifications of the Problem as Possible

Using the questions from Chapter 6, the problem solver developed Table 13.5.

Table 13.5 Questions/comments for Problem 13.2.

Question	Comment
Are all operating directives and procedures being followed?	All appeared to be correct and being followed
Are all instruments correct?	The instruments had allegedly been calibrated. The ethylene flowmeters and the analyzer for measuring the ethylene content of the unreacted gases were being used for the first time
Are laboratory results correct?	The technique for measuring the ethylene content of the polymer had been closely cross-checked against known standards
Were there any errors made in original design?	Since this was the first time that the ethylene meters and analyzers had been used, their design could be in error
Were there changes in operating conditions?	Not applicable
Is fluid leakage occurring?	Not applicable
Has there been mechanical wear that would explain the problem?	Not applicable
Is the reaction rate as anticipated?	Higher than anticipated ethylene conversion might explain the apparent discrepancies
Are there adverse reactions occurring?	Adverse reactions are highly unlikely
Were there errors made in the construction?	Since both the ethylene flowmeters and the ethylene analyzer were being used for the first time, this had to be considered

Based on the above questions, four hypotheses were developed:

1) There was a design mistake in sizing the ethylene flowmeters.
2) There was an error made in the installation of the ethylene flowmeters.
3) The unreacted gas analyzer had been incorrectly calibrated for ethylene, or else the installation or design of the analyzer made it unacceptable for ethylene monitoring.
4) The technique for analyzing for ethylene in the polymer was either incorrect or being done incorrectly.

While any of the four hypotheses could be selected as a "theoretically sound working hypothesis," the problem solver elected to do some further investigation prior to proposing a working hypothesis. Since the easiest and quickest thing to do was to review the instrument specification sheets, he elected this route. In addition, he recognized that the pretest instrument check by the

Table 13.6 Specification and actual data.

Variable	Specification Sheet	Actual Data
Pressure, psig	735	735
Temperature, °F	90	75
Critical pressure, psig	735	735
Critical temperature, °F	50	50
Compressibility (Z)	1.0	0.6
Density, lb/ft^3	3.56	6.11

instrument technicians did not involve a review of the instrument specification sheet. A review of the specification sheet indicated the values shown in Table 13.6.

The original instrumentation design used a density that was significantly less than the actual density because the gas compressibility that occurs at high pressures was not considered. There was also a difference in actual and design temperatures, but this was not significant. These calculations led to the following theoretically sound working hypothesis:

> "It is theorized that the discrepancies between the actual ethylene conversions and the indicated ethylene conversions are due to the fact that the density for the ethylene flowmeters was calculated assuming that the compressibility was 1. This resulted in a calculated density that is 40 percent below actual density. Since the measured flow rate is directly proportional to the square root of the density, this error results in a measured flow rate that is 30 percent below the actual flow rate. This single error explains the discrepancies observed."

Step 4: Provide a Mechanism to Test the Hypothesis

The mechanism for testing the hypothesis was very simple. Since the error was in the calculations, it was easily corrected by changing the range on the flow instrument to correct for the incorrect density. After this correction was made, the flow rates, ethylene content of the unreacted gases, and ethylene content of the polymer were all consistent with the anticipated results.

Step 5: *Recommend* Remedial Action to Eliminate the Problem without Creating Another Problem

In this case, the testing mechanism was the permanent solution. It should be noted that in this problem, there was a temptation to simply modify the process control computer with a fudge factor to adjust for the error. This was easy to do, but it brought with it the risk that at some future time point, the factor that was used in the computer would be removed because someone did not understand what it was. It was much more "fool-proof" to change the instrument range and revise the instrument specification sheet.

13.12 Lessons Learned

It should be noted that the error in the instrument specification sheet was only discovered after the problem solver realized that the pretest instrument check-out done by the instrument technicians only included a physical check of the instruments against the range given on the instrument specification sheet. They had no way of knowing that the density used in calculating this range was incorrect. The problem-solving lesson to be learned is that the problem solver needs to know the boundaries of the mechanics or operators when they are asked to check out a piece of equipment. If they are starting with an incorrect calculated setting, they can only confirm that the instrument or piece of equipment is set to that value.

While it could be argued that the effort to resolve this discrepancy was not justified, the discrepancy was a particular concern since the flowmeters were a key to changing grades. In addition, some grades produced at high ethylene contents could cause process shutdowns if control was lost. As a general rule, discrepancies in data or things that we do not understand will almost always cause future problems.

This example problem also illustrates again the value of doing the easiest calculation or data review first. While it seemed hard to believe that the design team would ignore the compressible nature of ethylene at high pressures, the simple review of the instrument specification sheet showed that this indeed was the case. Other hypotheses described earlier would have been more involved and taken longer to pursue.

13.13 Example Problem 13.3

This problem is included to show the calculation technique discussed earlier. The five-step problem-solving approach is not considered. In addition, the tables that show liquid volume as a function of level in a horizontal drum are not included. For simplicity, the head volume of the drum is not included. In order to illustrate the calculation technique, it is assumed that the bottom connection of the differential pressure-type level indicator is partially plugged so that the indicated level is 15% greater than the actual level.

Table 13.7 Summary of test basis.

Drum size	
Diameter, ft	11
Length, ft	30
Fluid density, lb/ft^3	30.6
Inlet flow—outlet flow, lb/hr	10,000
Test time, minutes	30
Volumetric flow, ft^3	163.3

A pump taking suction from a horizontal vessel continued to experience intermittent periods of operating below the pump curve. At times, it operated well. After an extensive problem-solving analysis, it appeared that the level instrument was incorrect. The level instrument was checked by instrument technicians, and the range and zero point appeared to be correct. The problem solver decided that the only way to confirm the accuracy of the level instrument was with a plant test using the inlet and outlet flowmeters. He planned to use the flowmeters to calculate the level change in the drum. He would then compare this change to the measured change to try to determine if the level instrument was correct.

A summary of the test is shown in Tables 13.7 and 13.8.

Table 13.8 Test run results.

	Test 1	**Test 2**
At test start		
Indicated level[a], %	25	65
Actual level, %	10	50
Actual contained volume, ft^3	123	1178
At test completion		
Actual contained volume, ft^3	286	1341
Actual level, %	17.9	55.5
Actual level change, %	7.9	5.5
Measured level change, %	6.1	5.8
Error, %	23	5

a) The problem description indicated that the level instrument showed 15% higher than the actual level.

An examination of Table 13.8 indicates the risk of running an accumulation test at only one level. If the test is run at a starting indicated level of 65%, it would be easy to conclude that the actual level change and calculated level change were very close. However, if the test runs were started at an indicated level of 25%, the conclusion would be that there is a significant difference between the actual level change and calculated level change.

13.14 Example Problem 13.4

Chemical production plants that consume fuel gas as utility in their processes often obtain their fuel gas from either self-production or through import from outside of the plant. This relationship typically holds:

$$\text{Fuel gas mass consumption} = \text{Fuel gas mass production} + \text{Fuel gas mass import} \quad (13.6)$$

This equation assumes that there is no fuel gas flaring or unintentional fuel gas loss from the process.

A chemical production plant experienced a fuel gas mass flow rate imbalance between consumption and production + import. The fuel gas used was methane. In the first 3 years, the fuel gas mass balance had been well balanced. However, after a turnaround, the fuel gas began to become imbalanced. Fuel gas consumption exceeds the sum of fuel gas production and import.

The problem solver's approach to using the five-step problem-solving technique was as follows.

Step 1: Verify that the Problem Actually Occurred

Just by looking at the graph of fuel gas consumption versus fuel gas production + import, the problem was very clear (Figure 13.4).

Step 2: *Write* Out an *Accurate* Statement of What Problem You Are Trying to *Solve*

The problem statement that was developed was as follows:

> "A chemical production plant had been experiencing a fuel gas flow rate imbalance between consumption and (production + import) for several years without knowing why. The problem solver was called in to solve this plant mysterious imbalance problem. Engineers in the plant were not sure which flow rate was the correct one and which one to believe. Moreover, the engineers wanted to find a way to correct this fuel gas flow rate imbalance and make it well balanced again."

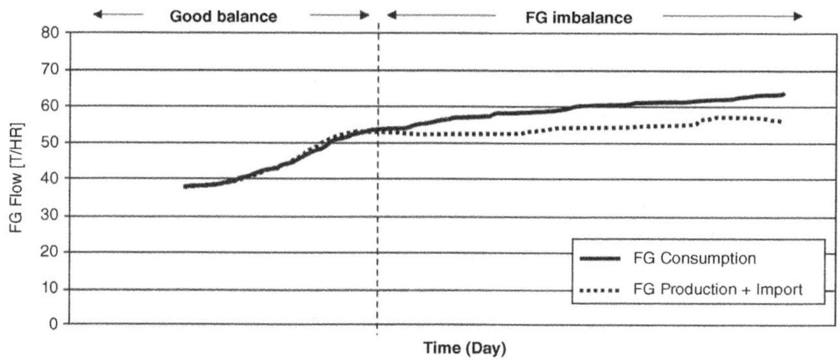

Figure 13.4 Fuel gas balance.

Step 3: *Develop* a Theoretically Sound Working Hypothesis that Explains as Many Specifications of the Problem as Possible

Using the questions from Chapter 6, the problem solver developed Table 13.9.

From Step 3, there were 4 hypotheses raised from the series of questions.

(1) Wrong flow instrument parameter settings in either consumption or (production + import).
(2) Fuel gas leakage to flare or atmosphere.
(3) Fuel gas flow instrument wear-and-tear or breakdown.
(4) Some errors in construction during the turnaround period.

Hypotheses (2) and (3) were eliminated because of the following reasons:

(1) The flow rate imbalance was so large that if the fuel gas leakage to flare had occurred, it would have resulted in very large flame or significant combustion difference seen at the flare tip. However, no such thing was observed. In addition, fuel gas leakage would make sense as a plausible root cause only if fuel gas consumption was smaller than production + import because there would be less fuel gas available for consumption. However, actual plant data showed that fuel gas consumption is larger than production + import, implying fuel gas addition into the system instead of leakage out of the system. Therefore, hypothesis 2 can be confidently eliminated.
(2) The fuel gas flow instrument received good preventive maintenance and showed no signs of wear-and-tear or breakdown. The calibration records were clearly available.

Therefore, only hypotheses (1) and (4) remained possible. However, hypothesis (4) was possible to occur simultaneously with hypothesis (1). Someone in the instrument and control team might have reset or changed parameters during the turnaround period. Therefore, we will focus the potential type of the flow meter error to only hypothesis (1) first.

Table 13.9 Questions/comments for Problem 13.4.

Question	Comment
Are all operating directives and procedures being followed?	Yes. Operating procedures were correctly followed
Are all instruments correct?	All instruments were assumed to have been calibrated. However, certain parameter settings in the flow instruments might have been altered unknowingly, causing a large change in the measured flow rate
Are laboratory results correct?	Yes. Fuel gas molecular weight remained unchanged
Were there any errors made in original design?	Not applicable because fuel gas had been well balanced in the first 3 years in operation. The problem began after the turnaround period
Were there changes in operating conditions?	No change in fuel gas operating conditions was found
Is fluid leakage occurring?	Fuel gas leaking to flare or atmosphere might have explained the problem
Has there been mechanical wear that would explain the problem?	Fuel gas flow instrument wear-and-tear or breakdown might have caused fuel gas consumption to incorrectly increase or decrease
Is the reaction rate as anticipated?	Not applicable
Are there adverse reactions occurring?	Not applicable
Were there errors made in the construction?	Yes. Some errors in construction might have been made during the turnaround period

Step 4: Provide a Mechanism to Test the Hypothesis

The most important thing is to understand how flow meters are typically set. Every flow meter will give a gas flow rate reading at "standard" or "normal" condition. Even if the flow meter measures "volumetric flow rate" and not "mass flow rate," the mass flow rate can be displayed by converting from "volumetric flow rate" to "mass flow rate" using "standard" or "normal" density.

$$\begin{aligned}(\text{Mass flow rate in kg/hr}) &= (\text{normal density in kg/Nm}^3) \\ &\quad \times (\text{volumetric flow rate in Nm}^3/\text{hr}) \qquad (13.7)\end{aligned}$$

However, the effect of the normal density value on the mass flow rate is nonlinear since both the orifice and pressure drop relationships include a density term.

$$
\begin{aligned}
&(\text{Furnace fuel gas mass flow reading})_{\text{new}} \\
&\quad = (\text{Furnace fuel gas reading})_{\text{old}} \\
&\qquad \times (\text{fuel gas density}_{\text{new}}/\text{fuel gas density}_{\text{old}})^{0.5}
\end{aligned} \tag{13.8}
$$

To prove hypothesis (1), the normal density parameter setting inside the flow meters needs to be checked. The historical data of the normal density setpoint is shown in Figure 13.5. We found that the normal density of the methane fuel gas was being set in the control room at 0.75 kg/Nm^3. The problem solver then used a simulation to calculate methane fuel gas normal density. The result was that the methane fuel gas density should have been around 0.6 kg/Nm^3 instead of 0.75 kg/Nm^3 according to the normal condition given.

However, we should not immediately tell the instrument and control engineer to change the normal density value back to 0.6 kg/Nm^3 unless we are confident

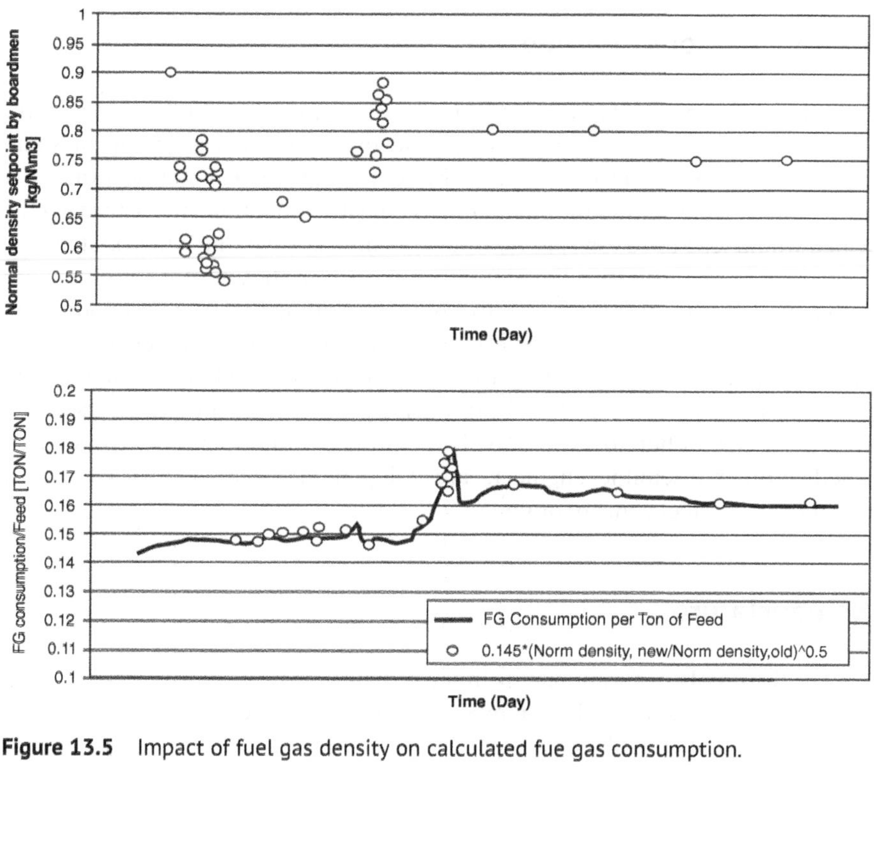

Figure 13.5 Impact of fuel gas density on calculated fue gas consumption.

that our root cause analysis is correct. A better way is to try to correct the flow meter reading on paper or offline first by applying the correction factor (fuel gas $\text{density}_{\text{new}}/\text{fuel gas density}_{\text{old}})^{0.5}$.

After applying the density correction factor, the fuel gas consumption became well balanced with fuel gas (production + import) on paper. Therefore, it decisively proved that hypothesis (1) was the main root cause.

Step 5: *Recommend* Remedial Action to Eliminate the Problem without Creating Another Problem

Remedial actions should be to change the normal density parameter setting inside all furnace consumption flow meters. The fuel gas production and import flow meters were also checked for normal density parameter settings, but were found to be already correct.

Therefore, this example shows that even a simple flow meter normal density setting could lead to a significant impact on the correctness of the mass flow meter reading.

13.15 Lessons Learned

Do not blindly believe that other disciplines of engineers are doing things right. Sometimes, people make mistakes because they lack understanding of some aspects of their work. In this case, instrument engineers may not be aware of a grave impact from setting the standard gas density in the flow meter incorrectly, which would lead to 10 tons per hour fuel gas imbalance in the whole system.

When we have a positive doubt and constructive criticism of what others are doing, it often leads to a more effective problem finding and troubleshooting. Do not just say that it is the mechanical engineer's problem or instrument engineer's problem. Take the responsibility to learn fundamentals of other engineering disciplines too. The objective is not to work details on behalf of other discipline engineers, but rather to have sufficient minimum knowledge to cross-check whether other discipline engineers are doing things right or at least with a theoretically correct concept or not.

Nomenclature

C	the number of components in the mixture
F	the degrees of freedom (temperature, pressure, and composition)
P	the number of phases present

14

Successful Plant Tests

14.1 Introduction

Step 4 of the five-step problem-solving procedure often involves a plant test of some kind. These tests can vary from very straightforward to very complicated. While some of these comments were included in Chapter 4, conducting a successful plant test is so important to solving many problems that it deserves a separate chapter.

As indicated earlier, a successful plant test is one that either confirms the hypothesis is correct or proves that the hypothesis is not correct. An unsuccessful plant test is one that fails to either confirm or negate the hypothesis. In addition, it is important that the plant test be conducted in such a fashion that it does not create a major plant operating problem. The concept of analyzing problem solutions for potential problems was discussed earlier. Before conducting any plant test, a well-thought-out potential problem analysis should be completed.

The Russian nuclear power plant disaster at Chernobyl was at least partially due to failure to complete a well-thought-out potential problem analysis ahead of the test. The test was designed to determine whether, in the event of a reactor shutdown, enough power to operate the emergency equipment and core cooling pumps until the diesel power supply came online was available. It was to be conducted during the planned shutdown of reactor 4. The following contributed to the disaster:

- Only six to eight control rods were used during the test despite there being a standard order stating that a minimum of 30 rods were required to maintain control.
- The reactor's emergency cooling system was disabled.
- The test was carried out without a proper exchange of information between the team in charge of the test and personnel responsible for the operation of the nuclear reactor.

Problem Solving Approaches for Maintaining Operational Excellence in Process Plants, Second Edition.
Joseph M. Bonem, Nattapong Pongboot, and Wiroon Tanthapanichakoon.
© 2025 the American Institute of Chemical Engineers, Inc. Published 2025 by John Wiley & Sons, Inc.

14.2 Ingredients for Successful Plant Tests

All successful plant tests will require the following:

- A full evaluation of instruments and laboratory procedures to be used for the test.
- A careful statement of what results are anticipated and how the anticipated results will be evaluated. This will almost always involve a significant amount of pretest calculations.
- A complete well-thought-out potential problem analysis including "trigger points," which, if violated, will cause the test to be terminated.
- A careful and detailed explanation to operating personnel about the test.
- A formal post-test evaluation and documentation.
- These items are discussed in more detail in the following paragraphs.

14.3 Pretest Instrument and Laboratory Procedure Evaluation

One might think of an evaluation of instrumentation prior to a plant test of involving only instrument technicians "zeroing" and calibrating instruments. In addition, the instrument specification sheets themselves should be reviewed if there is any doubt at all about the instrument validity. For example, as discussed in Chapter 13, during a plant test in a high-pressure process, it was difficult to get a good material balance using instrumentation. All instrumentation was checked and confirmed to be as indicated on the specification sheets. On a closer evaluation of the specification sheets, it was discovered that no allowance had been made for gas compressibility at the high pressures. This would result in a higher density than that calculated using the ideal gas relationships. There will probably be value in reviewing the specification sheets of all instruments to be used in a plant test.

While laboratory procedures may be thought of as the domain of the chemists, if things do not appear to make logical sense, the problem solver should not assume that the procedure is correct or is being correctly followed. As an example, during a start-up of a chloride removal system, strange laboratory results were being reported. The procedure developed to test for the level of the chloride ion called for acidification of the sample prior to running the test for chlorides. The laboratory technician used his most available acid (hydrochloric acid) for the acidification step. This was discovered only when a *tactful* process engineer asked if he could watch the laboratory technician perform the test. Upon a careful examination of the procedure, a paragraph was found that indicated that nitric acid should be used in the acidification step.

These two examples are provided to indicate how the process engineer or problem solver must be involved in every step of the problem-solving efforts and/or plant test preparations regardless of the discipline that has the primary responsibility. In these two cases, a more thorough review of things prior to the plant tests would have allowed discovering that the incorrect gas compressibility was used as the basis for flowmeter calculations and that the laboratory procedure was not as clear as it should have been.

14.4 Statement of Anticipated Results

Essentially all plant tests are directed at improving plant operations. However, in order to obtain an adequate evaluation of the plant test, it is imperative that the anticipated results be stated as quantitatively as possible. In addition, the variable or variables to be used to monitor the results of the plant test should be quantified. Both the positive and potential negative results should be spelled out. For example, a plant test on operating a fractionation column at a 15% increase in the reflux rate to improve the purity of the distillate might have anticipated results stated as follows:

> "It is expected that this plant test will provide an increase in distillate purity of 0.5 weight percent. The distillate purity will be measured by the standard laboratory test procedure which is accurate to within 0.1 weight percent. An increase in tower pressure drop of 0.5 psi with no indication of tower flooding is anticipated. Tower flooding will be monitored by deviations from anticipated values of tray efficiency, tower pressure drop and heat balance closure."

There are several things to notice in this statement. The fact that these are anticipated results means that calculations have been performed to determine the impact of the increase in the reflux rate. For these calculations to be meaningful, similar calculations must be performed on the "base case." That is, fractionation calculations should be performed for the normal operating conditions. These fractionation calculations are based on theoretical trays. The base case tray efficiency can be determined by varying the number of theoretical trays in the calculations until the calculated compositions match the laboratory results. At this point, the tray efficiency can be determined by dividing the number of theoretical trays by the actual number of trays. This same tray efficiency can be utilized to estimate the change in distillate purity with the increased reflux rate.

If the anticipated distillate purity was not obtained, it would be an indication that the tray efficiency had decreased. This could be the result of some kind of tray overloading condition. In a similar fashion, the tower heat balance closure can be determined for the "base case." Heat balance closure is the difference between the heat added including all heat sources (feeds and external heat source) and the heat removed with the products and cooling sources. In theory, this value should be 0. However, it is rarely 0 due to meter errors and heat losses. If the heat balance closure becomes worse as the reflux is increased, it is likely due to tower flooding. Tower flooding *can* cause liquid to be entrained from the top of the tower into the reflux accumulator. The heat balance calculations assume that anything that enters the reflux accumulator came out of the tower as vapor. If some of the material going to the reflux accumulator is in fact liquid entrainment, the heat balance calculations will show more heat being removed from the condenser than is the actual case. This will result in a change in heat balance closure.

The fact that an anticipated tower pressure drop increase of 0.5 psi is indicated means that tray pressure drops have been calculated. The value in doing tray pressure drop calculations is twofold. It is necessary in order to assess whether tray flooding at the higher reflux rates would be anticipated. In addition, it provides an anticipated pressure drop which can serve as a "trigger point" to abort the test if it is exceeded.

In addition to the pretest calculations, the anticipated results include a statement about the accuracy of the laboratory test. A plant test where the anticipated distillate purity increase is 0.5 wt% would be meaningless if the laboratory accuracy was only 0.5 wt%.

The example of a fractionation column is very straightforward and easy to quantify with both statements and calculations. A more difficult plant test might be one in which a new catalyst with anticipated higher reactivity was to be tested. A similar technique to that described in Chapter 9 could be utilized to determine the actual higher reactivity of the new catalyst. The simplified kinetic constant of the "base case" catalyst could be determined knowing the reactor residence time and the reactant concentrations. The simplified kinetic constant for the new catalyst could be determined using the same variables. The comparison of these two kinetic constants could be made to determine the increased reactivity of the new catalyst. While this is more involved than the less rigorous method of just comparing catalyst efficiencies (pounds of product produced per pound of the catalyst), it avoids the need for the plant test to be run at the exact same residence time and reactant concentration. Plant tests of new high-reactivity catalysts are often conducted at lower feed rate, production rate, and/or reactant concentration to avoid potential problems that might be associated with the higher reactivity. Catalyst efficiency will be impacted by these changes. Thus, a comparison based on catalyst efficiency only will not be valid. This type of a more complex plant test often involves considerations that on the surface seem difficult to assess. Many times, it is possible to assess these considerations using fundamental chemical engineering skills. Two example problems are discussed later.

14.5 Potential Problem Analysis

A complete well-thought-out potential problem analysis is mandatory. A "trigger point" should be developed for each major potential problem. If these "trigger points" are violated, the test will be terminated. Most of the variables considered in the statement of anticipated results will be considered in the potential problem analysis. While the concept of potential problem analysis is closely related to the previous discussion, there are some additional considerations. Some of these are related to safety. While the list below is not an all-inclusive list, it is a list of the type of safety-related questions that should be considered:

1) Are any new chemicals being used?
2) Have by-product reactions and by products been evaluated?
3) Are any new chemicals or reaction by products compatible with the existing materials of construction?
4) Are the test operating conditions outside acceptable ranges?
5) Are there any proposed conditions that will cause the safety release systems to be inadequate? For example, a higher-reactivity catalyst might cause safety release facilities to be undersized.
6) Will operating condition changes cause a highly reactive chemical to concentrate to an unsafe level or be at unsafe conditions? Some examples are as follows:
   - Oxygen concentrating in a vent to the point that the oxygen–hydrocarbon mixture is in the explosive range.
   - Operating conditions for systems handling ethylene or acetylene reaching a pressure and/or temperature that can result in thermal decomposition.
   - Operating temperatures for a thermally sensitive polymer being so high that decomposition with release of highly reactive monomers occurs.
7) Are there changes in operating conditions that seem innocuous but could, along with a single unexpected occurrence such as a utility or mechanical failure, lead to catastrophic results?

The above list as indicated is not inclusive but is only to serve as an example of the type of safety items that should be considered. Any list of this type must *not* be considered a "checklist." But it should be considered a guideline only. "Checklists" often have a way of defeating their purpose by allowing the person responsible to simply check off items. In the pretest work for a plant test, serious consideration and likely significant calculations should be completed to ensure that no safety-related problems will occur.

In addition to safety-related items, an analysis should be made of what kind of things could go wrong (potential problems) during the plant test. As indicated earlier, each potential problem should have a "trigger point." If this trigger point is violated, the plant test will be terminated. The value of having pretest trigger points is that they can be calculation-based when time is available for careful planning. This is opposed to waiting until unexpected events occur during the plant test and intuition becomes the mode of decision-making. Examples of these trigger points are as follows:

1) A new catalyst introduced into the reactor might lead to fouling of the heat transfer surface. In this example, the trigger point should be the heat transfer coefficient. A trigger point of reactor temperature only will not be sufficient to determine if the heat transfer surface is fouling. The trigger point for the heat transfer coefficient should be set high enough to avoid the possibility that a small increase in the catalyst rate will cause an uncontrolled increase in reactor temperature. Chapter 5 discussed a real-life example of this type of event.
2) A change in the reflux rate to a distillation column that should lead to an increased purity should include a trigger point to allow monitoring of unexpected tray performance deterioration. One possibility would be a trigger point specification on column pressure drop. The anticipated effect on column pressure drop could be calculated prior to the test and monitored during the test.
3) A plant test on adding a reagent to an exothermic batch reaction at a rapid rate might be monitored by a trigger point of temperature increase in the initial 2 or 3 minutes of reagent addition. Calculations prior to the test could determine the maximum temperature increase to avoid exceeding the maximum desired reactor temperature. If this calculated initial value was exceeded, the test would be terminated. Chapter 10 includes an example of this type of trigger point monitoring.

Hypothesis testing in a commercial plant may involve using results based on laboratory or pilot plant studies. Sometimes, these results are based on equipment that does not simulate the commercial process exactly. A potential problem analysis should include possibilities associated with differences in equipment such as

1) Did the reactor used in the studies simulate the commercial facilities? Considerations must be given to
   - Simulation of a continuous reactor with a batch reactor.
   - Differences in mixing patterns and regimes in a laboratory or pilot plant reactor and a commercial reactor.
   - The reduced heat removal capability in the commercial reactor caused by the smaller heat transfer area-to-reactor volume (A/V ratio).
2) Did the equipment used in the laboratory test adequately simulate commercial equipment? For example, chemical compound stability studies are often determined in a laboratory oven. If the material will be subject to high shear rates in the commercial equipment, laboratory oven studies may not be sufficient.

14.6 Explanation to Operating Personnel

An explanation of the plant test should be given to operating personnel (and perhaps mechanical personnel). This explanation can be conducted in a training session, with a written handout or by one-on-one discussions. The following items should be included:

1) Purpose of the test—The value of the test for the overall company objectives should be explained to the operating and mechanical personnel. The plant test could be directed at solving an operating problem that will make for an easier job or producing a superior product that will lead to increased sales. Regardless of the goal of the test, this is an opportunity to build enthusiasm for the test by explaining that the success of the enterprise leads to both job security and promotional growth for the individual.
2) How was safety evaluated?—Unfortunately, most hourly personnel have had experiences with plant tests or plant changes where safety was not adequately considered. This explanation should consider new chemicals, new laboratory procedures, and new operating or mechanical conditions/procedures. Many operating companies have safety and peer review committees that review new operating changes. The operations and mechanical workers should be informed of the details of these reviews. In addition, if new chemicals are involved, the MSDS sheets should be reviewed with the appropriate personnel and posted in visible places.
3) Why the test will work—This will be an opportunity to explain to the hourly worker the theory behind the proposed test. Most theoretical explanations can be expressed in terms that even people without engineering degrees can understand. This is an important step and should not be considered just a requirement, but it should be looked on as an opportunity to educate and obtain "buy-in" for the test.
4) How will the test be evaluated?—The purpose of this explanation is twofold. It is an opportunity to explain exactly what will be considered in evaluating the test. It may be desirable as part of the test to have the hourly worker fill out data sheets. The accuracy of this will be a function of how well he understands how the values will be used. In addition, this will be a time to explain that a test that does not prove the hypothesis is not a failed test. The only failed tests are those that are not conclusive.
5) If the test works, what does he get from it?—The hourly worker will often have questions such as:
   - If the test is successful and changes in operating or mechanical procedures are required, will it make my job easier or harder?
   - Will these changes reduce manning or limit future addition of jobs?
   - If the test is successful, how do I get any credit for doing all this extra work?

These are questions that are often not brought up. However, they are almost always in the minds of the hourly workers. Ignoring them will usually leave it to the operator or mechanic to assume the "worst case" answers.

14.7 Formal Post-Test Evaluation and Documentation

The formal post-test evaluation and documentation phase of the plant test is often one of the most overlooked areas of plant test execution. The desire to improve organization efficiency often creates pressures to move on to the next plant test to solve the current problem or to move on to the next problem to be solved. Whether the test proved or disproved the working hypothesis, the benefits of the documentation are to provide a lasting reference that will help avoid repeating the test or, even worse, avoid future changes that will cause the problem to reoccur. In addition, as indicated earlier, the writing process will often clarify the thinking process and improve problem-solving activity.

A plant test that disproves the hypothesis is often not documented because the problem solver does not see this as important. He feels the need to move on and work on the next hypothesis to get to a problem solution. The failure to document successful plant tests that disprove the hypothesis will often cause a similar hypothesis to be proposed at some future time point. Thus, the failure to document the work, because of the alleged need to be efficient, will cause a loss in organization efficiency. A comment often made by many experienced operating and/or mechanical personnel when presented with a proposed plant test is "We tried that before and it didn't work." This is an indication that formal documentation of previous tests is not being done.

Even more serious is the failure to document a plant test that proved a working hypothesis. In this scenario, changes that eliminate the problem are made. Several years later, a proposal is made to reverse the changes to try to solve another problem, increase production, or improve product quality. When the question is raised "Why are we operating at these particular conditions," no documentation exists to show the fact that changes were made to eliminate an operating problem. Thus, the assumption is made that the conditions can be returned to the previous conditions.

Another scenario is the case of a failed plant test, that is, one that neither proves nor disproves the working hypothesis. Rather than making another attempt with an improved plant test that will be successful (prove or disprove the hypothesis), the approach is just completely abandoned. No documentation is done to indicate why the test was not conclusive. At a later point in time, the hypothesis is reintroduced. When a similar test is proposed, the memories of both technical and operating personnel are "We tried it before and it didn't work." Actually, the validity of the hypothesis is unknown. What is known is that the test did not prove that the hypothesis was correct. The converse that the test proved that the hypothesis was incorrect is not true. Because the recollection is that the test did not work, the hypothesis is abandoned. A good documentation of the previous test would have indicated that what failed was the test itself rather than the hypothesis being incorrect.

The size of the actual document should be minimized. It should include items such as the objective of the test, the test procedure, the test results, and the conclusions. In addition, any comments concerning safety should be included. Because of the technical conclusions and the possible need to incorporate changes into operating conditions or procedures, the final document should be approved by both operating and technical management.

14.8 Examples of Plant Tests

Some plant tests to verify hypotheses will be simple, and some will be more complicated. The next few pages present two examples of involved plant tests that were directed at improving a product and/or process. While they are not directly related to solving plant problems, they are real-life examples that illustrate the concepts discussed above. These same concepts are applicable to plant tests directly associated with solving plant operating problems.

14.9 Example Plant Test 14.1

A synthetic rubber producer was pressured by customers to change stabilizers to one that would continue providing a product that did not discolor due to stabilizer oxidation but would have better product-related properties. Based on the customer request, a new stabilizer was identified by the technology organization. Tests to confirm the stabilizing and nondiscoloring properties were conducted at different temperatures in ovens. These tests were conducted by first mixing the stabilizer with the polymer at ambient temperatures in low-intensity mixers. The polymer containing the stabilizer was then heated and held at elevated temperatures in the oven. These tests confirmed that the product did not discolor and that the molecular weight of the polymer did not change even after several hours in an oven. Thus, the new stabilizer seemed to meet the goals. That is, the product did not discolor and the molecular weight did not decrease, indicating that the new additive was an effective stabilizer.

The actual synthetic rubber process consisted of a polymerization section where the stabilizer was added to a slurry of rubber and water. This slurry was then pumped to the finishing section where the rubber was dried to remove water, extruded, and rolled on hot mills to put the final product into a form that could be boxed and shipped to the customer.

A plant test was scheduled to assess the utilization of this new stabilizer in the plant. The hypothesis being tested was that the new stabilizer could be added effectively in the plant and that it would be as efficient as indicated in the laboratory results. Since the laboratory data indicated that the stabilizer was very effective, only some of the steps listed earlier were implemented. These steps and what was actually done are summarized below:

1) **Pretest Instrument and Laboratory Procedure Evaluation**—For the most part, this was done in a satisfactory manner. The laboratory was well prepared to analyze for the stabilizer, and in addition, instruments were adjusted for the slightly different density.
2) **Statement of Anticipated Results**—Since the technology experiments indicated that such good results were obtained, very little was done to prepare a statement of anticipated results. It was just assumed that the stabilization properties would be comparable to the those of the existing stabilizer or even better. That is, there would be no change in molecular weight during the finishing operation.
3) **Potential Problem Analysis**—Actually, very little was done to develop a potential problem analysis. The technology work was so convincing that this new stabilizer would meet the criteria of not discoloring and avoiding molecular weight change in the finishing operation that this phase was just ignored. It was anticipated to be a "boring test."
4) **Explanation to Operating Personnel**—This was not done.
5) **Formal Post-test Evaluation and Documentation**—As discussed later, the test disproved the hypothesis so conclusively that it was mandatory to document the results of the test.

Since, based on technology testing, it appeared that the new stabilizer had stabilizing properties equivalent to those of the current stabilizer, all operating conditions such as polymer molecular weight and amount of stabilizer added were held constant and operations were simply switched over to the new material. Very soon, the molecular weight of the product leaving the finishing operations began to decrease rapidly, even though the molecular weight in the polymerization operations did not change. This was completely the opposite of what would be expected if the stabilizer was performing as anticipated based on the technology tests. The lack of a well-thought-out potential problem analysis before the test was now causing a panic problem-solving attempt during the test. The two different groups involved (operations and technical) began pursuing different approaches.

The operating group increased the molecular weight in the polymerization section in an attempt to make "on-specification" product leaving the finishing section. Since the molecular weight in the polymerization process was inversely proportional to monomer conversion, it was necessary to reduce the monomer conversion in order to increase the molecular weight. This lowered monomer conversion tended to overload the monomer recycle system. The operations continued to deteriorate. The molecular weight in the finished product continued to decrease, and the monomer conversion continued to be decreased as a compensatory action. The test was aborted after only a few hours of operation.

The technical/technology group initiated problem-solving activities. An analysis of the final product indicated that there was very little stabilizer present. This brought up several potential hypotheses:

- An incorrect amount of the stabilizer was being added.
- The stabilizer was being washed off the polymer in the water slurry operation.
- The stabilizer was being vaporized in the finishing operation, even though long-term technology tests in the ovens indicated this would not occur.
- The laboratory results were wrong, and the stabilizer really was present at the desired concentration.
- There was the target amount of stabilizer in the polymer; it was just not an effective stabilizer. This would be in conflict with the technology data.

An evaluation of the data after the test was aborted indicated that the stabilizer was being added at a correct rate and was not being washed off the polymer in the water slurry operation. However, it was being vaporized in the high-shear zone of the extruder and hot mills. Not only were these zones at elevated temperatures, but also the high shear caused large amounts of the surface to be exposed to the atmosphere. This was a radically different condition than experienced in the low-shear environment of the ovens.

14.10 Lessons Learned

There are several lessons that can be learned from this plant test. They involve several areas such as the failure to conduct a potential problem analysis, the failure to provide an explanation to operating personnel, the blind acceptance of technology data without questioning the basis for conclusions, and the attitude that this is going to be a "boring plant test."

A potential problem analysis would have very likely uncovered that the shear rates in the extruder/mills were vastly different than those in the very low shear drying ovens. This consideration would have generated additional technology work to try to simulate the high shear rate. This additional work would have likely shown that large amounts of the stabilizer were lost in the high shear rate studies. If the proposed stabilizer still appeared to be the best available, the plant test could have been conducted at much higher stabilizer addition rates to compensate for the vaporization of the stabilizer in the extruders and mills. A potential problem analysis would have also led to consideration of the overloading of the monomer recycle system that occurred when the conversion had to be decreased to a low level to obtain the specification molecular weight in the finished product. Plant tests where technology data are accepted blindly and/or where the test is considered to be boring will almost always result in failure of some kind.

The lack of a careful and detailed explanation of the plant test to the operating personnel did not impact the results of this first test. However, when another test was scheduled, there was a considerable amount of tension caused by the failure to adequately explain the goals and purposes of the first test.

14.11 Example Plant Test 14.2

As a second illustration of a plant test, consider the following example. It was desirable to test the capacity of a single reactor in a plant that used two reactors in series. This would allow assessing the possible low-cost debottlenecking of the existing plant as opposed to the alternative of building a new plant. If the single reactor could be operated at the same capacity as the existing two reactors in series, it would be possible to increase the existing plant capacity by at least 50%. The hypothesis to be tested was that a single reactor could produce at the same production rate as two reactors in series, with no increase in the fouling rate.

The reactor was a boiling reactor where the heat of reaction was removed by the vaporization of the reactant, which was then condensed and returned to the reactor. The reaction product consisted of small insoluble solid particles. Because of the size of the particles, some of them were always entrained with the vapors leaving the reactor. The vertical condenser used to condense the reactant was provided with trays to allow washing the entrained solids back into the reactor. A sketch of this system is shown in Figure 14.1.

A plant test was deemed necessary since at the higher vapor rates associated with the higher reactor capacity, more entrainment might be encountered. The increased entrainment might lead to fouling of the exchanger, the vapor line, or the liquid return line. In addition, the liquid handling capacity of the return line under possible fouling conditions was uncertain. It was necessary to develop techniques for monitoring the possible fouling of the exchanger, as well as the vapor line and liquid return line.

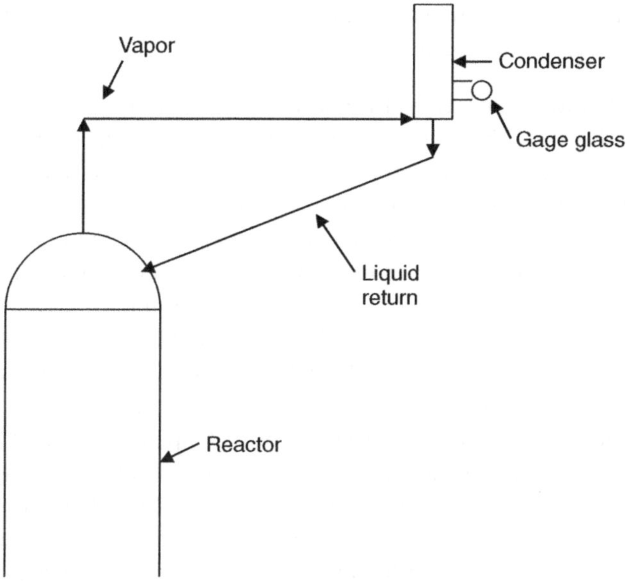

Figure 14.1 Reactor/overhead condenser schematic.

Three techniques were developed to monitor operations and fouling during the test of the single reactor. They were as follows:

1) The heat transfer coefficient on the overhead condenser was monitored hourly by the process control computer.
2) A local level gage on the bottom of the exchanger was monitored every 2 hours by visual observations. An increase in the liquid level in the bottom of the exchanger would indicate that the condensate was not flowing back into the reactor freely. That would mean that the liquid return line was likely partially plugged.
3) The vapor and liquid lines to and from the exchanger were "rung" every 2 hours. This "ringing" was a process of lightly tapping on the line with a hammer. If fouling was beginning to occur in the line, the pitch of the ring would have changed from a sharp clear ring to a dull thud. While this qualitative test was not described in textbooks, it had been proven by operating personnel in earlier experiences.

The manual monitoring (local level gauge and "ringing the lines") required the operating personnel to climb six flights of stairs in order to reach the desired physical area of the equipment.

As indicated earlier, there are five things that are essential to a successful plant test. For this plant test, these items were covered as follows:

1) **Pretest Instrument and Laboratory Procedure Evaluation**—This was relatively simple for this test. The calibration of instrumentation that would allow determining the exchanger heat transfer coefficient was part of a routine task order. However, as indicated later, it was necessary to re-range several instruments to allow operation of the reactor at higher capacity. The manual procedures for checking the level and condition of the vapor and liquid lines were well established by previous experiences with solid entrainment.
2) **Statement of Anticipated Results**—Prior to the plant test, the heat transfer coefficient on the condenser was determined from the existing temperatures and heat duty. Calculations were made to determine what the anticipated inlet and outlet water temperatures would be at the anticipated heat duty. These calculations assumed that the heat transfer coefficient would be slightly lower than that determined at the existing duty. This was done to allow for a higher thickness of the condensate flowing down the tubes. The actual change in condensate thickness would be very difficult to calculate with any accuracy. Thus, the reduction in the overall heat transfer coefficient caused by the thicker condensate layer was assumed to be 10%. Calculations were also made to confirm that the vapor and liquid lines had adequate capacity in their clean conditions. Operations personnel had a good intuitive feel for what a clean line sounded like when it was "rung." Thus, an overall statement of anticipated results would indicate the following evaluation of the reactor overhead circuit was expected:
   - Based on the anticipated heat transfer coefficient, the anticipated values of the water temperatures and flow rates around the condenser were estimated. These were included in the anticipated results.
   - Based on calculations, there was anticipated to be no level in the bottom of the exchanger. In this case, the calculations indicated that the liquid and vapor lines in a clean condition had sufficient capacity so that there should be no level buildup in the exchanger.

   The overall statement included a comment that the vapor and liquid return lines should "ring" with a clear sound. As with any process equipment or operating condition changes, safety must be considered. At the higher reaction rates, the safety valves would be required to release at a higher rate in an emergency. The required and actual capacities of the valves were reviewed, and it was found that the capacity of the valves was greater than the maximum anticipated release rate. After other potential safety problems had been evaluated, a statement was provided indicating that no safety problems were anticipated.

Other areas of the process were also considered to evaluate the effect of operating at full rates with only a single reactor. An evaluation of the catalyst feed system to ensure that adequate catalyst supply would be available was conducted. The catalyst efficiency was anticipated to decrease as reactor volume was removed from service. All of the instrumentation for the reactor to be operated at the higher rate was evaluated and meter ranges increased where necessary.

3) **Potential Problem Analysis**—In the case of this plant test, there were potential problems associated with the overhead condenser, as well as at other locations in the plant. Trigger points were developed for the heat transfer coefficient on the overhead condenser. This variable was routinely calculated by the process control computer, and a trigger point was set to ensure that the reactor temperature did not "run away." A reactor temperature "run away" could occur if the heat transfer coefficient on the condenser is so low that a slight increase in the reaction rate would cause the heat generated to increase at a faster rate than heat removal capabilities. This was explained in detail in Chapters 5 and 10. The trigger points for the liquid level in the condenser and for the "ringing" of the lines were more qualitative. Any accumulation of liquid in the condenser or "line ringing" that was not clear was set as a trigger point. Often when focusing on certain areas of the plant for a plant test, other areas are overlooked. As indicated above, the catalyst feed rate was anticipated to increase due to the reduced residence time associated with using a single reactor. Since the reactor was a boiling reactor, some of the reactor volume below the liquid level was bubbling vapor. As such, it would not be effective reaction volume. Correlations were used to estimate the effect of the higher "boil up" rate associated with the increase in production on the bubbling vapor volume of the reactor. As the "boil up" rate increased, the correlations predicted that the bubbling vapor volume below the liquid level increased and the effective reaction volume decreased. The potential problem that these correlations were inaccurate was considered. If the reduction in reactor volume, associated with the bubbling vapor, was greater than expected, the catalyst efficiency would be lower than anticipated. Thus, a trigger point was included for catalyst efficiency. Other trigger points were determined for other parts of the process where potential problems were anticipated. The key thing to consider is that having the well-thought-out potential problem analysis allows for reacting to situations in a predetermined way rather than in a reactionary mode in the "heat of battle."
4) **Explanation to Operating Personnel**—Most operating and mechanical personnel were happy to hear that their efforts to make a good product resulted in consideration of an increase in capacity. However, they had concerns that had to be addressed in order to obtain their cooperation for the plant test. Some of these comments and concerns were as follows:

- "We don't have time to climb stairs and do the checks every 2 hours."
- "You guys will get all the credit while we do all the work."
- "Won't running the plant at a higher rate make more work for me?"
- "If we expand by adding more plants that will mean more jobs for my friends and family. This 'debottlenecking' expansion doesn't add any jobs."
- "Is it really safe to operate the reactor in this fashion?"
- "I don't see how we can run this reactor at higher rates when we are having problems with carryover at the lower rates." Essentially, all of the questions that might come from the operating or mechanical personnel are in two categories. There are questions that can be answered based on calculations that have already been done as part of the pretest efforts or potential problem analysis. The value of doing the calculations is that the operator can be told that we have considered that and it is okay. The question of reactor safety is an example of this. It is always possible that a question will be raised, which has not been considered. *Serious consideration* should be given to any question of this nature.

  There will also be questions that cannot be answered based on calculations. Many of these will require an explanation that deals with company goals or the competitive situation. For example, the desire to expand by building more small plants could be answered by the fact that competitors are building larger plants and that being competitive with them provides job security. This job security for the employee is more important than creating jobs for friends and family.

5) **Formal Post-test Evaluation and Documentation**—While the test appeared to confirm the hypothesis, a formal test document was still prepared. The formal test evaluation and documentation included the conclusion that the heat transfer coefficient of the condenser at the higher rates was comparable to that at the normal rates. The slightly higher condensate thickness did not cause any decrease in the heat transfer coefficient. When the heat transfer coefficient was evaluated as a function of time, there was no indication that fouling was occurring during the test. In addition, there was no indication of fouling as gaged by the absence of any level in the condenser and the clear sound of the "pipe ringing" during the test.

   The catalyst efficiency anticipated at the higher rates was compared to the actual catalyst efficiency. Since these values were very close, it was concluded that no unanticipated loss in reactor volume associated with vapor bubbling occurred during the test.

   Perhaps one thing that is often overlooked in documenting a plant test is the comments from mechanics or operating personnel. Comments of the operating personnel were included. Their comments were that things ran very smoothly during the test.

14.12 Lessons Learned

This test was very successful from the standpoint that the results indicated very conclusively that it would be possible to operate the reactors at significantly higher rates. Even if the conclusions had been that it would not be possible to operate the reactors at high rates, the test would have been successful. The results would have been conclusive. As indicated earlier, a successful plant test is one where the test result is conclusive.

Several lessons were learned from this test. While the time to do the pretest calculations and potential problem analysis may seem to be inefficient, it provided a basis to convince management and operations personnel of the feasibility of the test. It also provided a basis for determining trigger points to allow a predetermined decision on aborting the test.

Using any short-term test to make decisions about fouling is difficult. The risk of concluding that fouling was not occurring was ameliorated by obtaining the maximum amount of data. In this case, three different techniques were used to monitor fouling.

There are always potential problems associated with any plant test. In the earlier example (Test 14.1), the anticipation that the test was going to be boring was radically different than the experience of this test.

Even in tests that appear to confirm a hypothesis, documentation of the results is important. In this test, it was obvious that the results associated with fouling should be documented. In addition, the analysis of the expected catalyst efficiency at the higher rates and the higher "boil up rate" proved very valuable. Without this analysis and documentation, erroneous conclusions regarding the reduced catalyst efficiency encountered as rates were increased might have been reached.

14.13 More Complicated Plant Tests

Frequently, it will be necessary to consider conducting a plant test where the cost of a plant shutdown is so great that "in-depth" theoretical *comparative* studies must be done as part of the potential problem analysis. These studies will allow determining the risk of such an event by a comparison to existing operations. These comparative studies will not provide absolute values but will allow determination of whether the plant test will increase or decrease the item of concern. An example of this might be a gas-phase reaction process where the presence of fines (small particles) could cause entrainment with a subsequent plant shutdown. Fines are known to be present with the current catalyst. However, the exact concentration is not known. The level of fines associated with the current catalyst results in minimal shutdowns. There are theoretical techniques available to predict the amount of fines produced by the catalyst knowing the catalyst attributes (reactivity, particle size, and particle size distribution) and reaction conditions.

A plant test using a new catalyst with radically different attributes in the gas-phase reactor was proposed. Concerns were expressed about the potential for this test causing a plant shutdown. In this case, the *absolute* level of fines was not as important as answering the question—"Will the amount of fines increase or decrease?" In order to assess the risk of a plant shutdown if a plant test is run with the new catalyst, calculations were done to estimate the amount of fines produced by both the existing and new catalysts. These calculations were then compared and an assessment made regarding the possibility of increased fines. If this assessment indicated that there will be a comparable or lower level of fines produced, then there is a high probability that the plant test can be conducted without shutting down the plant. Conversely, if the calculations indicate that the fines level will increase, a plant shutdown is a definite concern. In this case, it might be better to redesign the catalyst rather than risk a plant shutdown caused by the test of a new catalyst.

14.14 Other Uses for Plant Tests

While this chapter has been focused on the use of a plant test to test the validity of a working hypothesis, there are other utilizations of plant tests. Chapter 15 discusses the use of plant tests to validate and adjust digital twins that can be used for process optimization and training.

14.15 Key Plant Tests Considerations

The key thing to remember from this chapter is that if a plant test is used to confirm the hypothesis, it must be well planned with an aim to thoroughly prove or disprove the hypothesis. Successful plant tests are those that either confirm or reject the proposed hypothesis. Documentation of the plant test results regardless of the conclusion is imperative.

15

Utilization of Commercially Available Simulation Tools

15.1 Process Simulation and Modern Chemical Engineering

In the modern chemical engineering world, process or flowsheet simulation has become a prevalent tool for solving problems with varying complexities—be it process/equipment design, economic optimization, production planning, carbon footprint analysis, troubleshooting, etc.

Modern process simulators are equipped with sophisticated calculation algorithms for most unit operations with a user-friendly graphical interface, complemented with an extensive library of thermodynamic models, a comprehensive pure component databank, and physical property estimators. Thanks to advanced computing technologies, users have no longer been constrained by computation time, allowing them to perform complex and multiple case studies. With this capability, the user can check performances for multiple cases, conduct what-if analysis, perform sensitivity studies, and optimize process designs in no time.

As these process simulators are becoming more and more powerful, they are extensively utilized throughout the life cycle of process studies, from concept and feasibility to detail design, even extending to commissioning, operation, and revamps. In principle, process simulators perform calculations in such a way that thermodynamic models/rules are satisfied. The first outcome is a set of heat and mass balance data, which is a basis for all downstream process design activities including preparation of equipment and instrument specifications.

Some examples of commercially/open-source tools and their applications are listed in Table 15.1.

Problem Solving Approaches for Maintaining Operational Excellence in Process Plants, Second Edition.
Joseph M. Bonem, Nattapong Pongboot, and Wiroon Tanthapanichakoon.
© 2025 the American Institute of Chemical Engineers, Inc. Published 2025 by John Wiley & Sons, Inc.

Table 15.1 Examples of process simulators and their applications.

Simulator	Original Developer	Applications	License
Aspen Plus	AspenTech	Oil and Gas (upstream and downstream) Petroleum Refining Natural Gas Petrochemicals Chemical Industry Polymers Pharmaceuticals Life sciences Solid process Renewable	Closed-source
Aspen HYSYS	Hyprotech (now part of AspenTech)	Oil and Gas (upstream and downstream) Petroleum Refining Natural Gas	Closed-source
UniSim	Honeywell	Oil and Gas (upstream and downstream) Petroleum Refining Natural Gas Renewable	Closed-source
PetroSIM	KBC	Oil and Gas (upstream and downstream) Petroleum Refining Natural Gas Renewable	Closed-source
PRO/II	Schneider Electric (now part of AVEVA)	Oil and Gas (upstream and downstream) Petroleum Refining Natural Gas Petrochemicals Chemical Industry Polymers Pharmaceuticals	Closed-source
DWSIM	Daniel Wagner Oliveira de Medeiros	Oil and Gas (upstream and downstream) Petroleum Refining Natural Gas Petrochemicals Chemical Industry Polymers Pharmaceuticals Life sciences Solid process	Open-source

Despite their capability to solve and optimize virtually any process flowsheets and synthesis problems, process simulations at times fail to represent actual operating data for various reasons:

- Bad input – garbage in = garbage out
- Inappropriate selection of thermodynamic models or calculation methods—some specific systems require specific selections.
- Lack of fundamental process/equipment knowledge and experience—no inclusion of engineering judgment when performing the simulation, etc.

It is not uncommon to see that the same process flowsheet generates multiple and conflicting solutions for the same set of input—a better question is which solution is a valid one. This chapter discusses various issues and provides readers with guidelines to derive meaningful results from process simulation.

15.2 Breaking Down the Problem

There are numerous types of problems that can be encountered in process simulation. Nonetheless, most problems can be broken down into two major categories:

1) *Green field or design problem*: This usually means new designs, e.g., new process or equipment. Examples of deliverables are process configuration and equipment type/size/specification based on the process input/output. This type of problem allows more degree of freedom for the user but requires incorporation of design guidelines/rules and engineering justification to yield sensible results. This problem type is a crucial step to generate a heat and mass balance basis for downstream green field design activities.
2) *Brown field or simulation/rating problem*: This normally refers to existing process or equipment. In contrast to the green field problem, the process configuration and equipment type/size/specification are already available. The main objective of this problem type is to set up a process model based on the existing design data/parameter to represent the actual plant or laboratory data. A validated process model can then be used to perform what-if study, sensitivity analysis, or as a baseline for revamp study.

Each problem type requires different approaches/workflows when using a process simulator but share a common required process information, as summarized in Table 15.2.

In order to better understand how each problem can be solved using process simulators, they can be explained by block flow diagrams illustrated in Figure 15.1.

Table 15.2 Typical required process information.

Process information	Examples	Potential source—green field	Potential source—brown field	Note
Input parameter	• Temperature • Pressure • Flow • Composition • Bulk properties	• Design input, e.g., feed availability/ conditions/specifications • Research/literatures	• Actual plant data • Pilot-plant data • Lab data	
Design variables	• Process configuration • Equipment type/size/specification	• Design guidelines/standards • Past experiences of a similar process/equipment • Literatures	• PFD/P&ID • Datasheet • Drawing • Performance curve	
Output parameter	• Temperature • Pressure • Flow • Composition • Conversion • Yield • Product specifications	• Design objective, e.g., product demand/ conditions/specification • Research/literatures	• Actual plant data • Pilot-plant data • Lab data	
Tuning parameter	• Efficiency • Fouling factor • Reaction rate constant • Deactivation rate	Already included as design variables	Normally tuned to match the performances of the actual process/equipment	Thermodynamic model/product property estimators, and calculation method can also be considered a part of tuning parameters

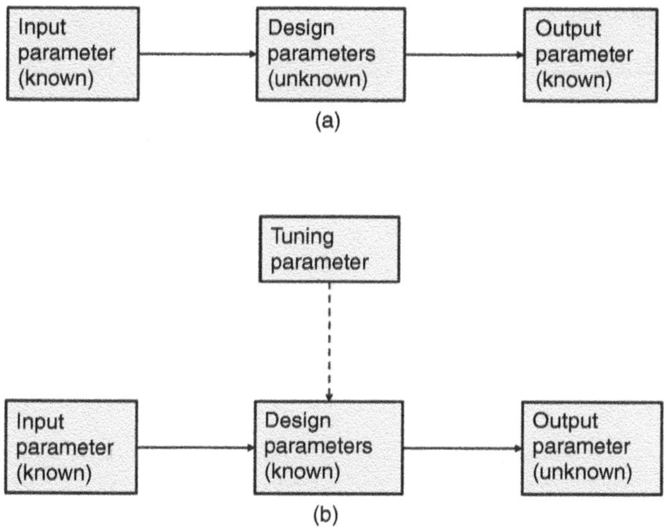

Figure 15.1 Definition of (a) green field problem (b) brown field problem.

The following examples provide a clear overview on how each type of problem is defined.

15.3 Green Field Problem Example

15.3.1 Situation

The gas mixture is to be compressed for injection into an offshore oil field as a means of enhancing well production. The objective here is to find the optimum process configuration and equipment sizes for this unit (jmcampbell.com 2006).

15.3.2 Input Parameter

- The dry basis gas composition is shown in Table 15.3
- The feed gas is saturated with water.
- The gas mixture is at 37 °C (98.6 °F) and 31 bar(g) (450 psig)
- The gas rate is 6.792×10^6 std. m^3/d (240 MMSCFD)

15.3.3 Output Parameter

The injection pressure is 131 bar (g) (1900 psig), and the temperature is 65 °C (149 °F).

Table 15.3 Feed gas composition—dry basis.

Component	mol%
Methane	73.2
Ethane	11.35
Propane	7.10
i-Butane	1.09
n-Butane	2.16
i-Pentane	0.63
n-Pentane	0.63
n-Octane	1.03
Nitrogen	0.06
Carbon dioxide	2.15
Hydrogen sulfide	0.60
Total	**100.00**

15.3.4 Design Parameters

The following design guidelines/considerations are incorporated to yield a reasonable system design:

- Soave–Redlich–Kwong (SRK) is selected as equation of state.
- The assumed isentropic efficiency is 72%, and the mechanical efficiency for each stage of compressor is 80%.
- The maximum discharge temperature is limited to 120 °C (for reasonable packing/seal life).
- Only the air cooler is available for interstage cooling, and no cooling water is available offshore. This limits the minimum achievable compressor suction temperature, especially in a hot climate like in this case where the average temperature of the warmest month is 42 °C.
- The inlet temperature of each compressor stage should not exceed 56 °C (132.8 °F).
- 5 psi (34 kPa) pressure drop is allowed between each compressor discharge and exit of the flash separator.

These guidelines/assumptions lead to the following design with a 2-stage gas injection compression unit, as illustrated in Figure 15.2.

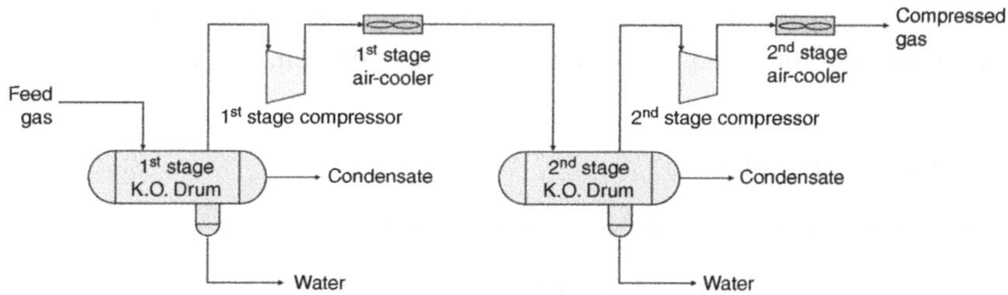

Figure 15.2 Process flow diagram for gas injection compression unit.

Table 15.4 Final summary of the gas injection compression unit design.

		Stage	
Design parameters Compressor	**Unit**	**1**	**2**
Suction pressure	bara	32.0	76.4
Discharge pressure	bara	76.7	132.3
Pressure ratio	–	2.40	1.73
Absorbed power	kW	9790.6	5804.1
Brake horse power	kW	12 238.3	7255.1
Actual discharge temperature	°C	112.3	103.4
Suction temperature	°C	37.0	56.0
Air cooler			
Inlet temperature	°C	112.3	103.4
Outlet temperature	°C	56.0	65.0
Cooling duty	kW	10 853.6	8407.5
Pressure drop	Bar	0.18	0.18
Total compression power	kW	19 493.4	
Total cooling duty	kW	19 261.1	

The interstage pressure (1st stage discharge pressure) is then further optimized to achieve the lowest total compression power and cooling duty (quick indicator of CAPEX/OPEX). The final results are summarized in Table 15.4.

15.4 Brown Field Problem Example

A gas mixture is to be compressed from one place to another by using an existing single-stage centrifugal compressor. With already defined gas composition, pressures, and temperature, a process engineer is tasked to estimate the maximum amount of gas that this compressor can deliver (jmcampbell.com 2011).

15.4.1 Input Parameter

- The wet basis gas composition is shown in Table 15.5
- The suction temperature and pressure are at 105 °F (40.6 °C) and 115 Psia (793 kPa), respectively.

15.4.2 Output Parameter

The destination pressure is 337 psia (2324 kPa).

Table 15.5 Feed gas composition—wet basis.

Component	mol%
Methane	83.099
Ethane	6.925
Propane	4.946
i-Butane	1.484
n-Butane	0.989
i-Pentane	0.495
n-Pentane	0.495
Hexane	0.198
Benzene	0.040
Heptane	0.148
Toluene	0.010
Octane	0.099
Ethyl-benzene	0.005
O-xylene	0.005
Nonane	0.050
Decane	0.030
Water	0.983
Total	**100.00**

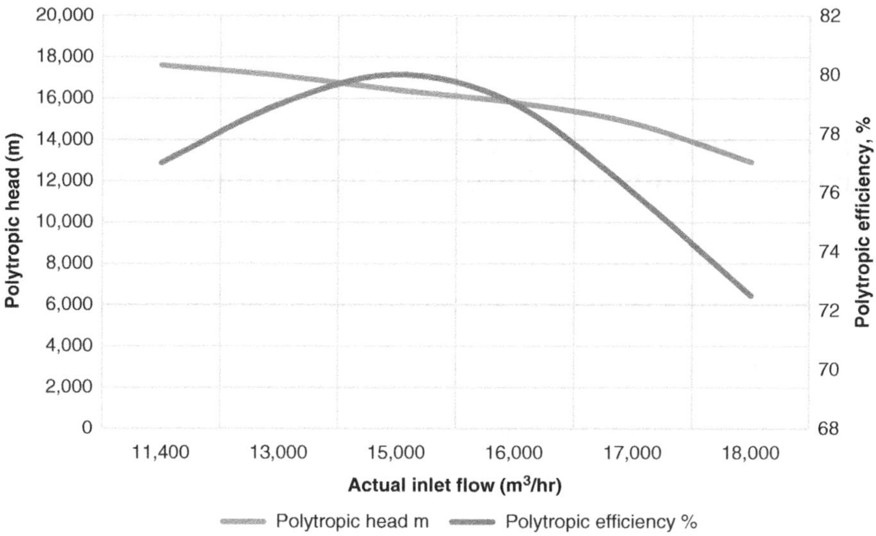

Figure 15.3 Centrifugal compressor performance curve.

15.4.3 Design Parameters

The existing centrifugal compressor has the characteristics (polytropic head and efficiency) described by Figure 15.3.

- Soave–Redlich–Kwong (SRK) as equation of state

This leads to the conclusion that this compressor can handle up to 101.00 MMSCFD (2.86×10^6 Sm^3/d).

15.5 Do Not Gamble with Physical Properties for Simulations

Regardless of the problem type, the thermodynamics model is critical for successful problem solving. As explained earlier in this chapter, all these process simulators perform calculations in such a way that thermodynamics are satisfied. Excellent guidelines are provided in the article with exactly the same headline (see Carlson 1996).

In order to clearly demonstrate the effects of the thermodynamics model on the final results, we will use the situation described in the Brown Field problem and compare the required compression power and cooling duty results from each equation of state.

15.6 Examples—Effects of Equation of State on the Required Compression Power and Cooling Duty

15.6.1 Input Parameter

- The total gas flow rate is 101.00 MMSCFD (2.86×10^6 Sm3/d) (jmcampbell.com 2011).
- The suction temperature and pressure are at 105 °F (40.6 °C) and 115 Psia (793 kPa), respectively.
- Same gas composition as in Table 15.5.

15.6.2 Output Parameter

- The final pressure target is 1015 psia (7000 kPa).

15.6.3 Design Parameters

- Three compression stages with the same pressure ratio of 3
- The stage efficiency is estimated to be 86%, 80%, and 79%.
- The suction pressure should always be kept at 40.6 °C.
- The pressure drop across each air cooler can be assumed as 5 psi (34 kPa).
- Results from the following equation of states will be compared:
  1) Soave–Redlich–Kwong (SRK)
  2) Peng–Robinson (PR)
  3) Lee–Kesler (LK)
  4) Benedict–Webb–Rubin-Starling (BWRS)

Table 15.6 shows that the maximum deviation in total compression power requirements is 1.4% and 2.0% for the total cooling load. For cases with large power requirement, these small differences in terms of total HP or kW could be

Table 15.6 Different compression power and cooling load obtained from different equations of state.

		Equation of State			
Design Parameters	**Unit**	**PR**	**SRK**	**LK**	**BWRS**
Total compression power	kW	15,528	15,655	15,520	15,422
Total cooling duty	kW	29,904	27,170	28,781	28,625

significant. Therefore, care should be taken to choose an appropriate equation of state (EOS) for handling vapor–liquid equilibrium (VLE) calculations and accurate predictions of enthalpy and entropies for the system under consideration. The deviation range could be different for other cases depending on the flow rates, condition composition, and pressure ratio.

15.7 Be Skeptical with your Initial Design Assumptions

In principle, process simulation software is a thermodynamics calculator. As such, some aspects, e.g., hydraulic, can be missing from the picture if not carefully addressed. In reality, hydraulic calculation is a critical part of real-world engineering design. A process designer must ensure that fluids/materials smoothly move from one place to another in the plant.

As an example, pressure drop always exists in an operating separation column due to hydraulic restrictions from column internals. However, it is actually possible to converge a separation column with zero tray pressure drop in the process simulator. A widely adopted inside-out algorithm is vapor–liquid equilibrium-based and does not take into account pressure losses, unless the user reasonably specifies the column pressure profile in the column model.

Therefore, the assumed column pressure profile should always be rechecked using the column internal design/rating software, e.g., SULCOL™ and KG-TOWER®, based on a fixed column internal design. Some design/rating iterations might be required to find a good match between the two calculations. A good separation column design should take into account different flooding (or weeping) mechanisms (e.g., jet flooding and downcomer backup) to ensure it can operate over a good range of operation. This design aspect can be checked by the mentioned column internal design/rating software. Modern process simulators already include column hydraulic calculation, so column design and rating can also be done interactively.

In addition to the separation itself, the pressure profile also affects the temperature profile. As an example, reboiler temperature is directly affected by the bottom column pressure (boiling liquid is at the bubble point), therefore affecting the reboiler design and sizing.

As a best practice, design assumptions used in the process simulator should always be cross-checked by dedicated equipment design software. This workflow is a feedback loop, design then rating, and so on. After the equipment design is fixed, the calculation results should be re-input into the process simulator to see how the equipment design affects the whole process flowsheet, as illustrated in the design flow path (Figure 15.4).

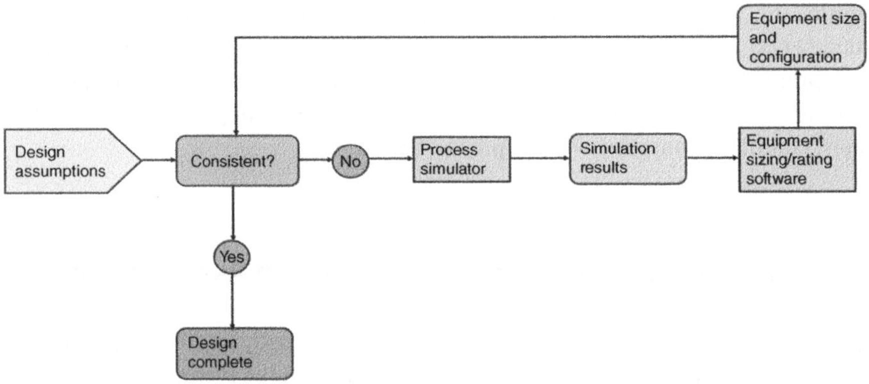

Figure 15.4 A feedback loop of process and equipment design.

Some examples of useful equipment design and rating software include the following:

- HTRI/Aspen EDR—for heat exchanger
- FRNC-5PC—for fired heater
- MYSEP—for phase separation devices
- PIPEPHASE—multiphase pipeline network

Some of the aforementioned applications can also be integrated and work interactively with modern process simulators providing the user with convenience by eliminating the extra steps required for data/information transfer.

15.8 Obtaining a High-Quality Plant Data for your Process Model

This step is a crucial part of creating a baseline process model from the existing plant (or pilot plant data) before starting to solve a Brown Field problem. As pointed out earlier, the calibrated process model can only be as good as the input. If the input data are of low quality, then it does not make sense to expect an accurate process model. Here are useful tips to obtain representative data from the existing plant:

- In general, a good process model should be able to fit/predict multiple sets of process conditions. As such, multiple test runs are preferred—1 as a bare minimum and ideally greater than 2. As an example, a hydrocracking unit should be tested at different feed conversions or feed compositions to gather all the information required to build a hydrocracker model which can cover a good range of operating scenarios.

- Some chemical processes might have special requirements. For example, the test run of the hydroprocessing unit should only be conducted at least few months after the start-up to let the catalyst stabilize. As a rule of thumb, minimum delay of 30 and 90 days for hydrotreaters and hydrocrackers, respectively, is required to obtain a stable start of run performance. For naphtha reforming, 7 days of oil before the start of a test run.
- Have a clear idea about the test run objective. If improved capacity/throughput is desired, the test run should be at the maximum documented throughput. If it is instead for improved reliability, better purities, or other reasons, then the test should not be at the maximum capacity limit.
- To ensure a steady state during the data collection, the test run should be conducted for 3 days minimum for each data set. Depending on the application, the data might need to be collected for multiple weeks for good statistical quality. As common sense, the data gathered during the process/equipment upset, e.g., emergency depressurization and thermal runaway, should not be used.
- A list of instrumentations to be calibrated should be generated. It should include thermocouples, pressure transmitters, and flow transmitters. Make sure that the critical flowmeters are ranged properly. Also, all important flowmeters are pressure-, temperature-, and density-corrected.
- Online analyzers should also be verified by off-line lab analysis to ensure that they work properly prior to the test run.
- For data reconciliation purposes, it is good to have redundant measurements when possible. There are many sources of inaccuracies in measured data. A good example is a heat duty of a shell and tube heat exchanger. In this case, the heat exchanger duty can be calculated by a flow rate and temperature range (difference between inlet and outlet temperature) of either the hot or cold side. A better question is which calculations are more representative. That is, the main reason why it is always better to have redundant measurements for data verification purposes.
- Flow data of reflux, pumparound, recycle streams are always helpful to verify the heat and material balance, although they tend to be less accurate due to composition and condition changes, and less often calibrated.
- Apart from the process measurement itself (flow, temperature, pressure, and composition). An accurate process/equipment configuration is necessary. It is not uncommon to see on-site modifications that conflict with the drawing/datasheet due to poor management of change. Always cross-check for consistencies between the paper and real life. The author has seen many undocumented bypass lines that appears in the field, but not in P&ID.
- It is always helpful to compare the obtained data with typical data for similar applications. Significant differences from the peers should be explored in detail.

- Regardless of problem types, a good representation of feed stream is essential for success. A simple method is to collect a feed sample and then analyze for composition/physical/chemical properties. Nonetheless, some complex problems might require a back-synthesis of the products to create a more representative feed. As such, the product rate, composition, or physical properties should be then gathered instead. This feed synthesis approach can be represented by Figure 15.5.

A good example is the FCC main fractionator, where the feed is in the vapor phase at very high temperatures >500 °C—accurately collecting sample and analyzing the properties is very difficult. Nevertheless, it is a best practice to gather both feed and product data as redundancy, so cross-checking can always be done.

- Pressures should be measured locally for all hydraulic systems to establish baseline performance data. This is especially critical for distillation towers where column pressure and temperature profiles are strongly inter-related from vapor–liquid equilibrium. The field pressure instrument should be selected to match the application, e.g., mano-meter for vacuum applications. This exercise also serves as a cross-check to already available pressure transmitters.
- Temperatures should be measured locally where remote indication does not exist. This can be accomplished by the use of a pyrometer gun. This exercise also serves as a cross-check to already available temperature transmitters.
- Control valves' travels and their bypasses that are partially open around control valves should be documented
- A suspectedly wrong flowmeter may be cross-checked by portable flowmeters, e.g., ultrasonic flowmeters. Feed/product flowmeters can also be cross-checked by the tank level.

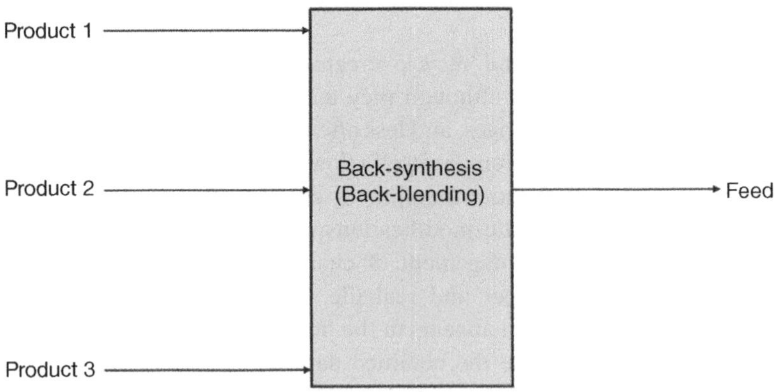

Figure 15.5 Feed synthesis from three product streams.

- Overhead vapor rate tends to be inaccurate due to varying gas composition.
- Product samples should be taken when it is certain that there is only one phase and from the well-mixed part of the process. Vapor samples to be taken where there is little chance of entrained liquid.
- Field measurements can also indicate equipment damages, e.g., damaged or dislodged trays have little or no pressure drop. This can also be checked before starting the test run if there are any known hardware damages/malfunctions, i.e., from inspection reports.
- Weather conditions should also be documented as performances of process equipment are also affected. For example, the colder the ambient temperature, the better the gas turbine efficiency.

15.9 Verifying your Plant Data

After obtaining the raw plant data, it should be first verified to eliminate obvious measurement errors. It is not uncommon to see the process unit/equipment that are not instrumented sufficiently to check the heat and material balance by independent measurements on all streams. Sometimes, even the major streams are not measured (i.e., recycle streams or vapor feeds) or measured but not calibrated often enough.

Some streams experience constant composition changes (e.g., overhead vapor from distillation column), making it even more difficult to obtain an accurate flow rate—as most of flowmeters still rely on the fluid composition to convert volumetric into mass flow rate (except Coriolis flowmeter).

A heat balance should be executed around any heat exchanger that impacts both the product and feed streams, like in a crude oil processing unit illustrated in Fig. 15.6. To verify the suspiciously wrong rates of any stream, a control valve flow rate can be estimated from the valve travel. Many times, a pressure drop across a flow element can be easily obtained and applied to a simple orifice equation to get an idea of the mass flow rate. This easy test is most often performed when the conversion factors in the control system are questionable or the system is operating far from original design.

In classical mechanics, no mass can disappear from the system, a component check can be performed for nonreactive (or minimally reactive) systems like distillation columns. As an example, a component balance can be performed across propane/propylene separation columns as negligible chemical reactions should occur inside the column.

For reactive processes (e.g., reactor), the reactant molecules are converted to form new product molecules so that component balances are no longer valid. As such, elemental balances offer a good validation check if the raw material data are

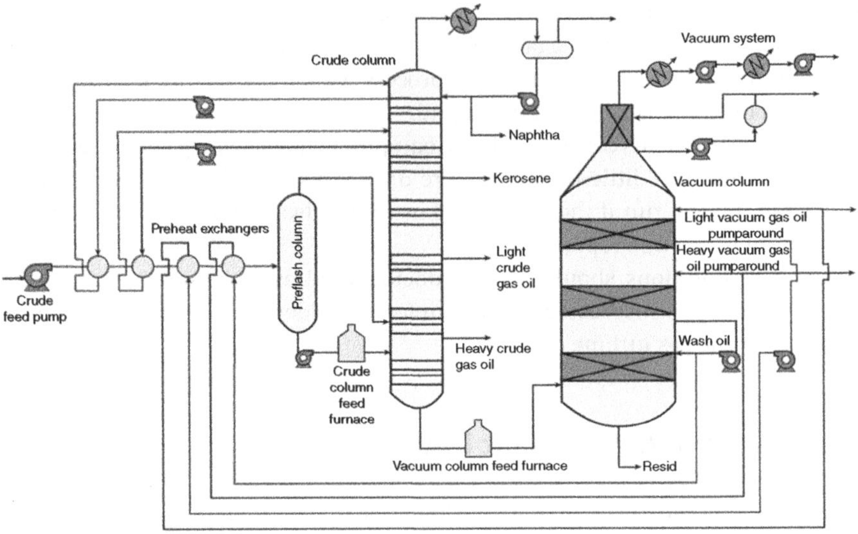

Figure 15.6 Crude oil distillation system.

valid or not. For complex systems in which feeds and products cannot be easily broken down into elements, e.g., crude oil processing, an estimation of elements can be done based on the analysis of bulk properties like distillation curve, density, aromatics content, bromine number and often already integrated as part of modern process simulation tools. The importance of this elemental balance concept will also be demonstrated in the subsequent section.

As a rule of thumb, if the overall flow rates, heat, and component/element balances are within 5%, these data can be considered a good data set already. Typically, flowmeters have a 2% accuracy; therefore, exact balances are almost impossible.

In order to give a clear idea of how a data verification can be accomplished, an example of heat balance calculation is given here.

15.10 Example—Heat Balance of Heavy Gas Oil Pumparound

Measured plant data are summarized in Table 15.7. In this example, an important assumption is that most heat is transferred as sensible heat.

In this case, the difference between hot and cold duty is less than 10%, but it can be considered acceptable, considering that there could be errors from heat of vaporization.

Table 15.7 Measured process data from a crude preheat exchanger.

Parameter	Unit	Heavy Gas Oil	Crude Oil
Flow rate	Lb/hr	492,333	642,292
Inlet temperature	°F	490	275
Outlet temperature	°F	375	370
Heat capacity	Btu/lb/°F	0.62	0.63
Duty	MMBtu/hr	35.1	38.4

15.11 Reconciling your Data

Despite an intensive data verification process, a certain level of measurement errors will always be there. Measured process data inherently contain some inaccurate information since measurements are obtained with imperfect instruments. Moreover, signal transmission, power fluctuation, improper instrument installation, and miscalibration can also contribute to measurement errors.

In order to further improve the raw process data set, a numerical technique called data reconciliation is widely employed and already embedded in modern process simulators. The use of data reconciliation allows for extracting accurate and reliable information about the state of processes from raw measurement data and produces a single consistent set of data representing the most likely process operation.

Data reconciliation is the estimation of process variables based on information contained in the process measurements and models. The process models used in the data reconciliation are usually mass and energy conservation equations. By principle, the data reconciliation technique allows the adjustment of the measurements so that the reconciled measurements are consistent with the corresponding mass and energy balances, as illustrated in Figure 15.7. This information from the reconciled data can then be used for process modeling purposes.

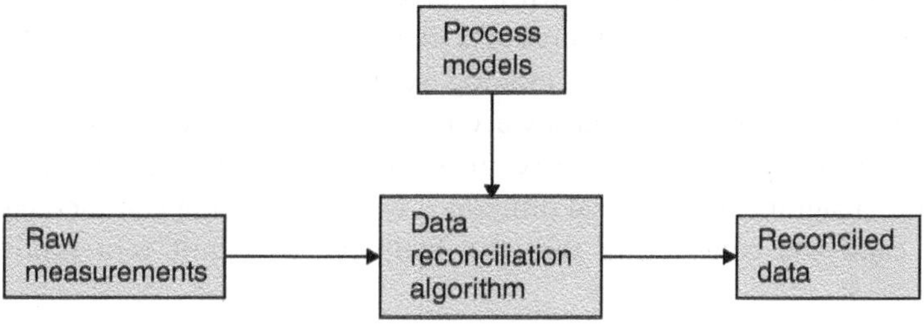

Figure 15.7 Data reconciliation process.

From the mathematical standpoint, the optimal estimates for process variables by data reconciliation are solutions to a constrained least-squares or maximum likelihood objective function, where the measurement errors are minimized with process model constraints. The in-depth details will not be re-iterated here but can be further explored in Romagnoli and Sanchez (1999).

The objective functions depend on the application. If you work with a power plant, then your reconciled data should well satisfy both mass and energy balance. However, if you work with reactor modeling, part of the energy balance (e.g., heat balance of heat exchanger/distillation column) may be compromised/relaxed to allow better overall and elemental mass balances, which are more important for tuning chemical reaction rate constants.

In order to illustrate how the data reconciliation process works, an example of the data reconciliation algorithm for lab-scale elemental balances will be given here.

15.12 Example—Hydrocracking Catalyst Testing

A hydrocracking catalyst testing was conducted to evaluate performances of commercial hydrocracking catalysts. The hydrocracking chemistry can be described by the following equation:

$$F_{\text{feed}} + F_{\text{chemical hydrogen}} = F_{H_2S} + F_{NH_3} + F_{\text{gas product}} + F_{\text{liquid product}} \quad (15.1)$$

Liquid hydrocarbon feed contains mostly C, H, S, and N with negligible metals in ppm or ppb level. As it undergoes hydrocracking reactions, reaction products including H_2S, NH_3, and shorter-chain hydrocarbon products are produced with a higher level of saturation. Some S and N atoms will remain in the shorter-chain hydrocarbon products. A certain amount of hydrogen is consumed to satisfy occurring chemical reactions, so-called "chemical hydrogen consumption."

"Raw" flow measurements and elemental compositions from this hydrocracking experiment are summarized in Table 15.8.

As pointed out earlier, there is no such thing as a perfect measurement, so the mass balances are imperfect, as shown in Table 15.9. It must be noted that the overall mass balance is defined by products divided by feed (equation 15.2), so this is an indication that something is missing from the product side (which is not uncommon as product losses can easily occur during handling of the sample or from error in gas analysis). This error seems small on the surface, but it can be translated into a significant financial difference when evaluating commercial hydrocracking catalysts, considering that the hydrocracking unit is a profit

Table 15.8 A raw data set from a hydrocracking catalyst test.

Category	Parameter	Unit	Test Scenario Conversion 1A	 Conversion 1B	 Conversion 2A	 Conversion 2B	Note
Feed data	Mass flow	g/hr	6.894	6.871	6.820	6.817	
	Composition-						
	Hydrogen	wt%	12.867	12.871	12.923	12.924	
	Sulfur	wt%	1.351	1.319	1.299	1.294	
	Nitrogen	ppm	883.4	901.8	848.0	830.7	
	Carbon	wt%	85.694	85.720	85.693	85.698	
Gas hydrocarbon product data	Mass flow	g/h	0.757	0.766	0.641	0.657	
	Composition-GC						
	Hydrogen	wt%	15.784	15.761	15.724	15.728	
	Sulfur	ppm	1.000	1.000	1.000	1.000	As organic compounds, typical value for demonstration
	Nitrogen	ppm	1.000	1.000	1.000	1.000	As organic compounds, typical value for demonstration
	Carbon	wt%	84.215	84.239	84.276	84.272	
Liquid hydrocarbon product data	Mass flow (g/hr)	g/h	6.103	6.046	6.113	6.105	
	Composition-						
	Hydrogen	wt%	14.81	14.86	14.83	14.79	
	Sulfur	ppm	10.00	10.00	10.00	10.00	Typical value for demonstration
	Nitrogen	ppm	2.00	2.00	2.00	2.00	Typical value for demonstration
	Carbon	wt%	85.19	85.14	85.17	85.21	

Table 15.9 Raw mass balance summary of a hydrocracking catalyst testing.

		Case			
Parameter	**Unit**	**Conversion 1A**	**Conversion 1B**	**Conversion 2A**	**Conversion 2B**
$\%_{\text{overall mass balance, raw}}$	wt%	98.99	98.62	98.60	98.76

generator in a typical refinery.

$$\%_{\text{overall mass balance,raw}} = (F_{\text{gas product}} + F_{\text{liquid product}} + F_{\text{H}_2\text{S,raw}} + F_{\text{NH}_3\text{,raw}}) \times \frac{100}{(F_{\text{feed}} + F_{\text{chemical hydrogen,raw}})} \tag{15.2}$$

Before this data set can be used for economic evaluation or as kinetic model input, it is always better to improve the data quality using data reconciliation. A rigorous universal data reconciliation method is described here by assuming that the feed composition/flow measurement is 100% trustable and instead reconciling the error-prone product flows.

This assumption is normally valid in a well-controlled environment like in lab-scale, where the liquid flow is normally measured by a high-accuracy weight-scale. The main benefit of this assumption is its simplification; therefore, the algorithm can be conveniently embedded in calculation spreadsheets without the need of least-square regression. The proposed method can be described by equation (15.3) to (15.11).

Reconciled overall mass balance equation

$$F_{\text{feed}} + F_{\text{chemical hydrogen,reconciled}} = k_{\text{scaling}} \times (F_{\text{gas product}} + F_{\text{liquid product}}) + F_{\text{H}_2\text{S,reconciled}} + F_{\text{NH}_3\text{,reconciled}} \tag{15.3}$$

Reconciled carbon mass balance equation

$$C_{\text{carbon,feed}} \times F_{\text{feed}} = k_{\text{scaling}} \times (C_{\text{carbon,gas product}} \times F_{\text{gas product}} + C_{\text{carbon,liquid product}} \times F_{\text{liquid product}}) \tag{15.4}$$

Reconciled hydrogen mass balance equation

$$F_{\text{chemical hydrogen,reconciled}} = k_{\text{scaling}} \times (C_{\text{hydrogen,gas product}} \times F_{\text{gas product}} + C_{\text{hydrogen,liquid product}} \times F_{\text{liquid product}}) + \frac{2 \times \text{AM}_{\text{H}}}{\text{MW}_{\text{H2S}}} \times F_{\text{H}_2\text{S,reconciled}} + \frac{3 \times \text{AM}_{\text{H}}}{\text{MW}_{\text{NH3}}} \times F_{\text{NH}_3\text{,reconciled}} - C_{\text{hydrogen,feed}} \times F_{\text{feed}} \tag{15.5}$$

Reconciled H_2S mass balance equation

$$F_{H_2S,\text{reconciled}} = \frac{\text{MW}_{\text{H2S}}}{\text{AM}_{\text{S}}} \times (C_{\text{sulfur,feed}} \times F_{\text{feed}} - k_{\text{scaling}} \times (C_{\text{sulfur,gas product}} \times F_{\text{gas product}} + C_{\text{sulfur,liquid product}} \times F_{\text{liquid product}})) \tag{15.6}$$

Reconciled NH_3 mass balance equation

$$F_{\text{NH}_3,\text{reconciled}} = \frac{\text{MW}_{\text{NH3}}}{\text{AM}_{\text{N}}} \times (C_{\text{nitrogen,feed}} \times F_{\text{feed}} - k_{\text{scaling}} \times (C_{\text{nitrogen,gas product}} \times F_{\text{gas product}} + C_{\text{nitrogen,liquid product}} \times F_{\text{liquid product}})) \tag{15.7}$$

Reconciled compositional balance equations

$$C_{\text{carbon,feed,reconciled}} = 100\% - C_{\text{hydrogen,feed}} - C_{\text{sulfur,feed}} - C_{\text{nitrogen,feed}} \tag{15.8}$$

$$C_{\text{carbon,gas product,reconciled}} = 100\% - C_{\text{hydrogen,gas product}} - C_{\text{sulfur,gas product}} - C_{\text{nitrogen,gas product}} \tag{15.9}$$

$$C_{\text{carbon,liquid product,reconciled}} = 100\% - C_{\text{hydrogen,liquid product}} - C_{\text{sulfur,liquid product}} - C_{\text{nitrogen,liquid product}} \tag{15.10}$$

From equation (15.4), k_{scaling} can be arithmetically determined by rearranging into equation (15.11).

$$k_{\text{scaling}} = C_{\text{carbon,feed}} \times F_{\text{feed}} / (C_{\text{carbon,gas product}} \times F_{\text{gas product}} + C_{\text{carbon,liquid product}} \times F_{\text{liquid product}}) \tag{15.11}$$

It must be noted that all these equations are mathematically rigorous, so every element will be reconciled to perfection.

In order to demonstrate the effectiveness of this data reconciliation technique in improving the data quality, a percentage chemical hydrogen consumption based on the liquid feed will be compared, as described in equation (15.12) (raw) and (15.13) (reconciled).

$$\%_{\text{chemical hydrogen,raw}} = \left(C_{\text{hydrogen,gas product}} \times F_{\text{gas product}} + C_{\text{hydrogen,liquid product}} \times F_{\text{liquid product}} + \frac{2 \times \text{AM}_{\text{H}}}{\text{MW}_{\text{H2S}}} \times F_{\text{H}_2\text{S,raw}} + \frac{3 \times \text{AM}_{\text{H}}}{\text{MW}_{\text{NH3}}} \times F_{\text{NH}_3,\text{raw}} - C_{\text{hydrogen,feed}} \times F_{\text{feed}} \right) \times \frac{100}{F_{\text{feed}}} \tag{15.12}$$

Table 15.10 Differences in chemical hydrogen consumption, raw versus reconciled.

Parameter	Unit	Case Conversion 1A	Conversion 1B	Conversion 2A	Conversion 2B
$\%_{\text{chemical hydrogen, raw}}$	wt%	2.08	2.07	1.95	1.93
$\%_{\text{chemical hydrogen, reconciled}}$	wt%	2.26	2.31	2.20	2.16
Δ,|raw-reconciled|	wt%	0.18	0.25	0.25	0.22

$$\begin{aligned}
&\%_{\text{chemical hydrogen,reconciled}} \\
&= \left(k_{\text{scaling}} \times (C_{\text{hydrogen,gas product}} \times F_{\text{gas product}} + C_{\text{hydrogen,liquid product}} \times F_{\text{liquid product}}\right) \\
&\quad + \frac{2 \times \text{AM}_{\text{H}}}{\text{MW}_{\text{H2S}}} \times F_{\text{H}_2\text{S,reconciled}} + \frac{3 \times \text{AM}_{\text{H}}}{\text{MW}_{\text{NH3}}} \times F_{\text{NH}_3\text{,reconciled}} - \text{C}_{\text{hydrogen,feed}} \\
&\quad \times F_{\text{feed}}\big) \times \frac{100}{F_{\text{feed}}}
\end{aligned} \tag{15.13}$$

Hydrogen is expensive, so small errors can cause a large financial difference. From Table 15.10, it can be seen that the chemical hydrogen consumption was missing by 0.18–0.25 wt% if the raw data sets are not reconciled. This is equal to approximately 10–13 million $ per year based on 2000 $/ton hydrogen price and 345 stream-days per year.

It must be noted here that there is no one-fits-all solution to data reconciliation. The data reconciliation strategy depends on the flow measurement accuracy and magnitude. Modern process simulators already include the data reconciliation module as part of the input data treatment. Here are some useful tips when performing the data reconciliation:

- Only adjust rates on streams with low accuracy levels:
  1) Streams with known metering issues, e.g., improper installation/measuring environment
  2) Uncorrected flowmeters
  3) Low pressure gas flowmeters—as the volumetric flow rate is more sensitive to pressure changes at low pressure
  4) Stripping steam
  5) Streams with varying composition, e.g., light naphtha
  6) Bottoms product rate—affected by swings in tower bottoms level

- Adjust the large feed or product streams for larger adjustments to the overall and carbon balances
- Adjust the low-pressure gas streams as they are inherently less accurate.
- Some streams might need to be specified as calculation results to facilitate mass and element closures. As an example, H_2S and NH_3 are not measured directly in hydroprocessing applications but reconciled/calculated as part of data reconciliation to close C/H/N/S balances.
- When properly performed, the reconciled flow should not deviate more than 2% from the raw flow (typical flow meter error).

15.13 Model Calibration

A first step is to reset your mind when working with model calibration. While there is more freedom when designing a process flowsheet from scratch, the degree of freedom is much more limited when trying to model the existing process/equipment. Most importantly, the workflow is reverse.

The main task here is to create a process model which can well predict the reality. Here are essential factors affecting the quality of calibrated process/equipment models:

- Correct hardware configuration/performance characteristics
- Accurate plant/lab data
- Suitable thermodynamics/physical model
- Proper calibration factors, e.g.,
  1) Distillation tower stage efficiencies
  2) Reaction rate constants/deactivation rate
  3) Heat exchanger fouling factors
  4) Fired heater heat flux imbalances: many heaters suffer from flux imbalances that result in a difference in pass flowrates as high as 50% between radiant passes.
  5) Pipe roughness/fouling

The process modeling workflow can be described by Figure 15.8. Once this is finished, the model can be considered complete.

In multiple cases, it is difficult to satisfy every aspect of the process model. For example, the calibrated crude distillation models in Figure 15.6 should be able to well predict the product yields and properties, but it is much more difficult to fully satisfy the heat balance, e.g., pumparound or reflux duty. Chances are that the model is accurate, but the measurements, which are used to compare with the model, are inaccurate. It is known among refinery operators that

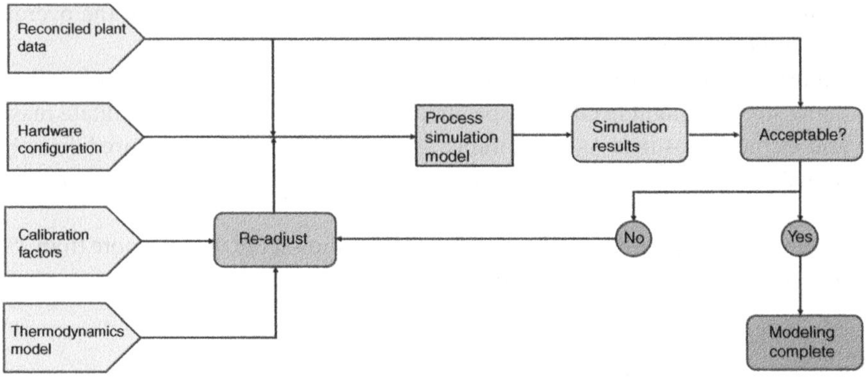

Figure 15.8 A typical process modeling workflow.

pumparound/reflux rates can be notoriously wrong due to lack of calibration or composition changes. Also, an erratic simulated column temperature profile can be just a result of an inaccurate pressure profile (or measurement)—as they strongly interrelate from the vapor–liquid equilibrium standpoint. If the product yield and property are good with the deviated temperature profile, then it is likely that there is a problem with the pressure profile. Apart from the already given examples, other inaccurate measurements include reflux, reboiler duty, and condenser duty.

As a best practice, adjust theoretical stages to fit the model to the plant and avoid tuning the column using tray efficiency—always leave tray efficiency at 100%. Inefficient stages (adjusted efficiency) have a vapor bypass, so the remaining vapor and liquid are in equilibrium, and then the vapor must recombine with the bypassed vapor. This can potentially cause the following problems:

- Tray sizing will have the wrong compositions and temperatures used for its calculations
- Temperature profiles can be significantly different
- Very heavy material can be carried up the column

Always compare the calibrated column model with typical industrial data; too much deviation might indicate that there might be something wrong with the calibration, e.g., wrong input. As a calibration guideline, typical column efficiencies for crude oil fractionation are provided here in Table 15.11.

Another interesting example is the hydroprocessing reactor model (Figure 15.9). A good hydroprocessing model should be well capable of simulating conversion, yields, product properties, and hydrogen consumption. These models at times cannot perfectly represent the overall heat balances of the system, e.g., heat recovery heat exchangers. The only heat balance that must be strictly satisfied

Table 15.11 Typical column efficiencies for crude oil fractionation.

Column	Zone	Theoretical Stages	Overall Efficiency
Crude distillation	Flash/wash zone	1–2	30
	Bottom section	1–2	30
	Heavy gas oil Section	3–5	50
	Kerosene/Light Gas Oil Section	3–5	55
	Naphtha/Kerosene Section	6–8	65
Vacuum distillation	Flash/Wash Zone	1	30
	Bottom section	1	30
	Heavy vacuum gas oil section	2	50
	Light vacuum gas oil section	3	55

in this case is the heat of reaction, which can be indicated by measured delta temperatures across the reactor. The matches in heat of reaction will dictate the calibrated reaction-rate constants. These models are usually used to generate input for refinery economic planning software; as such a compromise in heat balance is acceptable, as the product yields/properties and chemical hydrogen consumptions are the main focuses in refinery economic planning.

A good question here is whether the calibrated model is reasonable or not? As suggested earlier, it is always better to have data from at least two test scenarios to validate the model.

For example, a single-stage hydrocracker with two consecutive reactors is calibrated here using data from two conversion levels, 80 and 98%. Both test runs were conducted at approximately the same catalyst activity state, e.g., in less than 2 weeks. Intuitively, the number of active sites on the catalyst surface should be similar for both cases. Therefore, relatively close reaction-rate constants should be expected.

In order to evaluate if the two calibrations are reasonable or not, reaction-rate constants of key chemical reactions are compared in Table 15.12.

As shown in Table 15.12, the difference hydrocracking rate constant is only 6%, which is considered satisfactory in hydrocracking applications. While the differences are larger for hydro-denitrogenation and aromatic saturation, one should bear in mind that the magnitudes are much lower than that of hydrocracking reactions. In this case, these two calibrations can be considered acceptable.

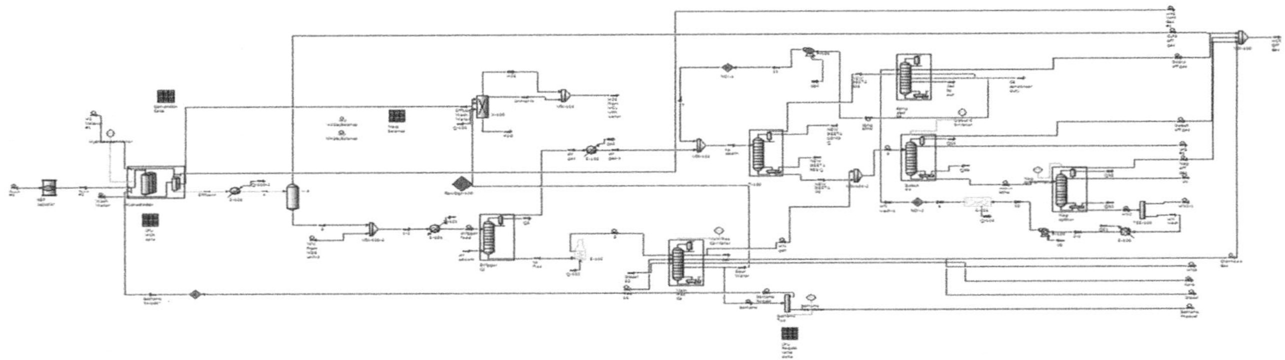

Figure 15.9 A hydroprocessing model consisting of reaction and separation sections in PetroSIM.

Table 15.12 Calibration results of a single-stage hydrocracker model.

Reactor	Key chemical reaction	Rate constant 80% Conversion	Rate constant 98% Conversion	%Difference
Pretreating	Hydro-denitrogenation	0.040	0.048	17%
	Aromatic saturation	2.318	2.022	13%
Hydrocracking	Hydrocracking	127.7	119.8	6%

There is no right or wrong answer here as to which model is more accurate. A model calibrated based on 98% conversion data should work better if the modeler tries to predict the conversion between 90 and 100%, while the other is more accurate at lower conversion ranges, e.g., 70–90%. Each calibrated model has a different validity range—that is why multiple test conditions are essential to generate a comprehensive model that covers the desired range of operations. For hydroprocessing reactors, the quality heat of reaction matching (calculated versus actual) can be significantly affected by a recycle gas flow, which is often not as accurate as it should be due to gas composition deviations. The position of the temperature sensor in the reactor can also play an important role as most of thermocouples in the field cannot entirely cover the whole catalyst bed length. This may make the heat of reaction matching difficult. Figure 15.10 shows the position of multipoint thermocouples, as indicated by the arrows above the ceramic support layer (granular ball) and deep into the catalyst bed.

Figure 15.10 Positions of multi-points thermocouples in a hydrocracking reactor.

There are two separate roles in evaluating a model: the modeler who constructed and calibrated the model and the user (plant engineer or operator) who will rely on the model to support decisions. Often, the user will ask the modeler, “Is the model calibrated?” when it should be the modeler who should ask, “Is the model calibrated well enough for your objectives?”

15.14 Process Simulation as a Training Tool

In addition to the more conventional design, rating, and optimization capabilities, process simulations can also provide a risk-free training environment for engineers and operators.

These days, process simulations (or digital twins) have advanced to the point where the outputs can be fed into an offline replica of the DCS/PLC, creating a training environment that allows the user to become familiar with actual plant SCADA. This allows users to practice startup or shutdown sequences, abnormal event management, and troubleshooting in a safe environment.

Furthermore, it can also help entry-level engineers develop their understanding in chemical engineering principles and process technologies, for example, learning how distillations or chemical reactors works. In a feedback loop fashion, process simulation helps the chemical engineering community to train less-experienced chemical engineers into more competent ones.

Nomenclature

$\%_{\text{chemical hydrogen, raw}}$	the raw amount of hydrogen being consumed by chemical reactions
$\%_{\text{chemical hydrogen, reconciled}}$	the reconciled amount of hydrogen being consumed by chemical reactions
$\%_{\text{overall mass balance, raw}}$	overall raw mass balance of the system
AM_C	atomic mass of C, 12.01 g/mol
AM_H	atomic mass of H, 1.01 g/mol
AM_N	atomic mass of N, 14.01 g/mol
AM_S	atomic mass of S, 32.06 g/mol
$C_{\text{carbon,feed,reconciled}}$	reconciled carbon content of the hydrocarbon liquid feed
$C_{\text{carbon,gas product, reconciled}}$	reconciled carbon content of the hydrocarbon gas product
$C_{\text{carbon,liquid product, reconciled}}$	reconciled carbon content of the hydrocarbon liquid product

$C_{\text{hydrogen,feed}}$	hydrogen content of the hydrocarbon liquid feed
$C_{\text{hydrogen,gas product}}$	hydrogen content of the hydrocarbon gas product
$C_{\text{hydrogen,liquid product}}$	hydrogen content of the hydrocarbon liquid product
$C_{\text{nitrogen,gas product}}$	nitrogen content of the hydrocarbon gas product
$C_{\text{nitrogen,feed}}$	nitrogen content of the hydrocarbon liquid feed
$C_{\text{nitrogen,liquid product}}$	nitrogen content of the hydrocarbon liquid product
$C_{\text{sulfur,feed}}$	sulfur content of the hydrocarbon liquid feed
$C_{\text{sulfur,gas product}}$	sulfur content of the hydrocarbon gas product
$C_{\text{sulfur,liquid product}}$	sulfur content of the hydrocarbon liquid product
$F_{\text{chemical hydrogen, reconciled}}$	reconciled mass flow rate of hydrogen consumed by chemical reactions or reconciled chemical hydrogen consumption
F_{feed}	mass flow rate of the hydrocarbon liquid feed
$F_{\text{gas product}}$	mass flow rate of the hydrocarbon gas product
$F_{\text{H2S,reconciled}}$	reconciled mass flow rate of hydrogen sulfide
$F_{\text{liquid product}}$	mass flow rate of the hydrocarbon liquid product
$F_{\text{NH3,reconciled}}$	reconciled mass flow rate of ammonia
k_{scaling}	scaling factor – greater than 1.00 in case of product loss or less than 1.00 in case of product gain
MW_{H2S}	molecular weight of H_2S, 34.08 g/mol
MW_{NH3}	molecular weight of NH_3, 17.03 g/mol

16

Putting it Altogether

16.1 Introduction

The previous chapters have been focused on various phases of problem-solving procedures and activities, as well as process engineering calculation techniques. The actual procedures and techniques, which are applicable for any problem-solving activity, have been discussed. In addition, Chapters 7 through 12 contain experience-based process engineering calculation techniques and guidelines. In each of these chapters, there are example problems that are directed toward the theme of the particular chapter. In real life, problems do not come packaged so neatly. For example, what is described as a problem associated with a prime mover may well be a problem associated with a reactor.

In order to see how these areas fit together, several real-life example problems are discussed in this chapter. The chapter shows how the five-step problem-solving procedure (Chapter 4) can be used, how working hypotheses can be formulated (Chapter 6), and how various process engineering calculations (Chapter 7 through 12) can be utilized to develop and confirm these hypotheses.

16.2 Do Not Forget to Use Fundamentals

Two of the most powerful tools that a process engineer can use in problem solving are heat and material balances. The material balance simply states that, with the exception of atomic power, mass cannot be created or destroyed. Thus, the total flow expressed in weight units into a process or a unit operation must be equal to the total flow out. A comparable truism applies to heat balances. In addition to the need for a process to be in balance, heat and material balances can be used to determine unknowns. For example, material balance principles can be used to determine the production rate of a desirable or undesirable component. If an undesirable by-product is being produced in a reactor and removed in a purge stream,

Problem Solving Approaches for Maintaining Operational Excellence in Process Plants, Second Edition.
Joseph M. Bonem, Nattapong Pongboot, and Wiroon Tanthapanichakoon.
© 2025 the American Institute of Chemical Engineers, Inc. Published 2025 by John Wiley & Sons, Inc.

the rate of production is simply the removal rate. Heat balances can be used to determine the boil-up rate from the steam rate. A problem solver might well be told that a tower is flooding because the trays are plugged. When he performs a heat balance, he might well conclude that the heat input to the tower is less than the heat being removed from the tower. If he makes a closer examination, he might find that the problem really is a steam meter that is indicating a flow rate much less than the actual flow. This is causing the tower to be flooded due to the excessive vapor rate being generated in the reboiler.

As will be noted in the example problems, knowing the flow rates of process or utility streams is a requirement for successful problem-solving. While this may seem very basic, it is amazing how many problem-solvers will start developing intricate hypotheses even though flow instruments indicate that the amount of material coming into the unit operation does not equal the amount of material leaving. Thus, one of the key things for a problem solver to remember is the idea of not forgetting the fundamentals of their discipline.

16.3 Example Problem 16.1: Do Fundamental Processes Developed in the United States Translate to Europe?

A chloride removal unit designed and operated in the United States was an integral part of a gas drying process that used triethylene glycol (TEG) as the circulating drying solvent. Water was removed from the gas in the absorber using a stream of TEG that had a very low water content. The properties of TEG are such that water has a very low partial pressure when dissolved in the TEG. Thus, it can be readily used to dry gases of all descriptions.

In the drying process, the TEG leaving the absorber and containing the water removed in the absorber flowed to a heated two-stage regeneration system. In this regeneration system, the dissolved water was stripped from the gas, and the dry TEG was recirculated to the absorber. The initial stage of the regeneration system was a heated flash step with about 5 minutes of residence time. The second stage was a vacuum distillation tower. The TEG from the vacuum distillation tower flowed back to the absorber.

The gas to be dried was a chlorine derivative. Some of the gas dissolved in the circulating TEG was converted to hydrogen chloride (HCl) as the solution was heated as part of the regeneration step. The reaction of the gas to form HCl was known to be first order with respect to the concentration of the chlorine derivative in the circulating TEG. However, the exact reaction rate constant and Arrhenius constant were not known. Essentially, all of this reaction occurred in the first stage flash drum of the two-stage regeneration system. This stage was operated at 175 °F and 5 psig. The level in the drum was controlled so that the residence time was limited to 5 minutes. The second stage of the regeneration system consisted of a small vacuum fractionating tower. HCl generation in this stage was very limited because of the short residence time and low pressure.

The HCl formed was neutralized using a soluble amine. However, the amine could not be added continuously since the HCl–amine complex would build up in the circulating TEG to an unacceptably high level. The HCl was removed from the HCl–amine complex by passing a small stream of dry TEG through a drum filled with ion-exchange resin. Ion-exchange resins are complex salts supported on a synthetic resin such as polystyrene. This particular ion-exchange resin had the capability of removing the Cl^- ion and replacing it with an OH^- ion. Thus, HCl was converted to water. The ion-exchange resin itself was converted from an OH^- form to a Cl^- salt. As this process continued, the bed would eventually become saturated with Cl^- ions and become ineffective. To restore the bed, it was removed from service and regenerated using water and a sodium hydroxide solution. This regeneration was done whenever the concentration of chlorides in the outlet increased above 50 ppm A simplified drawing of the process is shown in Figure 16.1.

The process was designed and installed in a location in the United States. After initial start-up problems, the operation of the process was very successful. The chlorides were controlled at a concentration of 500 ppm with a flow to the ion-exchange bed of 500 lb/hr. During the initial operations, high corrosion rates were encountered when the chloride level increased to 1000 ppm.

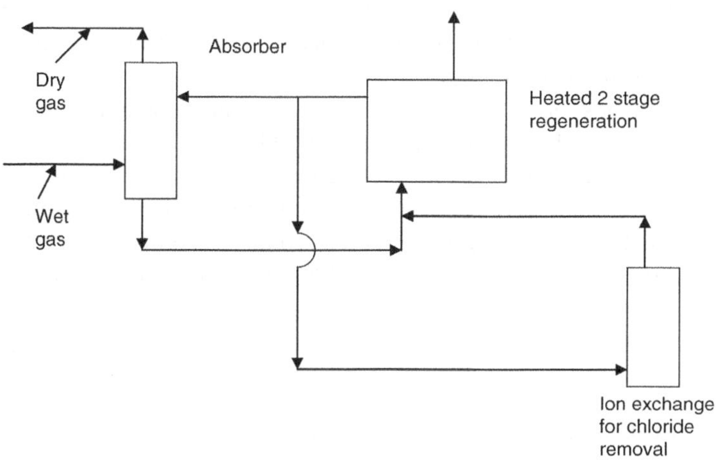

Figure 16.1 TEG system.

An identical unit was installed in Europe. Flow rates, temperatures, and pressures were all the same. While the gas was successfully dried, the concentrations of chlorides in the circulating TEG increased to 1500 ppm in a month after start-up. The ion-exchange bed appeared to be operating well since the outlet concentration from the bed was 0 until the ion-exchange resin became saturated with chlorides. However, the bed had to be regenerated more frequently than anticipated due to the heavy loading of chlorides. At the time, the plant in Europe was not using the problem-finding concepts discussed in Chapters 4 and 6. Because of the concern over corrosion, the European management shut the unit down. While significant corrosion had not yet been observed, they were convinced based on experience in the United States that high corrosion rates would be observed soon. After some preliminary problem-solving attempts on their own, they requested help from the original designer in the United States.

When the problem solver from the United States arrived on the scene, he began a methodical problem-solving activity using the approach discussed in the previous chapters. This methodical problem-solving approach was even more important in this case since there were significant geopolitical factors involved. There was a great deal of animosity between the European affiliate and the United States-based technical staff. The problem solver began the five-step approach as described in Chapter 4.

Step 1: *Verify* that the Problem Actually Occurred

Since operation at high chloride concentrations had not yet resulted in observable corrosion, the first problem to be verified was that the actual concentration of chlorides was as high as 1500 ppm The laboratory procedure was confirmed, and several samples were analyzed, confirming that indeed the chloride concentration was 1500 ppm Since the corrosion rate had been so severe in the United States at a chloride level of 1000 ppm, it did not seem wise to continue operation and simply monitor the corrosion rate.

The problem-solver's next step in verifying that the problem actually occurred was to compare the theoretical chloride buildup to the actual chloride buildup in Europe. Since the reaction rate constant was not known, the theoretical chloride buildup was determined based on operations in the United States. Based on material balance principles, the problem solver calculated the chloride production rates in the United States. This material balance principle is as follows:

$$RC = F^{*}X_{F} \tag{16.1}$$

where

RC = the rate of chloride production in the United States, lb/hr

F = the flow rate to the ion-exchange bed, lb/hr

X_F = the concentration of chlorides in the flow to the ion-exchange bed, weight fraction

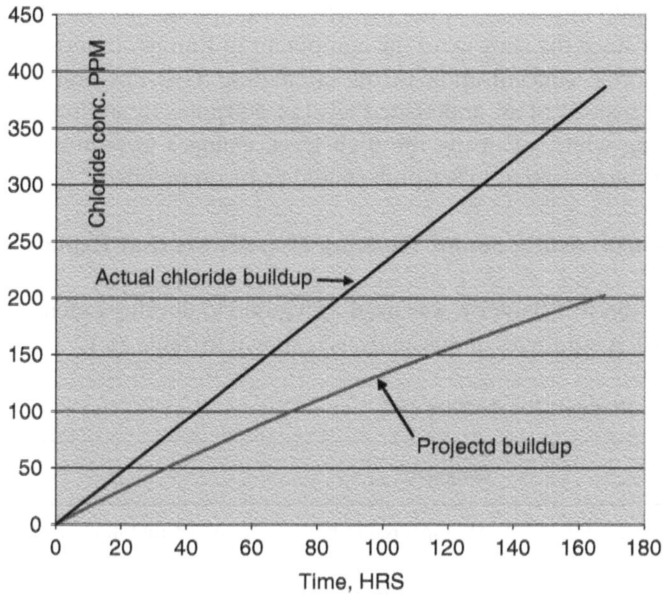

Figure 16.2 Actual and projected Cl buildup.

Note equation (16.1) assumes that the chloride concentration in the outlet from the ion-exchange bed was 0. This was true essentially all of the time.

Knowing the rate of chloride production in the United States and assuming that it was the same in Europe, the problem solver could then estimate the chloride concentration buildup rate in the European plant by knowing the TEG inventory. He then developed the relationships shown in Figure 16.2. In this figure, the projected chloride buildup based on kinetic relationships developed in the United States is shown, along with the actual chloride buildup. This provides a means of comparing the theoretical and actual chloride concentration buildup rates. Obviously, there was a significant difference between the two rates.

Step 2: *Write* Out an *Accurate* Statement of What Problem You Are Trying to *Solve*

The problem solver wrote out the problem statement as follows:

> "The chloride concentration in the circulating TEG in the European plant has built up to 1500 ppm instead of an anticipated level of 500 ppm. This is a major concern because experience in the United States indicated that significant corrosion would begin at levels above 1000 ppm. This high level of chlorides is being experienced in Europe at the apparent same identical process conditions used in the United States which resulted in a chloride level of only 500 ppm. An analysis of the rate of

chloride buildup indicates that the higher chloride production rate has been present since the start-up of the equipment in Europe. Determine why the chloride concentration in the circulating TEG has built up to 1500 ppm instead of an anticipated level of 500 ppm. In addition to understanding what is causing the high level, changes to reduce the steady-state concentration to 500 ppm should be recommended."

Step 3: Develop a *Theoretically Sound* Working Hypothesis that Explains the Problem

The question list given in Chapter 6 was used as a guide to develop potential working hypotheses. A summary of this analysis is shown in Table 16.1.

Table 16.1 Questions/comments for Problem 16.1.

Question	Comment
Are all operating directives and procedures being followed?	All appeared to be correct and being followed
Are all instruments correct?	The instruments had allegedly been calibrated. However, it was observed that a venturi meter was being used in Europe, whereas an orifice meter was used in the United States
Are laboratory results correct?	All appeared to be correct
Were there any errors made in the original design?	The design was essentially the same as that in the United States, with the exception of improvements in the regeneration section
Were there changes in operating conditions?	No. In fact, the operating conditions of temperature and pressure were identical to those used in the United States
Is fluid leakage occurring?	This would not explain the problem
Has there been mechanical wear that would explain the problem?	No
Is the reaction rate as anticipated?	Higher rates of HCl formation could explain part of the problem. On the other hand, the ion-exchange bed appeared to be performing as designed, except for the frequent regenerations. The frequent regenerations appeared to be associated with the higher-than-anticipated rate of HCl formation
Are there adverse reactions occurring?	There were no unusual reactions that could explain the problem
Were there errors made in the construction of the process?	Since the unit had only recently been built, this had to be considered

The value of using a list similar to Table 16.1 is that it helps eliminate superfluous hypotheses that might be suggested, and it allows focusing in on the likely areas for development of working hypotheses. For example, an examination of this table indicates that areas such as mechanical wear, fluid leakage, or changes in operating conditions are unlikely routes to pursue. In addition, the most obvious conclusion—"the ion-exchange resin is not working" is shown by Table 16.1 to be highly unlikely. On the other hand, meter errors, higher than expected HCl production rates and design improvements in the regeneration area, appear to be valuable areas to pursue. As is often the case in industrial problem solving, there may be more than a single item that is causing the problem. For this reason, the problem solver began looking in detail at the design improvements, the variables that might cause the reaction rate to form HCl to be higher than anticipated, and the difference in a venturi meter and orifice meter in the specific application. It was not possible to isolate the potential areas of flowmeter error and higher-than-anticipated HCl production rate because the technique for calculating the HCl production rate depended on the flow rate of TEG to the ion-exchange bed. If this flowmeter was in error, the calculated HCl production rate would be in error. Thus, both possibilities had to be considered.

Since the operating conditions (temperature and pressure) were identical to those in the United States, it would appear that the rate of HCl generation would be the same. However as indicated earlier, the possibility of higher rates of HCl generation could not be eliminated. Since essentially all of the HCl was produced in the first-stage flash drum, the problem solver began an analysis of that operation. Referring back to Chapter 9, the kinetic relationship can be expressed in terms of the reaction rate as shown below:

$$R^* = C^*\mathrm{DF} \tag{16.2}$$

where:

R^* = the rate of change with time per unit volume of the compound under study, mols of HCL/ft^3 -minute that are formed

C = a constant referred to earlier as the "lumped parameter constant"

In this case, the constant is the reaction rate for the formation of HCl from the chlorine derivative gas. This will be a function of temperature only. As indicated, the driving force will be the concentration of the gas in the TEG in the flash drum. This will depend on temperature and pressure only. Thus, the driving force and lumped parameter constant should be the same in Europe and the United States. This means that "*R*," which is the rate of HCl formation in mols per minute per unit of volume, should be the same. Thus, if the absolute value of HCl formation in mols/hr. is higher, it can only be due to an increase in drum volume or residence time at a constant flow rate. This possibility by itself does not provide enough data to formulate a hypothesis. In order to determine if there is a valid working hypothesis related to increased drum volume, the following items were reviewed:

- The dimensions of the first-stage flash drum were reviewed, and it was concluded that the dimensions were identical in Europe and the United States.
- While the measured liquid levels in the drums were both held at 20%, the ranges of the liquid level instruments were different. This was discovered only during a detailed review of the instrument specification sheets. The instrumentation philosophy in the United States was to only cover the planned range of operations with the liquid level instrument. This would provide a higher degree of accuracy. The range of the level instrument was 50 in. In Europe, the philosophy was to cover the entire height of the drum or 75 in. in this case. The Europeans considered this to be a significant design improvement.

Thus, the absolute liquid level in Europe was 15 in. versus 10 in. in the United States. Since the drums were of the same diameter and the flow rates were identical, this difference would provide 50% more reaction volume in Europe than in the United States. This would increase the HCl production by 50%.

In order to determine whether this hypothesis would explain all of the apparent increase in HCl production observed in Europe, an HCl material balance was developed for Europe in a similar fashion to that done for the United States. The premise of these balances was that the HCl removed was equal to the HCl produced. These balances are shown in Table 16.2.

Thus, it appeared that a working hypothesis that the increased residence time was responsible for an increased amount of chloride produced would be a valid working hypothesis. However, the calculated chloride production rate was only about half of the value obtained by material balance (0.38 lb/hr versus 0.75 lb/hr). The failure to get a good material balance calculation check indicated that there might be other problems. Thus, rather than proposing a plant test of lowering the level in the first-stage flash drum, the problem solver began considering other possible problems.

A review of the question guidelines and answers in Table 16.1 indicated that there was likely some construction mistake or mistake in the revised design. Several possible additional hypotheses were developed. Most of them tended to point to the flow measurement of the stream going to the ion-exchange bed. If the flow rate was significantly less than indicated, chlorides would still be removed, but the concentration would build up to a higher level than would be anticipated. For example, if the flow was actually 250 lb/hr, then the amount of chloride removed in the ion-exchange unit would be equivalent to that calculated based on the increased residence in the first-stage flash drum. The problem solver began developing a hypothesis associated with the flowmeter by selecting the simplest explanation possible. A review of the

Table 16.2 Chloride balances.

Variable	United States (Base Case)	Europe
Flow rate to ion exchange, lb/hr	500	500
Chloride concentration into ion exchange, ppm	500	1500
Chloride concentration out of ion exchange, ppm	0	0
Chloride production by material balance, lb/hr	0.25	0.75
Calculated chloride production[a], lb/hr	0.25	0.38

a) The calculated chloride production rate was set at the material balance level for the base case (United States) and was increased by 50% to allow for the increased residence time in Europe.

venturi meter calculations used for the flow to the ion-exchange bed indicated that it was selected to minimize pressure drop. The venturi meter discharge coefficient was assumed to be 1, which would have been true for a low-viscosity fluid. However, the TEG has a viscosity much higher than that of a typical hydrocarbon. While it appeared to the problem solver that a standard orifice meter would have provided a much more accurate installation, he refrained from indicating this to the Europeans.

The problem solver now developed the following hypothesis that actually contained two possible theoretically correct working hypotheses.

> "It is believed that the increased absolute level in the flash drum (15 inches versus 10 inches) is causing the chloride production to be 50 percent higher in Europe than in the United States. The level hypothesis does not explain that the concentration of chlorides in the TEG is three times that anticipated based on results in the United States. Thus, another problem must be present. It is believed that the measurement of flow to the ion-exchange bed is in error."

It should be noted that this problem illustrates that in industrial problem-solving, there are often more than one valid hypothesis. Using basic chemical engineering principles can often confirm whether one hypothesis can explain the entire problem.

Step 4: Provide a Mechanism to Test the Hypothesis

Two separate plant tests were developed to test both of these hypotheses. The tests were conducted concurrently since as conceived, there was no interaction between the two tests. In the first test, the level was reduced in the first-stage flash drum so that the residence time was reduced to 5 minutes. In the second test, the flow to the ion-exchange bed was diverted into a 5-gal bucket, and the actual flow rate was measured. The 5-gal bucket required about 11 minutes to fill as opposed to the 6 minutes that it would have required if the flow rate was really 500 lb/hr. In order to confirm that the problem was truly solved, the plant test on the flash drum was continued. That is, the operating directive for the level was set so that the residence time continued at 5 minutes. In addition, the flow rate to the ion-exchange bed was increased to a *measured* value of 1000 lb/hr. It was anticipated that a measured flow of 1000 lb/hr would give an actual flow of 500 lb/hr. The flow was again measured using the 5-gal bucket to confirm that a flow rate of about 500 lb/hr was achieved. After a few days at these conditions, the chloride concentration decreased to 500 ppm, the concentration experienced in the United States.

Step 5: Recommend Remedial Action to Eliminate the Problem without Creating Another Problem

The remedial action was relatively simple and consisted of only slight modifications to the changes made to conduct the plant test. These changes were designed to ensure that at a future time, comparisons made between the two plants did not cause changes in operations, which would recreate the start-up problems. The changes were as follows:

- The level instrument on the first-stage flash drum was re-ranged from 75 to 50 in. This would allow both flash drums in the United States and Europe to operate at the same apparent level (20%). This would also be the same absolute level of 10 in. If the level instrument had not been re-ranged, it would have been necessary to maintain the level in Europe at about 15%. There was a concern that this discrepancy between the United States and Europe might lead the European affiliate to raise the flash drum level at some point to be consistent with the level in the United States.
- The discharge coefficient for the European venturi flowmeter was changed to a value that was based on the actual viscosity rather than an assumed value of unity.

A detailed potential problem analysis did not reveal any significant new problems if these changes were made.

16.4 Lessons Learned

If the problem-solving concepts discussed earlier had been applied, the European technical staff might well have elected to predict the chloride buildup rate based on data from the United States. If they had been using this potential problem analysis concept, they would have been able to spot the problem much sooner than a month after start-up. As can be seen from Figure 16.2, it was readily apparent after 2–3 days that the chloride level was increasing much faster than would have been projected based on experience in the United States. The advantage of comparing the actual to projected chloride buildup is that problem-solving activity could have started 3 weeks earlier than actually experienced.

This problem illustrates the validity of calculations. There will always be a tendency to treat the first discovery as the root cause of the problem. Many industrial problems have more than a single root cause. For example, the discovery that the flash drum in Europe had more residence time than the comparable drum in the United States might have been considered the single root cause of the problem. If the problem solver had not concluded based on calculations that there must be another cause, the flash drum changes would have been made, but the problem of high chloride contents would have continued. In the case described, it was especially desirable to ensure that the problem solutions were complete since the problem solver had only a limited amount of time in Europe.

In our advanced age of electronic equipment, we often forget the more basic measurement techniques. The use of a 5-gal bucket to measure the low flow rate of a low toxicity and nonvolatile material is probably one of the best techniques available.

Any design change regardless of how small (use of a venturi meter and change in range of a level instrument) should receive a careful review including a potential problem analysis.

16.5 Example Problem 16.2: An Embarrassing Moment

A high vacuum system was designed as part of a new process. The vacuum system was required to achieve an absolute pressure of 15 mm of Hg. In order to do this, a three-stage steam ejector was selected. A schematic of the vacuum system is shown below in Figure 16.3.

The construction was relatively straightforward except for obtaining a steam supply for the ejector. In order to furnish steam and avoid a shutdown of the 200-psig steam supply line, it was necessary to "hot tap" the steam line. Hot tapping is a procedure where a valve with a flange is welded to the line. When the valve is fully secured and the welding is inspected, a cutting instrument is connected to the open end of the flanged valve. The valve is opened, and the cutting instrument is lowered into the valve opening until it touches the pipe. A small pilot drill is first used to cut a small (1/4 to 3/8 in.) hole. Following that, a full-size hole is cut. The tool used to cut the full-size hole is then used to pull the piece of the pipe that has been cut out back through the valve. The valve is closed as the tool is removed. The new steam piping to the process is then connected to the valve. The valve is then opened at the appropriate point in the start-up procedure. The hot tap crew will often save the section of the pipe cut from the main line as proof that the hot tap really had been made.

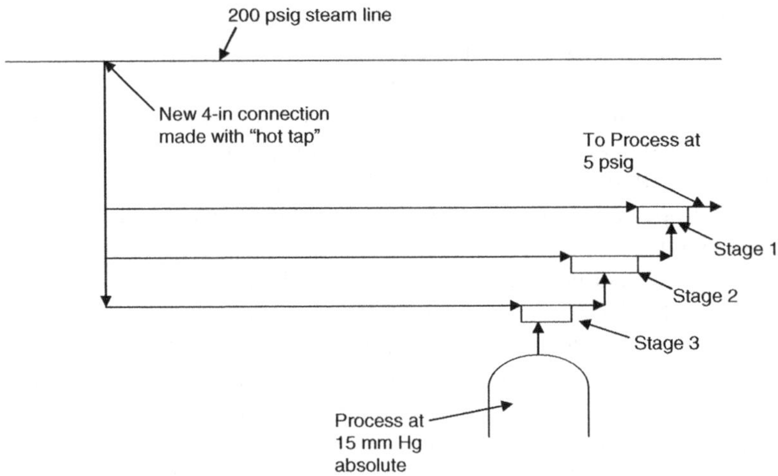

Figure 16.3 Schematic of three-stage steam ejector.

The start-up procedure for the three-stage ejector system called for starting the stages in the order shown on the schematic drawing. This allowed the lowest steam demand stages to be started first. The steam jet manufacturer's guaranteed steam consumptions were as follows:

- Stage 1 300 lb/hr
- Stage 2 500 lb/hr
- Stage 3 1700 lb/hr

The start-up procedure seemed to go well as the first and second stage jets were placed into service. The steam pressure as measured by a gage in the new 4-in. steam line remained relatively constant at about 200 psig, and the pressure in the vacuum drum decreased to the level anticipated with only two stages in service. However, when steam flow was started to the third stage, the measured steam pressure on the 4-in. line decreased rapidly to 105 psig, and the pressure on the vacuum drum increased as the first- and second-stage jets were no longer able to perform at the reduced steam pressure. The initial reaction of the problem solver was that there was obviously something wrong with the third-stage jet. After all, the system performed perfectly when only two stages were used. He called the sales representative and strongly suggested to him that maybe an orifice had been left out of the third-stage jet. The problem solver believed that this would cause a huge increase in steam flow and result in a large pressure drop in the 4-in. line. As time passed, the problem solver began a more methodical approach to analyzing the problem. He used the five-step approach as follows:

Step 1: *Verify* that the Problem Actually Occurred

There was little doubt that the problem occurred. The dropping in steam pressure and the loss of vacuum were also accompanied by loud noises inside the drum as material flowed backward through the ejectors. However, to satisfy the need to verify the problem and get a maximum amount of data, the start-up was repeated. The same results were observed. However, the increased attention to the steam pressure gage on the 4-in. line indicated that the steam pressure actually dropped slightly when the first- and second-stage ejectors were placed into service.

Step 2: *Write* Out an *Accurate* Statement of What Problem You Are Trying to *Solve*

The problem solver developed a problem statement as follows:

> "During start-up of the steam jet system, the steam pressure on the 4-inch steam supply line decreased rapidly to 105 psig when the third stage ejector was placed into service. The operation of the three-stage steam jet is impossible at this low pressure. Very small pressure drops were also observed as the first and second stage ejectors were placed into service. The same results occurred during both instances when the ejector system was being placed into service. The pressure on the 200-psig steam pressure header was normal during both trials. There are no steam meters available to measure the actual steam flow to the ejector system. Determine why the steam pressure on the 4-inch steam supply line decreased rapidly to 105 psig when the third stage ejector was placed into service. Recommendations for modifications to allow operating the steam ejector system are also to be provided."

The actual measured pressures from the second trial are shown in Table 16.3.

Step 3: Develop a *Theoretically Sound* Working Hypothesis that Explains the Problem

The question list given in Chapter 6 was used as a guide to develop potential working hypotheses. A summary of this analysis is shown in Table 16.4.

Table 16.3 Steam pressure measurements.

Ejectors in Service	Steam Pressure, psig
0	200
1	198.5
2	190.3
3	105

Table 16.4 Questions/comments for Problem 16.2.

Question	Comment
Are all operating directives and procedures being followed?	The vendor provided start-up procedure that was being followed exactly
Are all instruments correct?	The pressure gage that was initially used was replaced with a new gage before the second test
Are laboratory results correct?	Not applicable in this case
Were there any errors made in the original design?	The 4-in. steam line could be too small
Were there changes in operating conditions?	No. The process was being operated exactly as specified on the duty
	specification for the steam ejector.
Is fluid leakage occurring?	Not applicable
Has there been mechanical wear that would explain the problem?	Not applicable
Is the reaction rate as anticipated?	Not applicable
Are there adverse reactions occurring?	Not applicable
Were there errors made in the construction of the process?	Since the unit had only recently been built, this had to be considered

The approach of using the guidelines provided by Chapter 6 may seem trivial for this example, but they helped isolate the development of a hypothesis to two areas. The areas that the problem solver decided to investigate further were the sizing of the 4-in. steam supply line and some sort of construction error. The hypothesis of a construction error would also include the possibility of an error in the manufacturing of the steam ejector. Three potential working hypotheses were proposed as described below:

1. The process designer had made a mistake in the sizing of the 4-in. steam supply line.
2. There was an error made in the construction of the third-stage steam jet.
3. The hot tap crew had made an error and did not completely cut and remove the 4-in. piece of the line.

Initially the hot tap crew was contacted to ascertain that they did pull a piece of the pipe from their cut to determine that the hot tap had indeed been completed. Unfortunately, this was many weeks after the cut was made, and the problem solver was told that all cuts made during that time frame had been discarded.

In order to narrow down the possible number of working hypotheses to a minimum, the following actions were taken, and the indicated results were obtained:

- The steam supply line was resized, and it was concluded that the 4-in. line was more than adequate. The pressure drop should be less than 2 psi even if the steam rates were double the rates guaranteed by the steam jet manufacturer.
- The third-stage jet was removed and inspected, and the dimensions were compared to the factory-issued drawings. The dimensions of the ejector were as specified in the drawings.

The only remaining hypothesis was that the hot tap had not been completely cut through and that all of the steam was flowing through the pilot drill hole or another restriction in the piping.

Step 4: Provide a Mechanism to Test the Hypothesis

While the hypothesis could have been tested by insisting that the hot tap crew return and redo their hot tap, the problem solver decided to test the hypothesis by using calculations. He proceeded to consider the data shown in Table 16.5.

If the "hot tap" had not been cut completely, the pressure drop across the pilot drill hole should be proportional to the steam flow rate squared. Thus, he decided to plot the pressure drop versus the flow rate squared. The resulting plot is shown in Figure 16.4.

The resulting plot had a slope of 0.0000152 and an intercept of 0. Thus, it could be specified by the following relationship:

$$\Delta P = 0.0000152^{*}F^{2} \quad (16.3)$$

where

ΔP = the pressure drop across the restriction, psi

F = the steam rate, lb/hr

The intercept of 0 was as expected. That is, at no flow there should be no pressure drop. The slope of 0.0000152 was used to approximate the size of the opening

Table 16.5 Steam flow and pressure drops.

Stages in Service	Steam Flow (lb/hr)[a]	Pressure Drop (psi)
1	300	2
2	800	10
3	2500	95

a) The steam flows were taken from the steam jet manufacturer's specification sheet.

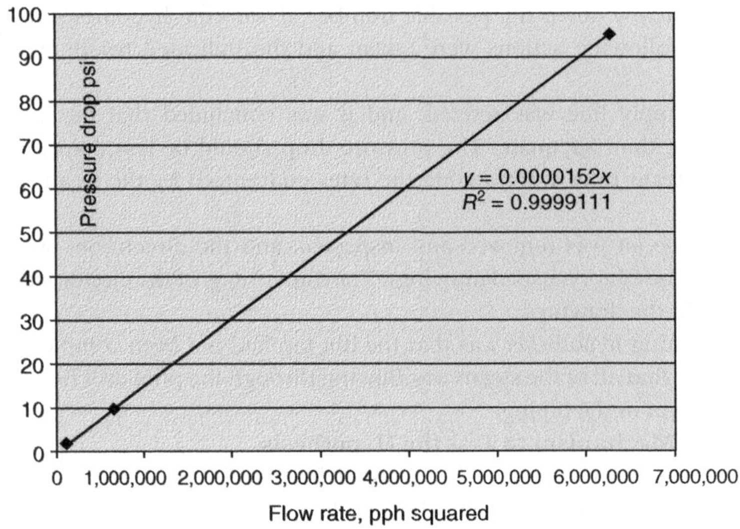

Figure 16.4 Pressure drop versus flow rate squared.

that steam was flowing through. To do this, the problem solver used the following relationship for fluid flowing through a restriction.

$$\Delta P = 0.5^* S^* U^2/148.2 \tag{16.4}$$

where

S = the density of the flowing fluid relative to water

U = the velocity of the flowing fluid through the orifice, fps

The constants represent conversion factors and the orifice discharge coefficient. Since there was now an experimental relationship between the flow rate and the pressure drop, as well as a similar theoretical relationship, these two equations could be used to estimate the diameter of the restriction. To make this estimate, the right-hand side (RHS) values of equations (16.3) and (16.4) were set equal to each other since they were both equal to the pressure drop in the restriction.

$$0.0000152^* R^2 = 0.5^* S^* U^2/148.2 \tag{16.5}$$

Knowing

$$U = R/(3600^* A^* \rho) \tag{16.6}$$

$$S = \rho/62.4 \tag{16.7}$$

where

A = area of restriction, ft^2

ρ = fluid density, lb/ft^3

Solving these two relationships for the area of the restriction and then calculating the resulting diameter yielded a value of 3/8 in. This was likely the diameter of the pilot drill used for the hot tap of the steam line.

It appeared likely that the hot tap had not been cut all the way through and that steam was only flowing through the opening that was cut for the pilot drill. When the hot tap crew was contacted, they agreed to return to recut the hot tap only after considerable discussion. When they did recut the hole, they were surprised to find that the piece of material that they removed was exactly as calculated by the problem-solver. That is, it was a 4-in. piece of metal with a 3/8-in. hole in it.

Step 5: Recommend Remedial Action to Eliminate the Problem without Creating Another Problem

No additional actions appeared to be required after the hot tap was recut. The steam ejector system was started up successfully, and the steam pressure on the 4-in. supply header remained constant at 200 psig throughout the start-up and operation.

16.6 Lessons Learned

This problem illustrates how "jumping to conclusions" can often lead to embarrassing moments. Rather than immediately confronting the sales representative, the problem solver should have made a careful study of the available data. A careful study of the data would have revealed that the steam supply pressure did not stay constant even when the smaller first- and second-stage jets were placed into service. The problem also illustrates the value of doing calculations to attempt to pinpoint the problem source. After the designer had rechecked the sizing calculations for the 4-in. steam line, the calculations described in Step 4 should have been done. If these additional calculations had been done, there would have been no reason to open and check the dimensions in the third-stage jet. The removal of this third-stage jet was a major effort since it was located three levels up in the structure and had large pipes connected to it.

The calculations described in Step 4 would have been sufficient to point out clearly that the "hot tap" was likely done incorrectly and thus would have eliminated the need to remove and inspect the third-stage jet. These calculations also provided a strong argument for redoing the "hot tap," as opposed to just a suspicion that it was not done correctly.

16.7 Example Problem 16.3: Prime Mover Problems are not Always What They Appear To Be

An ethylene refrigeration system was expanded by increasing the capacity of a blower. Prior to the expansion, this blower was used to boost the pressure on the system from 10 in. of vacuum to 8 psig. The system was expanded by increasing the blower discharge pressure (also the compressor suction pressure) to 10 psig. The increase in pressure to 10 psig was to provide an increase in capacity of about 10%. No other changes were required in the ethylene compression or condensation system. The increased capacity of the blower was to be obtained by replacing the existing impeller with a larger impeller. This increase in impeller size would allow an increased flow rate and an increase in discharge pressure to 10 psig. A short shutdown was required in order to install the new impeller. A schematic drawing of the process is shown in Figure 16.5.

Ethylene liquid flows from the compression and condensation block to provide refrigeration for a low-temperature process operating at approximately −150 °F. The ethylene liquid is vaporized at 10″ of mercury vacuum and flows as vapor to the ethylene blower. Prior to the blower, a series of economizers (heat exchangers) increase the temperature of the ethylene from about −156 to −40 °F. The blower boosts the pressure from 10″ vacuum to 10 psig. The ethylene gas at 10 psig flows to the reciprocating compression system where it is compressed to approximately 350 psig and condensed in heat exchangers that are cooled by vaporizing propane. The refrigeration load is not constant. The rate of vapor flowing to the ethylene

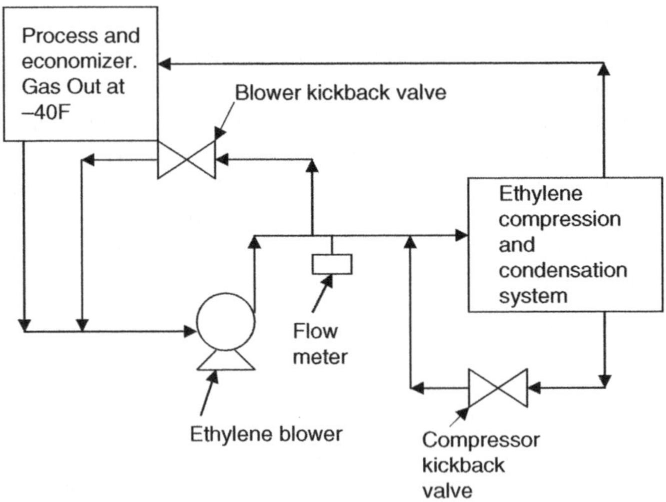

Figure 16.5 Ethylene refrigeration schematic.

blower and compressors varies significantly. To maintain the suction pressure of the blower and compressor constant, "kickback valves" are provided. As the refrigeration load decreases, the kickback valves open keeping the suction pressures constant.

The operation of the expanded blower was disappointing. While it appeared that additional ethylene flow was obtained, the pressure was well below the required discharge pressure of 10 psig. Since ethylene flowed from the blower discharge to the compressor suction, this lower discharge pressure also resulted in a lower compressor suction pressure. The lower suction pressure caused both a reduction in compressor capacity and an increase in the compression ratio. Operations personnel requested technical help because in their words "This new impeller is not as good as the one that we took out!! We never had this kind of trouble before and that kickback valve was always open at least 10 percent." The problem solver used the five-step approach to assess the situation.

Step 1: *Verify* that the Problem Actually Occurred

The verification that there was a problem was relatively easy. When attempts were made to increase the unit production and hence the refrigeration load above that possible before the new impeller was installed, the blower kickback valve would go all the way closed, and the blower suction pressure would increase above the operating value of 10 in. of vacuum. Since operating at 10 in. of vacuum was necessary in order to maintain the process temperatures, the blower discharge pressure was reduced, causing the reciprocating compressors to have less than the desired capacity.

Step 2: *Write* Out an *Accurate* Statement of What Problem You Are Trying to *Solve*

Prior to attempting to write out an accurate problem statement, the problem solver decided to look at the blower manufacturer's supplied compressor curve for the new impeller. As part of the preparation for this assessment of comparing the theoretical blower curve to actual performance, he had all the key meters checked. So he knew that the blower suction pressure, discharge pressure, and flow rate variables were as accurate as possible. In order to assess the blower performance, it was necessary to maintain the kickback valve in the closed position during the test. This was because the flowmeter was located outside the kickback valve line, as shown in Figure 16.5. The problem statement that he developed was as follows:

> "The performance of the ethylene blower seems to be less than anticipated with the new impeller. Rather than obtaining a 10% to 12% improvement in the plant capacity, operations since the start-up of the revised facilities have resulted in a capacity only slightly above the

previous capacity. While no test data exists for performance with the old impeller, operations personnel believe that the performance was adequate when the old impeller was being utilized. They also indicate that the kickback valve was normally open at least 10%. Currently, the valve is closed whenever the system is fully loaded. The ethylene compressors are operating as predicted. The problem is not related to instrumentation since all the meters have been checked. Determine the following:

- Is the blower operating as specified by the manufacturer's supplied blower curve?
- If it is not, determine why.
- Recommend changes to correct the problem or operating conditions that will allow operation at full capacity."

As indicated in Chapter 4, time is always an important component of a problem statement. In this problem, while no data were available from past operations, the problem solver still noted that performance seemed adequate with the old impeller. In addition, he indicated that the problem seemed to have been present since the start-up of the expanded facilities. This helps focus on the time period after the facilities were expanded. In parallel with developing working hypotheses, the problem solver decided to run a series of plant tests to assess the actual blower performance. A summary of these plant tests is shown in Table 16.6. In addition, the results are shown graphically in Figure 16.6.

Also, note in the problem specification that the problem solver takes into account the operator's observations that prior to the expansion, the performance was adequate and that the kickback valve was open at least 10% essentially all of the time. However, the problem statement did not include their conclusion that the blower was performing better with the old impeller.

The polytropic head is calculated using the following equation given in Chapter 7:

$$H = 1545^{*} T_S{}^{*} Z^{*} (R^{\sigma} - 1)/M\sigma \tag{16.8}$$

where:

H = the polytropic or adiabatic head, ft

T_S = the suction temperature, °R

Z = the average (suction and discharge) compressibility

R = the compression ratio

M = the gas molecular weight

σ = the polytropic or adiabatic compression exponent

Table 16.6 Blower capacity tests.

Variable	Design	Test 1	Test 2
Molecular weight	28	28	28
Specific heat ratio	1.25	1.25	1.25
Gas compressibility	1	1	1
Polytropic efficiency, %	70	TBD	TBD
Suction pressure, "Hg	10	7.5	8.7
Discharge pressure, psig	10	10	10
Gas density at suction, lb/ft^3	0.0609	0.0670	0.0633
Flow rate, lb/hr	35,000	42,590	36,560
ACFM	9580	10,590	9630
Temperatures			
Out economizer, °F	−40	−40	−40
Blower suction, °F	−40	−30.5	−29.3
Blower discharge, °F	87	81	91
Blower speed, rpm	10,000	10,000	10,000
Calculated polytropic head, ft	24,560	21,525	23,250
Projected polytropic head, ft		23,200	24,500
Performance deficiency, %		7.2	5.1

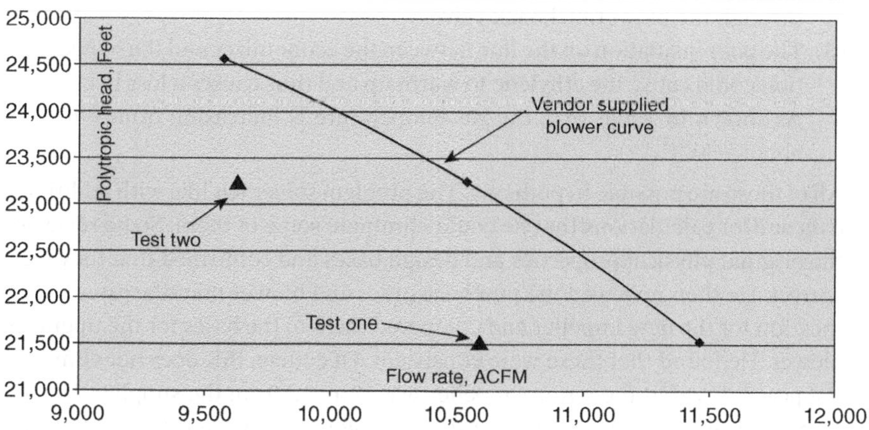

Figure 16.6 Blower capacity curve.

The projected polytropic head is taken from the blower manufacturer's supplied head curve shown in Figure 16.6. The performance deficiency is simply the deviation from the projected head curve expressed as a percentage.

Figure 16.6 clearly indicates that the blower does not appear to be performing as predicted by the performance curve. However, these data do not by themselves provide a working hypothesis. For example, if one simply presents a working hypothesis that says—"The blower is not performing as predicted by the blower manufacturer's supplied capacity curve," does this mean that the blower should be shut down for maintenance or is there another problem that is causing the blower to appear to be operating different than predicted by the performance curve?

Step 3: Develop a *Theoretically Sound* Working Hypothesis that Explains the Problem

The question list given in Chapter 6 was used as a guide to develop potential working hypotheses. A summary of this analysis is shown in Table 16.7.

Using these questions from Chapter 6, several hypotheses were developed as follows:

(1) There could have been errors in the design calculations for the new compressor impeller. This could consist of either errors in the data supplied to the manufacturer or errors made by the manufacturer.
(2) The wrong size impeller could have been installed.
(3) The increase in the blower discharge pressure could result in more leakage through wear rings.
(4) The increase in the blower discharge pressure could cause more leakage through the blower kickback valve.
(5) The poor insulation on the line between the economizer and the blower suction could cause the ethylene to warm up and thus causes a loss in capacity. As shown in Table 16.6, the gas temperature is increasing from −40 °F to about −30 °F.

All of these are possible hypotheses. The problem solver felt like with additional data and/or calculations that he could eliminate some of them. So he reviewed the original physical properties and design bases and confirmed that they were correct. He then reviewed the purchase order and blower manufacturer's specification for the new impeller and compared them to the bases for the upgraded blower. He found that these were consistent. Of course, this does not eliminate the possibility that the wrong impeller was shipped from the supplier. He also compared the old blower curve to the new blower curve and found that they were consistent. That is, when extrapolating from the old blower curve to the new blower curve using the appropriate diameter scaling factors, the extrapolated blower curve was essentially the same as the one supplied by the blower

Table 16.7 Questions/comments for Problem 16.3.

Question	Comment
Are all operating directives and procedures being followed?	Operating directives were being followed exactly
Are all instruments correct?	All instruments were checked
Are laboratory results correct?	Not applicable in this case
Were there any errors made in the original design?	There could be errors in the compressor impeller design calculations
Were there changes in operating conditions?	Yes. The blower discharge pressure was increased as part of the revised design
Is fluid leakage occurring?	Fluid leakage could be occurring through the kickback valve or through blower internals
Has there been mechanical wear that would explain the problem?	Blower wear rings are a potential problem
Is the reaction rate as anticipated?	Not applicable
Are there adverse reactions occurring?	Not applicable
Were there errors made in the construction of the process?	Since the unit had only recently been expanded, this had to be considered

manufacturer. Based on this work, he believed that he had done all that he could do except recommend a blower shutdown to eliminate hypothesis number 1. Before recommending that the blower be shut down to inspect the impeller to confirm that it was the correct diameter, he decided to consider the other hypotheses.

If there was internal leakage due to excessive clearance inside the blower, the internal gas recirculation would cause a decrease in the polytropic efficiency. This could be determined by the blower suction and discharge temperatures. The following equations from Chapter 7 were used to estimate the efficiency for the two tests. The results are shown in Table 16.8.

$$\sigma = (k - 1) * 100/(k^* E) \tag{16.9}$$

where

E = either the adiabatic or polytropic compression efficiency, %

k = the ratio of specific heats, C_p/C_v

$$T_d = T_s{}^* R^{\sigma}$$

where:

T_d and T_s = the absolute discharge and suction temperatures, respectively.

Table 16.8 *Calculated efficiencies.* for test runs.

	Test 1	Test 2
Suction temperature, °R	429.5	430.8
Discharge temperature, °R	541.0	551.1
Compression ratio	2.24	2.37
Compression exponent	0.286	0.285
Polytropic efficiency, %	70	70.2

Table 16.9 Calculation results.

	Test 1	Test 2
Gas rate, ACFM	10,590	9,630
Polytropic head at suction temperature, ft	21,525	23,250
Calculated polytropic head at −40 °F	21,050	22,700

Based on the test runs, there does not appear to be any indication of internal leakage since the calculated efficiencies from the suction and discharge temperatures appear to be essentially the same as the design. Thus, hypothesis 3 was eliminated.

As indicated earlier, one hypothesis was that the poor insulation was allowing a 10 °F increase in temperature between the economizer and the blower suction. This increased suction temperature would cause an increase in polytropic head. To determine if this was a theoretically sound working hypothesis, the problem solver calculated the blower head assuming that the gas temperature stayed at −40 °F. He got the results shown in Table 16.9.

If the gas temperature stayed at −40 °F, the required head would have been reduced slightly. The reduced head would decrease the horsepower requirements. However, the system does not appear to be limited by power requirements. As indicated in Table 16.6, the steam turbine driving the compressor remained at the design speed of 10,000 rpm throughout the tests. Thus, it appeared that the probability that this hypothesis was correct seemed very low.

Step 4: Provide a Mechanism to Test the Hypothesis

The elimination of these hypotheses left only the alternative hypothesis that there was excessive leakage across the kickback valve. If leakage was occurring through this valve when it was in the closed position, that would explain both the increase in temperature between the economizer and the blower suction, as well as the poor performance of the blower relative to the manufacturer's supplied curve. Rather than immediately recommending a shutdown to inspect the valve, the problem solver decided to test the hypothesis with calculations. He reviewed the specifications for the 14-in. kickback valve. When he reviewed the specifications, he found that the valve was not specified as a "tight shutoff valve." In addition, a review of the drawings indicated that the butterfly valve had a peripheral clearance of 0.05 in. That is, there was a clearance of 0.05 in. between the flap of the butterfly valve and the wall of the valve. He then estimated the leakage that could occur across the valve when it was completely closed. The flow rate through this small opening will be at sonic velocity. These calculations are as follows:

$$A = \pi^* \left(D_1^2 - D_2^2\right) /4 = \pi^*(14^2 - 13.9^2) = 2.19 \text{ in.}^2 = 0.0152 \text{ ft.}^2 \tag{16.10}$$

$$P = 0.55^*(14.7 + 10) = 13.58 \text{ psia} \tag{16.11}$$

$$\text{VS} = (P * g * k/\rho)^5(13.58 * 144 * 32.2 * 1.25/0.065)^{0.5} = 1100 \text{ ft./s} \tag{16.12}$$

$$F = \rho^* VS^* A = 0.065^*0.0152^*1100^*3600 = 3910 \text{ pounds/hr} \tag{16.13}$$

$$\text{ES} = F/(60^*\rho) = 3910/(60^*0.065) = 1000 \text{ ft}^3/\text{min} \tag{16.14}$$

where:

A = peripheral area with a clearance of 0.05 in
D_1 = approximate diameter of the valve, in
D_2 = approximate diameter of the butterfly wafer, in
P = pressure at restriction, psia
VS = sonic flow velocity, fps
g = gravity factor, fps^2
k = specific heat ratio
ρ = gas density, lb/ft^3
F = flow rate through peripheral area, lb/hr
ES = approximate volumetric flow, ft^3/minute

A brief explanation of the sonic flow conditions modeled by equations (16.11) through (16.13) may be appropriate. Essentially, all chemical engineering textbooks discuss this phenomenon in more detail than is possible in this book. The velocity across the peripheral opening will be at sonic flow velocity. This is because the pressure after the valve is only about 40% of the pressure before the valve. For a gas with a specific heat ratio (k) of 1.25, sonic flow properties occur if the pressure after a restriction is less than 55% of the pressure before the restriction. If sonic flow conditions are encountered, the maximum flow rate (sonic velocity) that will occur across any size opening with any amount of pressure drop is that which occurs when the outlet pressure is below 55% of the inlet pressure. Thus, the actual flow rate across the peripheral opening is evaluated at 55% of the absolute inlet pressure (equation (16.11)) and at the sonic velocity and density at these conditions (equations (16.12) and (16.13)).

As shown above, the estimated leakage through the butterfly valve could account for a capacity loss of approximately 1000 ft^3/minute. Referring to Figure 16.6, this difference in the suction flow rate could explain the deficiency in performance of the blower.

Step 5: Recommend Remedial Action to Eliminate the Problem without Creating Another Problem

The problem solver was faced with no other reasonable recommendation to make except to shut down the system and replace the kickback valve with one that had a "tight shutoff" rating. The potential problems that had to be considered were as follows:

(1) Was the replacement valve really a tight shutoff valve? Would it be possible to find a 14-in. valve that would fit into the space available and not have a peripheral opening similar to the existing valve?
(2) Would the new valve fit without making significant piping modifications? Could it be installed with a minimal amount of effort?
(3) Was there anything else that should be considered prior to a recommended shutdown? For example, should the efficiency of the steam turbine be determined to ensure that it is performing as designed?

16.8 Lessons Learned

This example problem indicates the value of doing a thorough problem analysis rather than just jumping to a conclusion that the blower is not performing as it was designed. If a complete analysis had not been done, the blower might have been shut down for an inspection or additional insulation might have been added to the blower suction lines in hopes that this would improve the performance. Either of these solutions that on the surface seemed to make sense would have delayed finding the fact that the "kickback" valve was leaking. If the blower had been shut down for an inspection and/or replacement of wear rings without knowledge that the kickback valve was leaking, another shutdown would have been required to replace the kickback valve. The analysis done illustrates the value of doing calculations to prove or disprove hypotheses.

As indicated in Chapter 4, a component of successful plant problem solving is a daily monitoring system that allows for the early detection of problems. This early detection will provide an earlier initiation of problem-solving activities than if the problem is allowed to continue to develop. If such a system had been in place in this example, a plot of "head curve deviation" as defined in Table 16.6 would have likely provided an early signal that there was a performance deficiency. An even better approach for a critical piece of revised equipment is to conduct a performance test as soon after start-up as possible.

The problem with the leaking "kickback" valve could have been detected even earlier than the start-up of the expanded facilities. If a plant test had been run prior to the shutdown to expand the plant or if the blower performance had been monitored on a daily basis, the problem with the valve could have been detected prior to the expansion. This would have eliminated the downtime required to replace the valve after the facilities start-up.

A conservative approach of inspecting the blower during the shutdown to replace the kickback valve could have been taken. However, this would have required major mechanical work and extended the time of the shutdown. Since the calculated polytropic efficiency was very close to the design value of 70%, it is highly unlikely that this would have been a "value-added" exercise.

16.9 Example Problem 16.4: The Value of a Potential Problem Analysis

While the utilization of a potential problem analysis was not emphasized in the previous problems, it would have been of great value for Example Problem 16.4.

A new fractionation process was designed to minimize cost by eliminating a reboiler, minimizing instrumentation, and maximizing heat integration. The fractionation tower products were a high-purity overhead and a high-purity bottoms stream.

A simplified sketch of the process is shown in Figure 16.7.

As shown in Figure 16.7, the feed to the tower is fractioned into a high-purity overhead methanol product and a high-purity xylene bottoms product. The heat integration is such that the heat input to the tower consists of a controlled vapor flow of xylene from a furnace. This vapor is the same material as the high-purity bottoms product so that no reboiler is required. That is, the vapor from the furnace is fed directly to the tower to provide heat input. The tower reflux is controlled to maintain a tower temperature profile. In addition to the controlled vapor flow to the bottom of the tower, the vapor output from the furnace is also used to heat the tower feed in exchangers. The xylene vapor condensed in the exchangers then flow to the accumulator as shown in Figure 16.7. The design of the exchangers was such that at *full capacity* and at the *design heat transfer coefficient* of the exchanger, the outlet material from the exchanger would be condensed xylene at the boiling point and pressure of the accumulator. That is, there would be no significant vapor flow from the accumulator to the tower. The fact that there was no meter on this

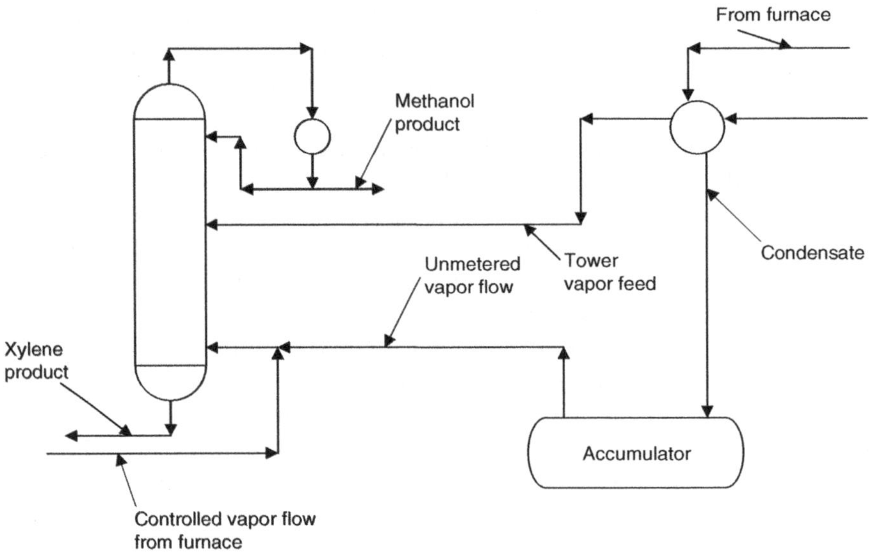

Figure 16.7 Simplified sketch of the fractionation system.

flow back to the tower did not seem like a significant problem since there would normally be no flow in the line. However, the designer recognized that there might be times when the conditions were such that the material leaving the exchanger would not be totally condensed and might contain vapor flowing into the accumulator, he provided a vent line to allow uncondensed vapor from the accumulator to flow to the tower. The designer believed that in cases where the vapor did vent out of the accumulator in an uncontrolled fashion, the control system would "take care of things" by adding more reflux to the tower.

As often happens, processes that are designed for steady state are rarely operated at steady state. The heat content of the material flowing into the accumulator was constantly changing. At times, there would be two phases (vapor and liquid) entering the drum. In this case, vapor would vent out of the drum into the tower. At other times, the xylene being condensed in the exchanger would be cooled to the point that the material entering the accumulator would be below the boiling point at accumulator pressure. If this occurred, vapor would flow back out of the tower. The predominant situation was the unsteady-state cycling of no vapor being vented back to the tower, transitioning to one where there were two phases in the flow to the accumulator. In this case, when there were two phases present in the flow to the accumulator, vapor would flow uncontrollably to the tower, creating an increase in the heat flow to the tower. The control system would respond *after* the temperature profile was disturbed and would cause more reflex to be added to the tower. During this transient condition, the purity of the overhead product stream would be less than desired since the increased vapor rate would cause more low volatility material to be carried overhead until the control system responded and increased the reflux rate. If the reflux rate had to be increased too much to compensate for the vapor venting out of the accumulator, it was possible that the tower would flood. As this situation transitioned to one where there was no vapor vent from the accumulator, the temperature profile in the tower would again be upset since there was now excessive reflux going to the tower. Again, the control system would correct the reflux rate, but only after the bottoms product was off-specification.

When the converse situation occurred, the vapor flowed back out of the tower due to the low pressure in the accumulator. In this case, the temperature profile would again be disturbed and the control system would again respond *after* the disturbance occurred. In this case, it was generally the bottoms stream that would be below specification during the transient.

The problem solver used the five-step procedure discussed earlier to begin solving the problem.

Step 1: *Verify* that the Problem Actually Occurred

The initial description of the problem was only that something was causing an upset in the tower, and the operations people believed that it was somehow associated with the accumulator. They often tried to compensate for these upsets by trying to adjust the controlled vapor rate. However, this was largely "guesswork" and often made things worse. The problem solver verified that upsets in the tower were being caused by changes (increases or decreases) in the fraction of vapor in the condensate flowing to the accumulator.

Step 2: *Write* Out an *Accurate* Statement of What Problem You Are Trying to *Solve*

The problem solver wrote out the following problem description:

> "Fractionation tower upsets are being caused by changes in an unmetered flow going to the bottom of the tower. These changes in the unmetered flow cause an increase or decrease in heat input to the bottom of the tower, the temperature profile in the tower to be upset, and the purity of the distillate and bottom products to be off specification. Determine how to eliminate the fractionation tower upsets caused by changes in an unmetered flow going to the bottom of the tower."

Step 3: Develop a *Theoretically Sound* Working Hypothesis that Explains the Problem

In this example, a start of the working hypothesis that explains the problem was included as part of the problem statement. However, it was not obvious how to

Table 16.10 Questions/comments for Problem 16.4.

Question	Comment
Are all operating directives and procedures being followed?	All operating directives and procedures were being followed. New ones were considered, but would not solve the problem
Are all instruments correct?	Yes
Are laboratory results correct?	Not applicable in this case
Were there any errors made in the original design?	The assumption of steady state was not valid
Were there changes in operating conditions?	No
Is fluid leakage occurring?	Not applicable
Has there been mechanical wear that would explain the problem?	No
Is the reaction rate as anticipated?	Not applicable
Are there adverse reactions occurring?	Not applicable
Were there errors made in the construction of the process?	Since the unit had only recently been built, this had to be considered

Table 16.11 Operating parameters before and after test runs.

	Before the problem	During the problem
Steam mass flow rate M, kg/hr	108,100	140,000
Steam inlet pressure, barg	42	42
Steam inlet temperature, °C	370	370
Surface condenser pressure, barg	−0.9	−0.3
Turbine exhaust temperature, °C	48.3	90.5
Power output, kcal/hr	20,000,000	20,000,000

solve the problem until the hypothesis was more fully developed. The questions given in Chapter 6 were used to fully develop a working hypothesis for obtaining a solution to the problem as shown in Table 16.10.

The only two reasonable hypotheses were that the assumption of steady-state operations was not valid and the possibility that a construction error had been made. No specific hypothesis was developed that would tie construction errors to the symptoms being observed. While it was possible that an error in the tray design, fabrication, or installation might be possible for fractionation upsets at the extreme conditions of high rates of uncontrolled vapor to the tower, it seemed unlikely that these tray errors would not show up at other times. Exploring the construction error hypothesis would likely require elaborate test equipment and/or a tower shutdown. It was decided to first consider the possibility that the original assumption of steady-state operation was the primary cause of the problem. It was clear that if the temperature of the condensate returning to the drum was not at the boiling point at the pressure in the drum that there would be an unmetered flow either to or from the tower. Developing the simplest solution that will work the problem as pointed out in Chapter 4 is always the best approach.

Thus, the problem solver developed the following hypothesis:

> "It is believed that the problems associated with the control of the tower are due to the fact that the heat content of the stream leaving the exchangers is not constant. At times there are large amounts of vapor in this stream which then vent to the tower as an uncontrolled heat input. At other times the stream leaving the exchangers is subcooled which causes vapors to flow from the tower to the accumulator. Tower control will be greatly improved if the vent or backflow from the accumulator can be measured."

Step 4: Provide a Mechanism to Test the Hypothesis

As what was thought to be a permanent solution to the problem, a venturi meter was installed in the vapor line. The venturi meter was selected because it would have a low-pressure drop and because it had the inherent capability to measure flow in both directions. If the enthalpy of the flow to the accumulator was such that some flashing occurred in the accumulator, the venturi would measure flow from the accumulator into the tower, and the controlled vapor rate from the furnace would be reduced to compensate for this vapor flow from the accumulator. Thus, the vapor rate in the tower would remain constant. Conversely, if the enthalpy of the flow into the accumulator was such that the liquid in the accumulator was subcooled, creating backflow from the tower, the controlled vapor rate would be increased to compensate for this backflow. It was believed that the installation of the venturi meter would maintain the vapor flow in the tower constant and thus avoid tower upsets. The control algorithm for the system is described as shown below:

$$F = V - Y + ZF \tag{16.15}$$

where:

V = the tower internal vapor rate which should be held constant

F = the controlled vapor rate from the furnace

Y = the flow rate from the accumulator to the tower

ZF = the flow rate from the tower to the accumulator

A simplified sketch of the venturi meter design is shown in Figure 16.8. A typical venturi pressure profile is also shown for the case where there is flow from the accumulator to the tower. It was recognized that the accuracy of the flow from the accumulator to the tower (measured by the pressure drop from point 1 to 2) would be more accurate than the backflow from the tower to the accumulator (measured by the pressure drop from point 3 to 2).

Unfortunately, no potential problem analysis as suggested in Chapter 4 for this problem solution was done. The fact that there is pressure recovery with any type of meter was not considered. This pressure recovery is represented in the sketch as the pressure increases from point 2 to point 3 when flow is from the accumulator to the tower. Since the pressure at point 3 is greater than the pressure at point 2, the control system would assume that this was backflow from the tower. The control system would then have values for both "Y" and "ZF," Of course, when the flow was from the accumulator to the tower, the actual value of "ZF" was 0. However, the pressure recovery made the control system think that "ZF" had a nonzero value. A potential problem analysis that included a detailed understanding of the venturi meter would have discovered this problem and allowed an engineering solution well before start-up of the revised facilities.

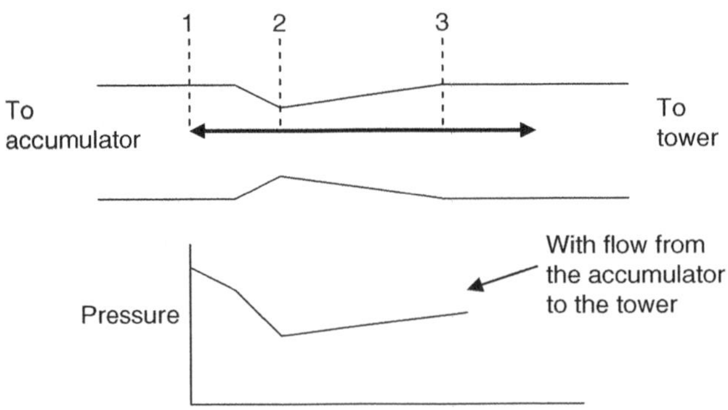

Figure 16.8 Venturi sketch.

Step 5: Recommend Remedial Action to Eliminate the Problem without Creating Another Problem
After the initial start-up difficulties discussed above, a selector switch was installed to allow the control scheme to select the greater value of "*Y*" or "ZF" and set the other variable to 0. The system performed flawlessly after that minor modification.

16.10 Lessons Learned

There are several lessons that can be learned from this problem. While the skill of process design involves the assumption of steady state, consideration should always be given to the question of "How does unsteady state impact the design?" If the process designer had considered the unsteady state, it is likely that he would have provided flow-measuring devices as part of the original design. The problem solver can also use the question of "How does unsteady state impact things?" as a problem-solving tool by questioning the validity of the steady-state assumption.

There is great value in both understanding the equipment involved in a problem solution and in performing a potential problem analysis prior to making a recommendation. In the example given here, the fact that pressure recovery would impact the results was blatantly obvious to anyone with a minimal knowledge of flow instruments. However, in the rush to get the facilities designed and installed, it was overlooked. A potential problem analysis would have pinpointed this problem before the venturi meter was installed. Potential problem analyses are often not done, except if they are required as part of a disciplined procedure.

16.11 Example Problem 16.5

A multi-billion dollar petrochemical plant has a condensing turbine for driving a gas compressor in the process. The condensing turbine section shown in Figure 16.9 had been operating normally at around 100 mbara (−0.9 barg). The reason for operating at vacuum pressure is to extract more work or power output, *W*, from the turbine as a result of higher value of (H1 − H2) from lower turbine exhaust enthalpy H2 according to the following equation (16.16).

$$W = M^{*}(\mathrm{H1} - \mathrm{H2}) \tag{16.16}$$

where:

W = turbine extracted work (or power output) (kcal/hr)

M = steam mass flow rate (kg/hr)

H1 = turbine steam inlet enthalpy (kcal/kg)

H2 = turbine steam outlet enthalpy (kcal/kg)

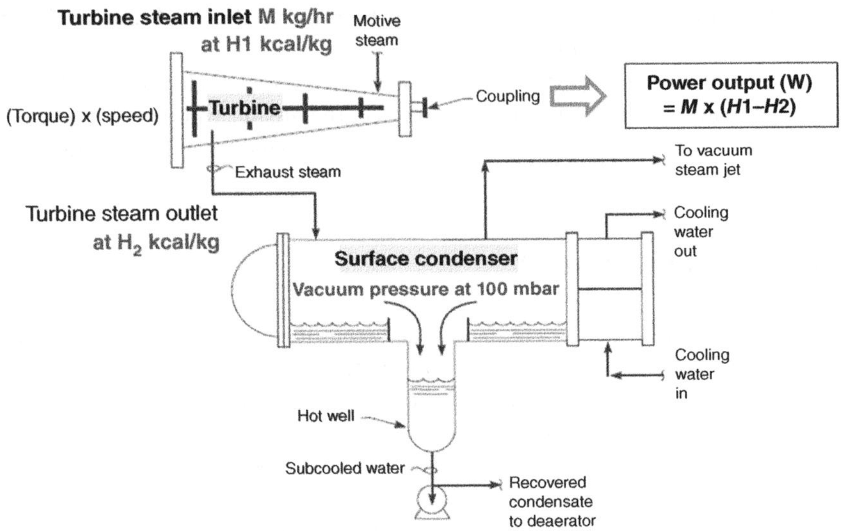

Figure 16.9 Simplified sketch of the condensing turbine surface condenser system.

At the outlet of condensing turbine exhaust is the surface condenser which helps condense turbine exhaust steam into the condensate and collect at the hot well for recycling back into the process by a condensate pump. One day, the steam side of the surface condenser started losing vacuum pressure as shown in Figure 16.10. In other words, the pressure at the surface condenser started to gradually increase to be closer to atmospheric pressure, from 100 mbara to 200, 300, 400 mbara, etc. This caused H_2 to go up, reducing the value of (H1 – H2) and thus work obtained from 1 ton of steam.

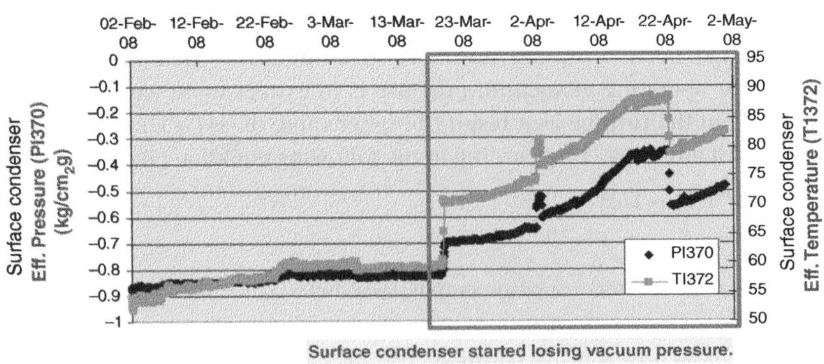

Figure 16.10 The pressure at the surface condenser started to gradually increase to be closer to atmospheric pressure, from 100 mbar to 200, 300, 400 mbar.

In order to obtain the same work to drive the compressor, steam mass flow rate, M, needed to be increased, increasing compressor operating cost significantly. The work obtained from the condensing turbine was monitored by using a power meter with a dynamometer concept to measure torque and speed (rpm) and convert the values to power output based on the following equation (16.17).

$$\text{(Turbine Power Output)} = \text{(Torque)} \times \text{(Speed)} \tag{16.17}$$

There was urgency to solve this problem as this caused an incremental operating cost debit of thousands of USD per day. The problem solver was called in to find the root cause and solve the problem as quickly as possible.

The equipment used to create vacuum was steam jet ejectors. There is a large hogging ejector used to achieve desired vacuum pressure in the commissioning stage where the air evacuation rate reaches a peak load. Then, after the condensing turbine and surface condenser have been fully started up and the target vacuum pressure has been achieved, the hogging ejector would be stopped, and smaller ejectors would be used to maintain vacuum instead.

The problem solver used the five-step procedure discussed earlier to begin solving the problem.

Step 1: *Verify* that the Problem Actually Occurred

The initial description of the problem was only that something was causing a loss of vacuum at the steam side of the surface condenser. There are several things that could be used to verify that the problem actually occurred.

(1) The vacuum pressure indicator should show clear signs of pressure increase.

(2) The condensing turbine exhaust temperature should also increase, as a result of saturated steam temperature corresponding to saturated steam pressure.

(3) The turbine steam flowmeter should show higher mass flow rate at the same power required from the compressor. The steam control valve should also show an increase in the opening percentage.

All three signs above needed to be consistent so that it confirmed that the problem really occurred.

Step 2: *Write* Out an *Accurate* Statement of What Problem You Are Trying to *Solve*

The problem solver wrote out the following problem description:

> "One day during normal operation, the vacuum pressure at the surface condenser of the condensing turbine gradually increased, causing higher steam consumption for the same turbine power output. The steam mass flow meter showed an increase, and the condensing turbine exhaust temperature also increased. The steam inlet control valve also opened more than previously observed by the operator. The problem statement is shown graphically in Figure 16.11."

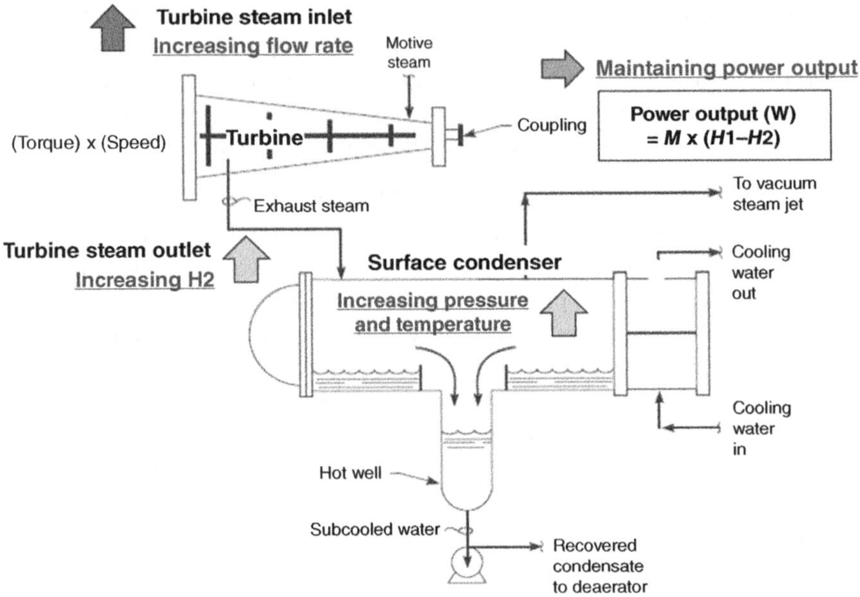

Figure 16.11 The problem statement.

Step 3: Develop a *Theoretically Sound* Working Hypothesis that Explains the Problem

The questions given in Chapter 6 were used to fully develop a working hypothesis for obtaining a solution to the problem as shown in Table 16.12.

The only two reasonable hypotheses were

(1) Air leakage into the system

(2) Loss of turbine isentropic efficiency

Hypothesis 2 was eliminated because calculations in Table 16.12 proved that turbine efficiency essentially the same, but the lower (H1 – H2) due to surface condenser loss of vacuum pressure was the key reason for this problem. Moreover, hypothesis 2 would not lead to the simultaneous increase in both turbine exhaust pressure and temperature. Therefore, only Hypothesis 1 was considered to be valid.

It is important to note that the turbine exhaust needs to have some moisture or liquid condensation (e.g., 5–7 wt%) for the power output to match the actual power output monitored from the dynamometer.

Since only the turbine side power output was considered in this example problem, we did not include the mechanical transmission loss between the turbine and compressor (Table 16.13).

Table 16.12 Calculated parameters before and after test runs.

	Before the Problem	During the Problem
Steam mass flow rate M, kg/hr	108,100	121,500
Steam inlet pressure, barg	42	42
Steam inlet temperature, °C	370	370
Surface condenser pressure, barg	−0.9	−0.3
Turbine exhaust temperature, °C	48.3	90.5
Turbine exhaust liquid wt%	6.2	6.2
Turbine inlet enthalpy H1, kcal/kg	−3046	−3046
Turbine outlet enthalpy H2, kcal/kg	−3231	−3211
Power output per steam feed, kcal/kg	185	164.6
Actual power output, kcal/hr	20,000,000	20,000,000
Calculated power output, kcal/hr	20,000,000	20,000,000
Turbine isentropic efficiency, %	72	73

During the hypothesis development, there was a mechanical engineer who proposed with confidence in a brainstorming meeting that there was likely no air leakage into the surface condenser system. He insisted on finding another root cause, even though he could not propose another theoretically correct hypothesis that could explain the problem and all symptoms.

The reason he believed that there was no air leakage was because the condensate in the hot well was sampled to be analyzed in the lab for dissolved oxygen. He hypothesized that if there was an air leak, the dissolved oxygen in the hot well condensate should have increased. However, the dissolved oxygen value during the problem was lower than before the problem.

Is this a correct theory and assumption? The answer is NO. Since the surface condenser is operated at an equilibrium or saturated condition, the higher the temperature in the hot well condensate, the lower the dissolved oxygen equilibrium limit as shown in Figure 16.12. This is the same philosophy as the deaerator where dissolved oxygen was reduced by boiling water inside the deaerator. To explain in a more daily-life example, making ice with boiled water will eliminate bubble voids in the frozen ice because of less dissolved oxygen in the boiled water. Thus, the theoretically correct hypothesis should imply air leakage into the system when dissolved oxygen value from lab analysis decreased due to higher operating temperature and pressure of the surface condenser hot well.

Table 16.13 Questions/comments for Problem 16.5.

Question	Comment
Are all operating directives and procedures being followed?	Yes. The operator was following the same procedure as before
Are all instruments correct?	The flow and pressure and temperature instruments have allegedly been calibrated correctly. The instruments were deemed correct also by process analysis
Are laboratory results correct?	Steam quality was found to be the same. The condensate in the hot well showed a decrease in the dissolved oxygen value
Were there any errors made in the original design?	No. The surface condenser worked fine from startup until the day of problem occurrence
Were there changes in operating conditions?	No intended operating condition changes were made, but the vacuum pressure continued to increase
Is fluid leakage occurring?	Air leakage into the surface condenser system might explain the problem
Has there been mechanical wear that would explain the problem?	Turbine isentropic efficiency loss could have led to more steam consumption at the same turbine power output
Is the reaction rate as anticipated?	Not applicable
Are there adverse reactions occurring?	Not applicable
Were there errors made in the construction of the process?	Not applicable because the surface condenser had been working well after startup until the day of the problem occurrence

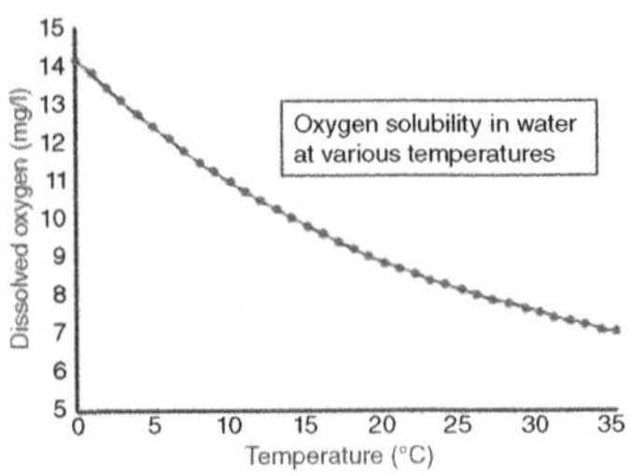

The dissolved oxygen value in the hot well condensate.

	Dissolved 02 (mg/l)		
	2-Apr-08	18-Apr-08	
E-300	4	1.53	Problem now
E-660	4.25	2.18	Another Surface condenser
E-2400	3.8	1.62	Another Surface condenser

Remark : Checked by compared with TSC from another surface condensor which use steam at the same source.

Figure 16.12 The dissolved oxygen analysis.

Thus, the problem solver developed the following hypothesis:

> "It is believed that the problems associated with the loss of surface condenser vacuum are due to air leakage into the surface condenser system."

Step 4: Provide a Mechanism to Test the Hypothesis

The mechanism to prove the hypothesis was to find the point of leakage. This is easier said-than-done. For internal pressure leakage into the atmosphere, it is much easier to detect the leakage. We can use a surfactant solution like soap liquid to spray onto the leak point and observe bubbles on the surface. However, for vacuum pressure leakage, the air will be sucked from the atmosphere into the process, also known as "air in-leakage." It would be very difficult to observe points of leak by observing surface bubbles. Some suggest using shaving cream, but in reality, it is still quite difficult to detect the air in-leakage.

The problem solver went out into the field for days, trying to find the leakage point, but to no avail. Then, a maintenance engineer proposed a much better solution, though much more expensive. The solution was to use a helium tracer injection and detection system to identify the point of leak. The concept was to expose helium gas from the outside location to be sucked into the process if there is a leak point and then put a helium sensor downstream of the helium exposed location. By this method, the problem solver finally managed to find the point of air-in leakage. To no surprise, the location was in a very difficult-to-access location behind a concrete column at the turbine exhaust bellows expansion joint. The bellows joint was used to accommodate thermal expansion when the turbine system operates at higher temperatures where metal would expand and cause excessive mechanical stresses if not using bellows joint.

Step 5: Recommend Remedial Action to Eliminate the Problem without Creating Another Problem

The remedial action was to stop the air-in leakage at the turbine exhaust bellows. This is done by shutting down the system and replacing the whole bellows or applying a temporary pad to seal the leakage. However, it needed to be done with care, and scaffolding was necessary to access the point of leakage in this case due to high point and difficult-to-access location.

16.12 Lessons Learned

There are several lessons that can be learned from this problem. We should not use a theoretically incorrect hypothesis to analyze the problem. In this case, a mechanical engineer clearly misunderstood the phenomena and determined the root cause incorrectly. Another lesson learned was the need to have a valid skillset and toolset to find air leakage points in a vacuum system.

Nomenclature

A	Area. In this chapter, it is used to represent the area of a restriction or a peripheral area with a clearance of 0.05 in. The value is in ft^2
C	a constant referred to as the "lumped parameter constant"
D_1	the approximate diameter of valve, inches
D_2	the approximate diameter of the butterfly wafer, inches
DF	the driving force or incentive for reaction to occur, mols of chlorine derivative gas absorbed/ft^3 of TEG
E	either the adiabatic or polytropic compression efficiency, %
ES	the approximate volumetric flow, ft^3/minute
F	flow rate. In this chapter, it is used to represent the ion-exchange bed feed rate, flow rate through a peripheral area, the flow of steam through the restriction, or the controlled vapor rate from the furnace all in lb/hr
G	gravity factor, fps^2
H	the polytropic or adiabatic head, ft
H1	turbine steam inlet enthalpy, kcal/kg
H2	turbine steam outlet enthalpy, kcal/kg
K	the ratio of specific heats, C_p/C_v
M	steam inlet mass flow rate to turbine, kg/hr
M	the gas molecular weight, g/mol
P	the pressure at restriction, psia
R	the compression ratio
R^*	the rate of change with time per unit volume of the compound under study, mols of HCl/ft^3-minute that are formed
RC	the rate of chloride production in the United States, lb/hr
S	the density of the flowing fluid relative to water
T_D	the absolute discharge temperature
T_S	the suction temperature, °R
U	the velocity of the flowing fluid through the orifice, fps
V	the tower internal vapor rate which should be held constant
VS	the sonic flow velocity, fps
W	turbine extracted work (or power output), kcal/hr
X_F	the concentration of chlorides in the flow to the ion-exchange bed, weight fraction
Y	the flow rate from the accumulator to the tower
Z	the average (suction and discharge) compressibility
ZF	the flow rate from the tower to the accumulator
ΔP	the pressure drop across the restriction, psi
ρ	the fluid density, lb/ft^3
σ	the polytropic or adiabatic compression exponent

17

A Final Note

Since problem solving as described in this book is not an intuitive process, there will be a learning period or an induction period before it can be done efficiently. This learning period will result in the first few problems or first few calculations that are done using this process being slow and labor-intensive. A personal assessment is that the first time that a new calculation technique is utilized that it often takes five to 10 times as long to do the calculation as after the calculation technique becomes "second nature." This is because the problem solver often will not have a feel for what the magnitude of the value that he is calculating should be, and he will conservatively check his calculations against some known standard. However, as experience is gained with this technique, speed will increase, and some of the labor-intensive steps will become almost automatic. Best of all using this approach means that, problems will be solved one time for all time. In addition, as indicated in earlier chapters, this approach is not to be used for every plant problem-solving activity. There will always be questions of "optimum technical depth" to consider.

There are some guidelines that should be considered prior to beginning to either implement these techniques as an individual or as a manager to implement these techniques throughout an organization. Some of these guidelines have been mentioned in earlier chapters. The purpose of this chapter is to present some of these guidelines together in one location.

The problem solver and his management should realize that the first time these techniques are used that more time will be required. The process of using a list of questions, such as those in Chapter 6, to develop a working hypothesis will be cumbersome until it is done a few times. There will be calculation techniques that the problem solver is not familiar with. He may well want to check his calculations because he has not yet developed an intuitive feel for the particular calculation. The management can be of great assistance in this area by insisting that they want the problem solver to take the time to do it right and insist on using the techniques described in this book. Our culture demands quick answers in almost all areas. From video games to interactive computer-aided learning, we have become

Problem Solving Approaches for Maintaining Operational Excellence in Process Plants, Second Edition.
Joseph M. Bonem, Nattapong Pongboot, and Wiroon Tanthapanichakoon.
© 2025 the American Institute of Chemical Engineers, Inc. Published 2025 by John Wiley & Sons, Inc.

an iterative society. The high school student who is using the computer to study often can keep trying to get the answer without any thought or calculation until he "gets it right." It will take strong management action to insist on replacing this "quick answer culture" with a "do it right culture."

The problem solver should recognize the role that the operator or mechanic can play in problem solving. The data that they have are invaluable in developing a theoretically sound working hypothesis. They have first-hand experiences and have made observations that are not available via any other means such as process control computers, instrumentation, or laboratory results. They may at times have problems putting their observations or experiences into quantitative descriptions. The problem-solver can help them express themselves by patiently asking questions. However, regardless of how much experience and valuable input they may have, they may not have sufficient training to formulate a correct working hypothesis. Their input and relationships should be cultivated by the professional.

There will be times when it is necessary to make assumptions. Many times in industrial problem solving, the direction of the change and the order of magnitude of the impact are all that is required. An exact value may not be required. For example, it may be necessary to assume the constant in the relationship between the rate and the driving force. The problem solver should not be discouraged by the need to make an assumption. It is often more important to get the form of the driving force correct than it is to get the constant correct. As described in Example Problem 16-1 given in Section 16.3, knowing the relationship between chloride production and residence time was much more important than the actual rate constant.

Bureaucracy should be avoided at all costs. There will be a tendency for management to micromanage the exact wording of the document described in Table 4-3. This problem specification is meant to be used by the problem solver to provide the best description possible of the problem that he is trying to solve. Wording changes will no doubt end up diluting and confusing his efforts. On the other hand, a review of means to test the working hypothesis and the final conclusion is mandatory. These reviews should focus on the technical accuracy of the calculations and the preparation for any proposed plant tests. Safety should be a major component of this review.

The concept of "one riot, one ranger," as described in Section 6.6, should be utilized. While committees are of value, the management should make it absolutely clear as to who has responsibility for developing a problem solution.

There may be value in initially having an individual assigned as the sponsor for utilization of these problem-solving procedures. This individual would be available to consult in the utilization of each of the procedural steps. He would not necessarily be an expert in the process of interest or in the utilization of equipment calculations. However, he would have sufficient know-how in the techniques to ensure that all the steps in the process were adequately considered.

Appendix

Conversion Factors from English Units to CGS Units

English Units	CGS Units	Multiply by to Go to CGS
Feet (ft)	Meter (m)	0.3048
Pounds (lb)	Kilograms (kg)	0.454
Pounds/in.^{2} (psi)	Bar (bar)	0.069
Gallons (g)	Liters (L)	3.785
BTU	Calories (cal)	252
Centipoise (cP)	Poise (P)	0.01
Horsepower (BHP)	Kilowatt (kw)	0.746
Degrees Fahrenheit (°F)	Degrees Centigrade (°C)	(°F – 32) * 0.5556
Degrees Rankin (°R)	Degrees Kelvin (°K)	0.5556

Problem Solving Approaches for Maintaining Operational Excellence in Process Plants, Second Edition.
Joseph M. Bonem, Nattapong Pongboot, and Wiroon Tanthapanichakoon.
© 2025 the American Institute of Chemical Engineers, Inc. Published 2025 by John Wiley & Sons, Inc.

References

AWWA Research Foundation (n.d.). *Filter Maintenance and Operations Guidance Manual | Water Treatment Plant Operators*. AWWA Research Foundation. thewaternetwork.com.

Carlson, E.C. (1996). Don't gamble with physical properties for simulations. *Chemical Engineering Progress* 92: 35–46.

Chaware, D. and Joshi, A. (2014). Refinery specs: the user's view. *World Pumps* 2014: 38–41.

Crane Company (2018). *New Flow of Fluids TP-410 2018 Edition*. Crane Company. cranebsu.com.

Efficient Petrochemical Processes: Technology, Design and Operation, Wiley, October 2019 Publisher: WILEYNew Jersey John Wiley & Sons, Inc. Corporate Headquarters 111 River Street Hoboken, NJ 07030-5774 Telephone: 201.748.6000 Facsimile: 201.748.6088 www.support.wiley.com Telephone: 1.800.567.4797Facsimile: 1.800.565.6802.

Flynn, D. (1979). *The NALCO Water Handbook*, 4e. Google Books.

Golden, S.W., Fulton, S.A., and Hanson, D.W. (2002). Understanding centrifugal compressor performance. *Petroleum Technology Quarterly* Quarter 2: 1–10.

GPSA (n.d.). *GPSA Engineering Databook*, 13e. gpsamidstreamsuppliers.org.

Henshaw, T.L. (1981). *Chemical Engineering*, 105. McGraw-Hill.

jmcampbell.com (2006). Finding the Optimum Compressor Interstage Pressure | Campbell Tip of the Month. jmcampbell.com.

jmcampbell.com (2011). Compressor Calculations: Rigorous Using Equation of State vs Shortcut Method | Campbell Tip of the Month. jmcampbell.com.

jmcampbell.com (2012). Which Equations of State Should be Used for Compressor Calculations? | Campbell Tip of the Month. jmcampbell.com.

Kaes, G.L. (2000). *Refinery Process Modeling. A Practical Guide to Steady State Modeling of Petroleum Processes*. Athens, GA: The Athens Printing Company.

Problem Solving Approaches for Maintaining Operational Excellence in Process Plants, Second Edition.
Joseph M. Bonem, Nattapong Pongboot, and Wiroon Tanthapanichakoon.
© 2025 the American Institute of Chemical Engineers, Inc. Published 2025 by John Wiley & Sons, Inc.

Mokhtar, A.A. and Zulkifli, A. (2018). Effects of flow velocity on RBI analysis of cooling water heat exchangers. *Journal of Engineering and Applied Sciences* 13 (14): 4265–4271.

NALCO (2004). *Corrosion in the Petroleum Industry*. NALCO.

NALCO (2015). *The NALCO Guide to Cooling Water Systems Failure Analysis*, 2e. New York: McGraw-Hill Education.

Narsimhan, G. (1981). Prediction of Intrinsic kinetics in diffusion disguised immobilized enzyme systems. *Chemical Engineering Journal* 22: 105.

Romagnoli, J.A. and Sanchez, M.C. (1999). *Data Processing and Reconciliation for Chemical Process Operations*, 1e, vol. 2. elsevier.com.

Scheel, L.F. (1968). New ideas on centrifugal compressors part 1. *Hydrocarbon Publishing* 47 (9): 253.

Scheel, L.F. (1972). *Gas Machinery*. Gulf Publishing.

Seymour, R.B. and Carraher, C.E. Jr., (2003). *Polymer Chemistry: An Introduction*. Marcel Dekker Inc.

Smith, J.M. (1981). *Chemical Engineering Kinetics*, 3e, 413. McGraw-Hill.

The NALCO Water Handbook, 3rd edition Publisher: McGraw-Hill New York 1325 Avenue of the Americas New York, NY 10019.

Veolia (n.d.). *Handbook of Industrial Water Treatment*. Veolia. watertechnologies.com.

Zhu, F., Johnson, J.A., Ablin, D.W., and Ernst, G.A. (2019). *Efficient Petrochemical Processes: Technology, Design and Operation*. Wiley.

Index

a

anticipated plant test results 41, 283, 292, 293–294, 304–305

b

batch operations 12, 187
Bernoulli's equation 76
best efficiency point (BEP) 77
black box approach 158
blowing 138

c

calorimeter 145
catalyst efficiency 33, 35, 197–198, 215, 294, 305–307
cavitation 78
Chernobyl 291
chloride removal unit 340
compressibility factor (Z) 104
compression
  exponent evaluation 81–82
  head 81
  thermodynamic assumptions 80
compression leakage 70, 89, 105
computer calculations
  checks with manual calculations 141, 199
  sources of errors 267

d

daily monitoring system guidelines 35–36
data verification 37, 265–269, 273, 275, 321, 324, 325
determining optimum technical depth guidelines 16
diffusion limited drying 156, 159, 162
directionally correct hypothesis approach 47–48
Disciplined Learned Problem Solving–5 step approach 36–42
display element 266–268
downcomer filling 138
downcomers 41, 136–140, 319
driving force 39, 143–144, 155–157, 159, 162, 167, 213, 345, 380

e

engineering problem solving components 21, 24, 29, 145
  determining optimum technical depth 16, 42–47
  developing a daily monitoring system 29–36

Problem Solving Approaches for Maintaining Operational Excellence in Process Plants, Second Edition.
Joseph M. Bonem, Nattapong Pongboot, and Wiroon Tanthapanichakoon.
© 2025 the American Institute of Chemical Engineers, Inc. Published 2025 by John Wiley & Sons, Inc.

engineering problem solving components (*contd.*)
utilizing a disciplined and learned problem solving approach 36–42
engineering problem solving definition 29

f
fast response model 189, 190, 199, 200, 205
first time calculations 379
flooding 137–138, 148, 150–153, 293, 294, 319, 340
flow meter errors 126, 287, 331
formulating hypotheses *via* key questions 68–71

g
Geiger counter 273

h
head curve 76–78, 80, 98, 230, 231, 360, 365
heat balance 148, 149, 151, 189, 198, 210, 212, 271, 293, 294, 323–326, 331, 333, 339, 340
heat transfer 2, 3, 8, 34, 40, 53, 55, 68, 72, 144, 155–159, 171, 172, 189, 193, 194, 210–215, 279, 280, 296, 303–306
high vacuum system 349
historical data bases 27, 29–31, 97, 108, 115, 172, 175, 289
horsepower 39, 77, 79, 82, 84, 87–89, 92, 95, 96, 106, 107, 109, 113, 115, 116, 118, 123–125, 128, 362
horsepower load point 84, 88
hot tap 70, 349, 352, 353, 355
human resources 265, 275
hydraulic gradient 137
hypothesis verification 6, 71–72

i
industrial examples 49
polymerization reactor 49–51
rotary filter 56–57
synthetic rubber discoloration 299–300
instrument specification sheet 69, 265, 266, 282–284, 292, 346
interpersonal skills 275
ion exchange bed 341–344, 347, 348

k
kinetically limited process 68, 72, 155–185

l
leaking "kickback valve" 365
limitations to problem solving 24–27
lower explosive limit (LEL) 180–183
lumped parameter constant 33, 155–157, 159, 189, 194, 345
lumped parameter limitations 158–159

m
material balance 2, 69, 148–151, 164, 270, 292, 321, 323, 339, 342, 346, 347
mean time between failures 35, 60, 61

n
net positive suction head (NPSH) 24, 79, 91, 92, 102–104, 110, 111, 219
novel process engineering 255–264
relocating tanks 257–260
tank roof 260–262

o
off the head curve 78

p
percent of range 266
phase rule 272, 274, 277, 279, 280

phase separation 135, 320
pitot tube 270
plant test evaluation 292–293, 298–299
pocket clearance 105
polymer-solvent equilibrium calculations 168–170
post test evaluation and documentation 292, 298–300, 306
  failed tests 297
  successful tests 103, 182, 204
pretest communication 40
pretest instrument evaluation 282, 284, 292–294, 300, 304
pretest laboratory evaluation 292–293, 300, 304
pretest potential problem analysis 295–296, 306, 307
pretest safety evaluation 295, 306
primary element 266, 267
prime movers 75–133
  displacement 75, 82–83, 85–87
  kinetic or dynamic 75–78
problem solving considerations/guidelines/approaches for
  fractioning towers 139–140
  hydraulic limitations 209, 227
  kinetically limited processes 68, 72, 155–185
  prime movers 72, 75, 78, 82, 83
  reactor expansions 212
  reboiler circuits 142–145
  sand filters 233
  unsteady state problems 188

r

rate of accumulation 188, 200, 206, 271
reaction 70, 72, 155–156, 189–199, 211, 213, 334
reciprocating pump calculation of 78–80, 85–87
  acceleration head 86, 111
  volumetric flow 85, 105
reciprocating compressor calculation of 80–82
  horsepower load point 84, 88
  leakage 89
  volumetric efficiency 84, 88
risks of "jumping to conclusions" 25 355

s

sieve trays 135–140
simplified approach 71, 214, 216
simplified kinetic constant 294
slow response model 189
stability limit 74, 77
startups 211, 336
steam pressure modulation 144
steam reforming 125
steam traps 39, 144, 145
steam turbines 90
  efficiency calculation 77, 90, 96, 113–118, 364
  steam consumption 350, 373
successful plant tests 5, 36, 40, 167, 291–308
  definition 38
  ingredients 292

t

technical resources 24, 265–269, 276
temperature driving force 39, 144, 213
theoretical chloride buildup 342
theoretical stages 140, 141, 332
thermal gravitmetric analyzer (TGA) 167
thermosiphon reboiler 142, 143
thermowell 272, 277, 279, 280
three stage ejector 349
tray efficiency 141, 293, 332
tray loading diagram 154
tray restrictions 138, 139

tray slope 139
tray stability diagram 137–140, 147–150, 152
tri ethylene glycol (TEG) system 340–345, 347, 378
trigger point definition 32–35, 156, 292, 294, 296, 305, 307
trigger point setting based on
  plant demonstrated 33
  theoretical/laboratory/pilot plant demonstrated 33
  vendor demonstrated or guaranteed 33
trigger points–statistical use 33
two way flow in venturi meter 41

u
unmetered heat input 367, 369
unsteady state operations 187–189

v
volumetric efficiency 84, 85, 88, 105

w
weeping/dumping 138, 139

x
X-rays 97–99, 135, 141, 142, 152

z
zero point 266, 273–275, 279, 285

Printed and bound by CPI Group (UK) Ltd, Croydon, CR0 4YY

17/09/2026

14964311-0004